Biology of
BLOOD-SUCKING
INSECTS

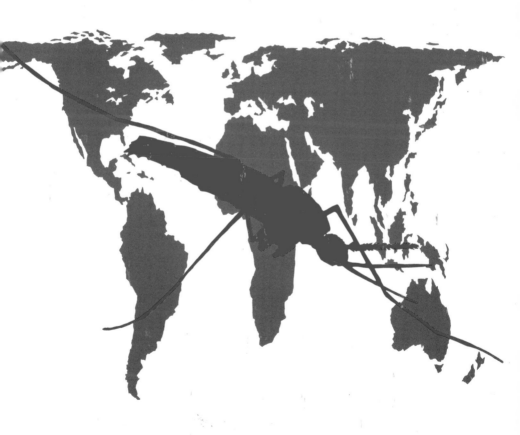

M. J. Lehane

BIOLOGY OF
BLOOD-SUCKING
INSECTS

TITLES OF RELATED INTEREST

Bacterial control of mosquitoes and black flies
H. de Barjac & D. Sutherland (eds)

The biology of Echinococcus and hydatid disease
R.C.A. Thompson (ed.)

A biologist's advanced mathematics
D.R. Causton

Cell movement and cell behaviour
J.M. Lackie

Parasitic protozoa
J.B. Kreier & J.R. Baker

Parasitic helminths and zoonoses in Africa
C.N.L. Macpherson & P. Craig (eds)

BIOLOGY OF
BLOOD-SUCKING
INSECTS

M.J. Lehane

School of Biological Sciences,
University of Wales

HarperCollins *Academic*
An imprint of HarperCollins *Publishers*

Published by
HarperCollins_Academic_
77–85 Fulham Palace Road
Hammersmith
London W6 8JB
UK

First published in 1991

British Library Cataloguing in Publication Data

Lehane, M.J.
Biology of blood-sucking insects.
1. Blood-Sucking insects
I. Title
595.77
ISBN 0–04–445409–0 (H/B)
ISBN 0–04–445410–4 (P/B)

Library of Congress Cataloging-in-Publication Data

Lehane, M.J.
Biology of blood-sucking insects / M. J. Lehane.
    p.   cm.
Includes bibliographical references and index.
ISBN 0–04–445409–0 (alk. paper).
ISBN 0–04–445410–4 (P/B).
1. Bloodsucking insects – Anatomy. 2. Bloodsucking
insects – Physiology. 3. Bloodsucking insects –
Digestive organs.   I. Title.
QL494.L34   1991
595.7'05249–dc20   90–22268
                                                              CIP

Typeset in 10 on 12 point Palatino by
Computape (Pickering) Ltd, North Yorkshire
and printed in Great Britain by
The University Press, Cambridge

Contents

List of tables   *page* ix

List of boxes   xi

Preface   xiii

Acknowledgements   xv

1   The importance of blood-sucking insects   1

    1.1   The importance of blood-sucking insects   1

2   The evolution of the blood-sucking habit   6

    2.1   Prolonged close association with vertebrates   6
    2.2   Morphological pre-adaptation for piercing   11

3   Feeding preferences of blood-sucking insects   14

    3.1   Host choice   14
    3.2   Host choice and species complexes   23

4   Location of the host   25

    4.1   The behavioural framework of host location   25
    4.2   Appetitive searching   27
    4.3   Activation and orientation   30
    4.4   Attraction   46
    4.5   Movement between hosts   49

5   Ingestion of the blood meal   52

    5.1   Vertebrate haemostasis   52
    5.2   Insect anti-haemostatic factors   56
    5.3   Probing stimulants   61
    5.4   Phagostimulants   63
    5.5   Mouthparts   64
    5.6   Blood intake   72

6 Managing the blood meal 79

6.1 Midgut anatomy 79
6.2 The blood meal 82
6.3 Gonotrophic concordance 92
6.4 Nutrition 94
6.5 Host hormones in the blood meal 99
6.6 Partitioning of resources from the blood meal 101
6.7 Autogeny 105

7 Host–insect interactions 111

7.1 Insect distribution on the surface of the host 112
7.2 Morphological specializations for life on the host 116
7.3 Host immune responses to insect salivary secretions 121
7.4 Behavioural defences of the host 128
7.5 Density-dependent effects on feeding success 135

8 Transmission of parasites by blood-sucking insects 143

8.1 Transmission routes 143
8.2 Specificity in vector–parasite relationships 156
8.3 Origin of vector–parasite relationships 160
8.4 Parasite strategies for contacting a vector 163
8.5 Parasite strategies for contacting a host 170
8.6 Vector pathology caused by parasites 173
8.7 Insect defence mechanisms 178

9 The blood-sucking insect groups 193

9.1 Insect classification 194
9.2 Phthiraptera 195
9.3 Hemiptera 199
9.4 Siphonaptera 204
9.5 Diptera 208
9.6 Other groups 246

References 248

Index 280

List of tables

1.1 An outline of the investigations which laid the foundations of medical and veterinary entomology. *page* 2

1.2 Estimates for the number of people infected and the number at risk from various major insect-transmitted diseases (WHO 1976). 3

1.3 Estimated losses in agricultural production caused by blood-sucking insects. (Information largely from Steelman 1976.) 5

4.1 The responses of various blood-sucking insects to spectral information; a range of flies are generally attracted to blue/black targets but are repelled by yellow ones. 42

5.1 Anticoagulants are found in the saliva of many (+) but not all (−) blood-sucking species. (A more extensive list can be found in Gooding 1972.) 57

5.2 Adaptations of mouthpart components for different purposes in various haematophagous insect groups. (Information largely from Marshall 1981 and Smith 1984.) 66

5.3 The size of red blood corpuscles in various host species. (Data from Altman & Dittmer 1971.) 76

6.1 A rough guideline to the 'average' meal size and the 'average' digestion time in various haematophagous insects. 84

6.2 The major blood constituents for three host species. Proteins are the most abundant of the nutrients which are unevenly distributed between whole blood (B), red blood cells alone (E), and plasma alone (P). (Data drawn from Albritton 1952 and Altman & Dittmer 1971.) 86

6.3 An outline of the anatomical location and the means of transmission of insect symbionts between generations in various insect groups. (Information from Buchner 1965.) 95

6.4 The influence of larval feeding success on the reproductive pattern of three types of female *Aedes taeniorrhyncus* identified in terms of egg development: 1, autogenous; 2, autogenous if mated; 3, anautogenous. (From O'Meara 1985.) 108

6.5 Results of an experiment to show that some mosquitoes can use sugar meals (10% sucrose in this experiment) to increase the number of autogenously produced eggs. (Adapted from O'Meara 1985.) 109

7.1 The choice of feeding site of *Aedes triseriatus* on eastern chipmunks and grey squirrels is influenced by body hair length and density. The different feeding patterns on the two hosts reflects the differences in hair cover between them. (From Edman *et al.* 1985.)   114

7.2 The anti-mosquito behaviour of a range of ciconiiform birds showing that different host species display various types and degrees of defensive behaviour against blood-sucking insects. (From Edman & Kale 1971.)   131

8.1 Some of the most important disease-causing organisms carried to man and other animals by blood-sucking insects: (a) viruses, (b) rickettsia and bacteria, (c) protozoa, and (d) nematodes.   144

8.2 Relationship between the total blood volume of the host and the insect blood meal volume.   163

8.3 The periodicity of the microfilariae of filarial worms, the major host, and the major vector species which has a peak biting time synchronized with the periodicity.   166

8.4 Feeding differences between tsetse flies infected with trypanosomes and uninfected flies. (From Jenni *et al.* 1980.)   171

8.5 Comparison of the rate of formation of the peritrophic membrane among various mosquito species.   180

8.6 The correlation between the infection rate of tsetse flies with trypanosomes and the development of the peritrophic membrane. Whether this is cause and effect or coincidental is still unknown. (Information from Lehane & Msangi (forthcoming) and Wijers 1958.)   181

8.7 The melanization response to subsequent challenge of infected and uninfected *Aedes aegypti* as shown by the intrathoracic injection of specific microfilariae (mff) which normally induce a strong melanization reaction. (From Christensen & LaFond 1986.)   192

9.1 An outline classification of the class Insecta.   194

9.2 The geographical distribution of triatomine species which have become highly adapted to the domestic–peridomestic environment of man and so represent a particular threat as vectors of Chagas' disease. (Information largely from Schofield 1988.)   201

9.3 The divisions of the order Diptera and the major families in each division. Families containing blood-sucking species are in bold type.   211

9.4 Characteristics of the three tsetse fly (*Glossina*) groups: *fusca*, *palpalis* and *morsitans*. (Information from several sources notably Jordan 1986 and Weitz 1963.)   236

List of boxes

3.1 The precipitin ring test   *page* 19
3.2 The importance of man-biting rates for the transmission of
     malaria   21
4.1 Hormonally controlled sensitivity changes in mosquito lactic
     acid receptors   32
7.1 Histopathology of the various stages in the sequence of host
     response to insect bites   126
8.1 An outline of the properties of the six types of haemocyte
     commonly found in insect blood   184

Preface

Blood-sucking insects are the vectors of many of the most debilitating parasites of man and his domesticated animals. In addition they are of considerable direct cost to the agricultural industry through losses in milk and meat yields, and through damage to hides and wool, etc. So, not surprisingly, many books of medical and veterinary entomology have been written. Most of these texts are organized taxonomically giving the details of the life-cycles, bionomics, relationship to disease and economic importance of each of the insect groups in turn. I have taken a different approach. This book is topic led and aims to discuss the biological themes which are common in the lives of blood-sucking insects. To do this I have concentrated on those aspects of the biology of these fascinating insects which have been clearly modified in some way to suit the blood-sucking habit. For example, I have discussed feeding and digestion in some detail because feeding on blood presents insects with special problems, but I have not discussed respiration because it is not affected in any particular way by haematophagy. Naturally there is a subjective element in the choice of topics for discussion and the weight given to each. I hope that I have not let my enthusiasm for particular subjects get the better of me on too many occasions and that the subject material achieves an overall balance.

Although the book is not designed as a conventional text on medical and veterinary entomology, in Chapter 9 I have given a brief outline of each of the blood-sucking insect groups. This chapter is intended as a quick introduction for those entirely new to the subject, or as a refresher on particular groups for those already familiar with most divisions of blood-sucking insects. There are several 'conventional' textbooks of medical and veterinary entomology available to those requiring more information.

The book is primarily intended for advanced undergraduate and postgraduate students, but because it looks at topics which cut across the normal research boundaries of physiology and ecology, behaviour and cell biology, I hope it may also be useful to the more established scientist who wants to look outside his own specialism. I have tried to distil this broad spectrum of information, much of which is not readily available to the non-specialist, into a brief synthesis. For those who want to look further into a particular area I have included some of the references which I found most useful in writing the text, and these will provide an entry into the literature. Clearly the subjects covered by the book encompass a vast number of publications and I am sure to have missed many important and interesting references, for which I apologize in advance to both the reader and my fellow scientists. Many of the topics discussed in the different chapters are interrelated. To avoid repetition, and still give the broadest picture possible, I have given cross-references in the text which I hope the reader will find useful.

From a comparative point of view it is an unfortunate fact that most of the work on blood-sucking insects has been carried out on a few species. Consequently, tsetse flies and mosquitoes pop up on every other page. In many instances it remains to be seen how widely the lessons we have learnt from these well-studied models can be applied. Where possible I have tried to point to general patterns which fit whole groups of blood-sucking insects. To help me in this I have divided the blood-sucking insects into three convenient but artificial categories: temporary ectoparasites, periodic ectoparasites, and permanent ectoparasites. These categories are based solely on the behavioural biology of the blood-feeding stadia in the lives of these insects. Temporary ectoparasites are considered to be those largely free-living insects which only visit the host for long enough to take a blood meal, such as the tabanids, mosquitoes, blood-feeding bugs and blackflies. I also include insects such as the tsetse here even though the adult male may be found in following swarms closely associated with the host for large parts of its life. Permanent ectoparasites are considered to be those insects which live almost constantly on the host, such as lice, the sheep ked, and tungid fleas. Finally, periodic ectoparasites are considered to be those insects which spend considerably longer on the hosts than is required merely to obtain a blood meal, but which nevertheless spend a significant amount of time away from the host. Insects which fall into this category include many of the fleas and Pupipara. These categories are no more than a useful generalization in the text; I make no claims for their rigour and I realize that it could be argued in several instances that an insect will sit as easily in one category as another.

Acknowledgements

I gladly and gratefully acknowledge the help I have received from many people during the writing of this book, particularly J.D. Edman, P. Billingsley, J. Brady, H. Briegel, E. Bursell, R. Galun, A.G. Gatehouse, M. Gillies, R.H. Gooding, C. Green, A.M. Jordan, K.C. Kim, J. Kingsolver, A.M. Lackie, B.R. Laurence, A.G. Marshall, D. Molyneux, P. Morrison, W.A. Nelson, G. O'Meara, G. Port, N.A. Ratcliffe, J.M. Ribeiro, P. Rossignol, M. Rothschild, W. Rudin, C.J. Schofield, M.W. Service, J.J.B. Smith, G.A. Vale and J. Waage. I also thank Paula Hynes, Maria Turton, Paula Dwyer and Daffyd Roberts for help with the illustrations. Finally most thanks go to my family, particularly to Stella, without whose encouragement, support and practical help this book would never have been finished.

CHAPTER ONE

The importance of blood-sucking insects

1.1 THE IMPORTANCE OF BLOOD-SUCKING INSECTS

Insects are the pre-eminent form of metazoan life on land. The class Insecta contains over three-quarters of a million described species. Estimates for the total number of extant species vary between 1 and 10 million, and it has been calculated that as many as 10^{18} individual insects are alive at any given instant – 200 million for each man, woman and child on earth! Thankfully only relatively few species feed on blood; approximately 300–400 species regularly attract our attention. These blood-sucking insects are of immense importance to man.

Man evolved into a world already stocked with blood-sucking insects. From his earliest days they will have annoyed him with their bites and sickened him with the parasites they transmit. As man evolved from hunter to herdsman, blood-sucking insects had a further impact on his well-being by lowering the productivity of his animals. It is reasonable to assume that, because of their annoyance value, man has been in battle with blood-sucking insects from the very beginning. In recent years this battle has intensified because of an increasing intolerance of the discomfort they cause, our fuller understanding of their role in disease transmission and the demand for increasing agricultural productivity. But despite considerable advances in our knowledge of the insects and improvements in the weapons we have to use against them, there is still no sign of an eventual winner in this age-old battle.

Many keen observers of nature suspected that insects were in some way involved with many of the febrile illnesses of man and his animals well before confirmatory scientific evidence was available. The explorer Humboldt recorded such a belief amongst the tribes of the Orinoco region of South America. The great German bacteriologist Robert Koch reported the belief of the tribes of the Usambara mountains of East Africa, that the mosquitoes they encountered when they descended to the plains were the cause of malaria (Nuttal 1899). Burton, in his travels in East Africa, recorded the similar belief of Somaliland tribes that mosquitoes were responsible for febrile illnesses (Burton 1860). Many of the peoples living near the tsetse fly belts of East and West Africa associated tsetse flies with sleeping sickness of humans and nagana of animals. In our own western tradition North American stock ranchers held the belief that Texas cattle fever was transmitted by ticks (in the class Arachnida, not Insecta) well before this was confirmed experimentally.

The fact that insects are vectors of disease was only confirmed scientifically at the end of the 19th century. The key discovery was made in 1877 (reported in 1878) by a Scottish doctor, Patrick Manson, working for the customs and excise service in China. He found that larval stages of the filarial worm, *Wuchereria bancrofti*, developed in the body of a mosquito, *Culex pipiens quinquefasciatus* (Manson 1878). This was the start of an avalanche of investigations which laid the foundations of medical and veterinary entomology. Some of the key discoveries of this era are outlined in Table 1.1.

The main insects involved in the transmission of all the most important vector transmitted diseases (Table 1.2) are now well known. The list of diseases transmitted is an impressive one including the medical scourges

Table 1.1 An outline of the investigations which laid the foundations of medical and veterinary entomology.

1878	Manson	development of *Wuchereria bancrofti* in a mosquito
1893	Smith & Kilbourne	*Babesia bigemina*, the causative agent of Texas cattle fever, transmitted by the tick, *Boophilus annulatus*
1895	Bruce	transmission of nagana by tsetse fly
1897	Ross	malaria parasites seen to develop in mosquitoes
1898	Ross	transmission of avian malaria by mosquitoes
1898	Simond	transmission of plague from rat to rat by fleas
1899	Grassi, Bignami and Bastianelli	*Anopheles* spp. are the vectors of human malaria
1901	Reed, Carroll, Lazear and Agramonte	transmission of yellow fever by the mosquito, *Aedes aegypti*
1902	Graham	transmission of dengue by mosquitoes
1903	Bruce & Nabarro	sleeping sickness in humans transmitted by tsetse fly
1903	Marchoux & Salimbeni	transmission of fowl spirochaetes, *Borrelia conserina*, by the tick, *Argus persicus*
1907	Mackie	spirochaete of relapsing fever transmitted by lice
1909	Chagas	*Trypanosoma cruzi*, causative agent of Chagas disease, transmitted by reduviid bugs

Table 1.2 Estimates for the number of people infected and the number at risk from various major insect-transmitted diseases (WHO 1976).

Disease	Estimated number of people infected	Estimated number of people at risk	Geographical distribution	Major vectors
malaria	365 million (in highly endemic areas)	2217 million	tropics and subtropics	anopheline mosquitoes (*Anopheles* spp.)
onchocerciasis (river blindness)	40 million		tropical, Africa, Yemen, Central and South America	blackflies (*Simulium* spp.)
lymphatic filariasis (elephantiasis)	90 million	905 million	Africa, Asia and South America	mosquitoes (*Culex* spp., *Aedes* spp., *Anopheles* spp., *Mansonia* spp.)
African trypanosomiasis (sleeping sickness)	>20 000	50 million	sub-Saharan Africa	tsetse flies
American trypanosomiasis (Chagas disease)	15–20 million	65 million	Central and South America	reduviid bugs (*Rhodnius* spp., *Triatoma* spp.)

malaria, sleeping sickness, leishmaniasis, river blindness, elephantiasis, yellow fever and dengue and the veterinary diseases nagana, surra, souma, bluetongue, African horse sickness and Rift Valley Fever (the piroplasms being tick borne); gauging the extent of these diseases is much more problematical, even for human disease. One reason is that health statistics are a moving target, particularly for those organisms such as yellow fever which occur as epidemics. But the greatest problem is that the heartland of these vector-borne diseases is the under-developed world where, for a variety of reasons, accurate statistical data is often difficult or impossible to gather. For this reason figures given for the extent of a disease are often not based on hard data, but are an estimate founded largely upon the experience of an expert. For example, a statistic that is repeated in classrooms and texts all over the world is that malaria kills about a million children a year in Africa. The figure is just such an informed estimate, originally made over 30 years ago and repeated regularly ever since. Table 1.2 gives estimates of the extent of some of the major vector transmitted diseases of humans, but obviously, as just indicated, care needs to be taken in the interpretation of the figures.

Blood-sucking insects cause very serious losses to agriculture. One way this happens is through the transmission of parasites. Indeed in the most celebrated case it has been argued that trypanosomes transmitted by tsetse flies preclude the use of about 10 million square kilometres of Africa for cattle rearing (Steelman 1976). In 1973 it was estimated that trypanosomiasis in cattle cost the agriculture industry $5 000 000 000 a year! (Goodwin 1973). The counter argument has also been proposed, that the tsetse has prevented desertification of large areas of land by overgrazing, and has been the saviour of Africa's game animals. The debate has been clearly outlined by Jordan (1986). Other examples of spectacular losses caused by insect transmitted disease are the death in 1960 of 200 000–300 000 horses in Turkey, Cyprus and India caused by African horse sickness, transmitted by *Culicoides* spp. (Huq 1961), and the estimated deaths in the USA, between 1930 and 1945, of up to 300 000 equines from Western and Eastern equine encephalitis transmitted by mosquitoes (Shahan & Giltner 1945).

In the developed countries it is usually direct losses caused by the insects themselves which are of greatest concern (Table 1.3). In exceptional circumstances the insects may be present in such numbers that stock are killed; for example, 16 000 animals died in Romania in 1923 and 13 900 in Yugoslavia in 1934 because of outbreaks of the blackfly *Simulium colombaschense* (Ciurea & Dinulescu 1924, Baranov 1935). More normally losses are caused not by death but by distress to the animal. Good examples are the reductions in stock of milk yields, weight gains or feed efficiencies which are commonly caused by the painful bites of the tabanids and biting flies (Table 1.3).

The sheer annoyance that blood-sucking insects cause to man can easily be overshadowed by their importance in medical and veterinary medicine. In some parts of the world, at certain times of the year, there may be so many blood-sucking insects that any activity outside is difficult or impossible without protective clothing. This is seen during the summer blooms of insects at higher latitudes, and also in many of the wetter areas of the tropics. These levels of annoyance are still rare for most people and for this reason the concept of nuisance insects is much more difficult to grasp than that of a vector or an agricultural pest causing economic damage. Perhaps the best way to view annoyance caused by insects is as a tolerance threshold. It can then be viewed as a variable with widely separated upper and lower limits; a handful of mosquitoes may be a minor inconvenience to the beggar in the street but intolerable to the prince in the palace.

I suggest that, in the developed world at least, we are increasingly intolerant of nuisance insects. There are several underlying reasons: the

4

Table 1.3 Estimated losses in agricultural production caused by blood-sucking insects. (Information largely from Steelman 1976.)

Insect	Year	Animal affected	Estimated losses US$ (millions)	Geographical region
Haematobia irritans (horn fly)	1965	cattle	179	USA
Stomoxys calcitrans (stable fly)	1965	cattle	142	USA
tabanids	1965	cattle	40	USA
mosquitoes	1965	cattle	25	USA
Melophagus ovinus (sheep ked)	1965	sheep	9.4	USA
lice (including non-blood-feeding species)	1965	cattle sheep swine goats	47 47 3 0.8	USA
tsetse fly	1973	cattle	5000	sub-Saharan Africa

increased awareness in the general population of the importance of insects in the spread of disease (sometimes over-exaggerated); the increasing stress placed on hygiene and cleanliness (almost manic at times); and increasing urbanization, so that for many people blood-sucking insects are not familiar, everyday things that they were once to our fathers and grandfathers working in a rural economy. This decrease in our tolerance of nuisance insects causes problems. The extended leisure time and mobility of many people in the developed world means they spend more and more time in increasingly distant places. The countries involved are often anxious to promote and develop their tourist industries and this has led to pressures to control nuisance insects and can be seen in places like the Camargue in southern France, the Scottish Highlands, the Bahamas, Florida and many parts of the Caribbean (Linley & Davies 1971). In addition population growth has put increased pressure on marginal land which in the past may have been left alone because of nuisance insect problems. Development of this land for leisure, commerce or housing with no insect control input can be disastrous for the developer, user or purchaser.

CHAPTER TWO
The evolution of the blood-sucking habit

Because of the very patchy nature of the insect fossil record, any discussion of the evolution of the blood-sucking habit among insects must rely heavily on detective work. The major clues lie in the diversity of forms and life-styles seen in modern day insects, and in some cases in the details of their relationships with vertebrates. From careful interpretation of this evidence quite credible accounts of the likely evolution of the blood-sucking habit can be made. From this starting point it has been convincingly argued that the evolution of the blood-sucking habit in insects has occurred along two main routes (Waage 1979).

2.1 PROLONGED CLOSE ASSOCIATION WITH VERTEBRATES

In the first route it is suggested that haematophagous forms may have developed subsequent to the prolonged association between the vertebrate and insects which had no specializations immediately suiting them to the blood-sucking way of life. The most common association of this type is likely to have centred around the attraction of insects to the nest or burrow of the vertebrate host. Insects may have been attracted to nests for several reasons. The humid, warm environment would be very favourable to a great many insects. In some circumstances, such as the location of the nest in a semi-arid or arid area, the protected habitat offered by the nest may have been essential to the insect's survival. For many insects the nest would also have proved attractive for the abundant supply of food to be found there. Certainly many current-day insects such as the psocids are attracted to the high concentrations of organic matter to be found in nests. Indeed, psocids may become so intimately associated with this habitat that they develop a phoretic association with birds and mammals, climbing into fur and feathers, to be translocated from one nest site to another (Pearman 1960, Mockford 1967, 1971).

Initially feeding on dung, fungus or other organic debris, the insects attracted to the nest will also have encountered considerable quantities of sloughed skin, hair or feathers. The regular, accidental ingestion of this sloughed body covering probably led to the selection of individuals possessing physiological systems capable of the efficient use of this material. Behavioural adaptations may then have permitted occasional feeding directly from the host itself. It is easy to see how this may have gone hand in hand with the adoption of a phoretic habit. Morphological and further behavioural adaptations would have allowed the insect to remain with the host for longer and longer periods with increasingly efficient feeding on skin and feathers.

The mouthparts developed for this life-style, in which the insect is feeding primarily on skin and feathers, were almost certainly of the chewing type such as those seen in the current-day Mallophaga. While these mouthparts are not primarily designed to pierce skin, some mallophagans do feed on blood. *Menacanthus stramineus*, a current-day mallophagan, feeds at the base of feathers or on the skin of the chicken. It often breaks through to the dermis giving the insect access to blood on which it will feed (Emmerson *et al.* 1973). Blood has a higher nutritional value than skin and is far easier to digest. This is reflected in the increased fecundity of blood-feeding Anoplura compared to skin-feeding Mallophaga (Marshall 1981). Once blood was regularly encountered by insects it is likely that its high nutritional value favoured the development of a group of insects which regularly exploited blood as a resource. This would have developed progressively, through physiological, behavioural and morphological adaptations, first to facultative haematophagy and eventually, in some insects, to obligate haematophagy. One way in which the progression from skin feeding to blood feeding may have occurred is seen in members of the mallophagan suborder the Rhynchophthirina, such as the elephant louse, *Haematomyzus elephantis*. This insect possesses typical mallophagan biting-type mouthparts (Ferris 1931, Mukerji & Sen-Sarma 1955) which are not primarily adapted for obtaining blood. By holding the mouthparts at the end of an extended rostrum (Fig. 2.1) the insect manages to use them to penetrate the thick epidermal skin layers of the host to get to the blood in the dermis.

It is thought that haematophagous lice developed from an original nest-dwelling, free-living ancestor (Kim 1985) along the pathway described above. It is impossible to know when the change occurred from free-living nest dweller to parasite, but it may well go right back to the appearance of nesting or communal living in land-dwelling vertebrates, which is thought to have happened during the Mesozoic (65–225 million

Figure 2.1 Despite having the chewing mouthparts typical of mallophagans, *Haematomyzus elephantis* is unusual in feeding on blood. The chewing mouthparts are held on the end of an unusual, elongate rostrum which may well be an adaptation helping the insect reach the dermis-containing blood through the thick skin of its host.

years ago). So lice may have pre-dated the emergence of mammals and birds and been parasitic on their reptilian ancestors (Hopkins 1949, Rothschild & Clay 1952). Others suggest a later date for the emergence of the lice (Fig. 2.2). Whichever educated guess is correct, it is certainly highly likely that ancestral forms were parasitic on primordial mammals and that from there they radiated along the lines of mammalian evolution. Coevolution of the host and permanent, and to a lesser extent temporary, ectoparasite stock probably led to rapid speciation in lice and other ectoparasitic forms. Helping drive this rapid speciation of the permanent ectoparasites was the reproductive isolation they suffer from being confined on specific vertebrate hosts, which may well have enhanced the effects of classical geographic reproductive isolation.

Some beetles also appear to be developing along this evolutionary highway. Several hundred species have been reported from nests and burrows (Barrera & Machado-Allison 1965, Medvedev & Skylar 1974). Most of these are probably free-living, feeding on the high levels of organic debris to be found at these sites. Some of these beetles have developed a phoretic association with the mammal which allows them to transfer efficiently between nest sites. Many of these phoretic forms also feed on the host by scraping skin and hair, and some have progressed to

8

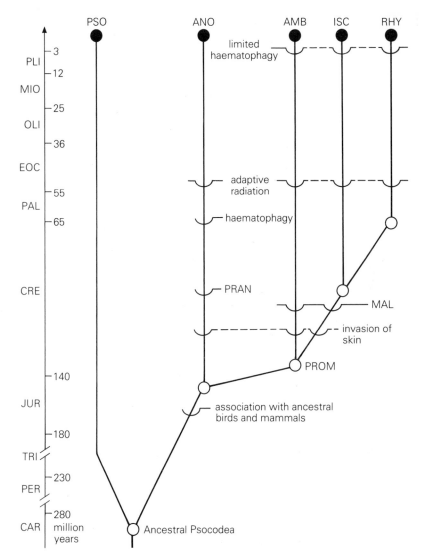

Figure 2.2 An outline of the possible origins of lice from the Psocodea. ANO, Anoplura; AMB, Amblycera; ISC, Ischnocera; RHY, Rhynchophthirina; MAL, Mallophaga; PSO, Psocodea; PRAN, Protanoplura; CAR, Carboniferous; PER, Permian; TRI, Triassic; JUR, Jurassic; CRE, Cretaceous; PAL, Palaeocene; EOC, Eocene; OLI, Oligocene; MIO, Miocene; PLI, Pliocene. (Redrawn from Kim & Ludwig 1982.)

9

the stage where they will occasionally take blood (Wood 1964, Barrera 1966).

The prolonged association of the insect with the vertebrate, which is the corner-stone of this first route for the evolution of the blood-feeding habit, may not have always relied on encounters in the nest habitat. Free-living ancestral forms with few, if any, clear adaptations for the blood-sucking way of life, may have also developed prolonged associations with the vertebrate at some point distant from the nest. This type of association may have had several different underlying reasons such as attraction to feed on vertebrate secretions, or the use of the vertebrate as a basking or swarming site. But probably the most important factor was the use of the host's dung as a larval habitat.

The vertebrate may live in a harsh environment where dung rapidly dries up or it may bury its dung. In either case closely associating with the vertebrate would help the insect in locating usable dung. Non-blood-feeding associations arising for these reasons are seen between some scarabaeid beetles and vertebrates, but competition to be the first to exploit dung was probably the commonest reason driving the insects to an ever closer association with the vertebrate. Dung is a limited resource and there is often intense competition to utilize it as a larval site. Consequently there is pressure on the insect to be the first to introduce its eggs into newly deposited dung. One way for the insect to achieve this is for an increasingly close association with the vertebrate providing the dung. Such a close association can be seen in the horn fly, *Haematobia irritans*, which as an adult is permanently associated with large vertebrates, only leaving them to oviposit. It will lay eggs in dung within 15 s of defecation by the vertebrate and, within its distribution range, is almost always the first colonizer of freshly deposited dung (Mohr 1943). If such an insect can feed on the vertebrate, it minimizes the time it will have to spend away from the host and therefore maximizes the advantage to be gained from the close association and this may have led the ancestral adult female to feed on body secretions, open wounds and sores. As was argued above, because blood is such a nutritious substance, once it was encountered in the diet of the insect, selection is likely to have favoured the progressive development of physical and behavioural adaptations leading to full haematophagy. Selection will have led to the progressive development of the mouthparts, allowing the insect to dislodge scabs, open up old sores and eventually to penetrate unbroken skin. The evolution of organisms capable of breaking the skin and obtaining a blood meal may well have been an accelerating process. The wounds produced by the first blood-suckers will have provided regular blood feeding opportunities for other organisms which could not as yet break the skin by their own efforts. Once given access to blood these too may well have followed the evolutionary pathway outlined above.

The spasmodic appearance of blood feeding among male insects is a difficult issue to explain, but the close or permanent association of females with the vertebrate as discussed above is one factor which may explain it in some insects. Under these circumstances it may become advantageous for the male to become associated with the vertebrate because of the greater likelihood of success in mate finding. This may then lead to blood-feeding in the male as this will minimize the time spent away from its host and therefore maximize the advantage to be gained from the close association. Similarly, if blood-sucking females are irregularly and/or widely dispersed in a habitat, then the male may gain a mating advantage by staying with the vertebrate and waiting for the female to arrive to feed. This is seen in tsetse flies and in the mosquito *Aedes aegypti*, for example (Teesdale 1955).

2.2 MORPHOLOGICAL PRE-ADAPTATION FOR PIERCING

The second route for the evolution of the blood-sucking habit suggests that blood feeding developed in some insect lineages from ancestral insects which were morphologically pre-adapted for piercing surfaces (Beklemishev 1957, Downes 1970, Waage 1979). Entomophagous insects are strong candidates for such a conversion. The Rhagionidae are a good example: most of the group are predacious on other insects, but a few species have turned to blood feeding. How could this changeover have come about? Entomophagous insects would have been attracted to nests and burrows by the accumulations of insects to be found there and in so doing would also have encountered vertebrates. Away from nests they may have been attracted to vertebrates by the accumulations of insects around them, or the vertebrates may have regularly congregated in the wet areas which are the breeding sites for many of the 'lower' Diptera. The vertebrates involved may have been permanently resident amphibians or reptiles or larger vertebrates who regularly visited such sites for drinking or bathing purposes. In each of these cases it is easy to see how entomophagous insects made repeated and possibly prolonged contact with vertebrates. These predatory insects would have physiological and morphological adaptations (such as efficient protein digesting enzymes and piercing mouthparts) facilitating the switch to haematophagy. Haematophagy in these individuals was at first probably an occasional, chance event which led to full haematophagy through continued close association with the vertebrate host. It is thought that haematophagy developed along these lines in the ancestors of the blood-feeding bugs and in blood-feeding rhagionids and possibly in some blood-feeding Diptera. In other Diptera which may have evolved along this line, there is

11

some doubt about which came first, larval feeding on nest debris or adult feeding on other insects in the nest habitat. Fleas may also have evolved along this pathway from free-living mecopteran stock (Tillyard 1935, Hinton 1958). The Mecoptera, or scorpion flies, contain a modern day group, the Boreidae, which are apterous and are capable of jumping. They live in moss and feed on insects. Similar insects may well have been the ancestors of the fleas.

The life-styles of several current day insects support the idea that entomophagous insects gave rise to some blood-sucking insect groups (Waage 1979). Many personal experiences in Britain (as many as three in one day) with the flower bug, *Anthocoris nemorum*, show that this insect is willing and able to pierce human skin. This insect is entomophagous, living on and around flowers where it pounces on small insects visiting the flowers to feed. While its probings of my skin cause a sharp pain, I have still to find one which has obviously ingested any blood; however, it still establishes the fact that entomophagous insects will often show an interest in vertebrates as potential sources of a meal. The hemipteran bug *Lyctocoris campestris* is also an entomophagous species but it takes matters further. It lives in birds' nests where it feeds on other insects but it will also take blood meals from vertebrates (Stys & Daniel 1957). Evidence from blood-sucking insects themselves also points to the close links between entomophagy and the blood-feeding habit. The mosquitoes *Aedes aegypti* and *Culex tarsalis* will take body fluids from insect larvae presented to them under laboratory conditions (Harris *et al.* 1969). Indeed this form of feeding is so successful that these mosquitoes go on to produce viable eggs as a result, which opens up the intriguing possibility that this may occur naturally in the field.

It has also been argued that haematophagy may have arisen in some insect groups (including the mosquitoes; Mattingley 1965) from plant-feeding ancestors. This is certainly a possibility as many plant-feeding insects possess piercing and sucking mouthparts which would pre-adapt them for haematophagy. This is seen in the moth *Calpe eustrigata* which is one of a group of noctuiids which possess an unusually modified, sharp proboscis used in most species for the penetration of fruit rinds. But *C. eustrigata* uses it to penetrate vertebrate skin for the purposes of blood-feeding. It is probable that plant feeding ancestors of modern day blood feeders will only have developed haematophagy if they were in a position where they continually associated with the vertebrate host. This may have occurred through mechanisms similar to those already out-lined. Attraction to free-living vertebrates may have occurred in order to feed on bodily secretions or to use dung as a larval medium. Or insects may have been attracted to nests to feed on fruits or seeds stored there by the vertebrate. In this context it is interesting to note that the hemipteran bugs are exceptional in using cathepsin-like digestive proteinases. This is

12

consistent with a proposed evolutionary path for bugs from sap-sucking ancestors (Houseman *et al.* 1985, Billingsley & Downe 1988, Terra 1988). Sap feeders, not needing proteases, may have lost their trypsins. If they then moved to blood feeding they would need to re-acquire proteolytic activity. It is argued that having lost their trypsins they had to make use of the cathepsins which are contained in the lysosomes of all cells, re-routing them for extracellular digestion.

CHAPTER THREE
Feeding preferences of blood-sucking insects

3.1 HOST CHOICE

Blood-sucking insects feed from a range of different host animals. For obvious and often painful reasons, we are aware of the fact that many of them feed from mammals, but many other host animals are also exploited including birds, reptiles, amphibians and fish, and even insects, arachnids and annelids (Hocking 1971). Any one insect does not feed equally well from all of these potential resources; it displays host choice. For some insects, particularly some permanent ectoparasites, the host chosen may be very specific; occasionally, for example human lice, just a single species. For other blood-sucking insects host choice is clearly not so restricted as introduced exotic hosts quickly become incorporated into the diet of local blood-sucking insects. Let us consider what is meant by host choice. In its main sense it denotes the species of host animal or animals from which a blood-sucking insect obtains its blood meals, but host choice can go beyond the particular species of host chosen. Insects often choose to feed on particular individuals from among the preferred species, which may well have implications for disease transmission (Burkot 1988).

Although most blood-sucking insects in their undisturbed, natural surroundings show a preference for feeding from a particular group or species or even a particular cohort of its chosen species, the degree of host specificity shown varies greatly from one insect to the next. Some are entirely dependent on a single species of host while others are willing to feed from a wide range of hosts. As a rough rule of thumb, it has often been said that there is a direct relationship between the locomotory capabilities of an insect and the number of different hosts that it utilizes. Thus, the permanent ectoparasites, with their limited capacity for movement away from the host, contain most cases of precise dependence. For example, the louse *Haematomyzus elephantis* is confined to elephants and

H. hopkinsi to the wart hog (Clay 1963). Considering the more mobile periodic ectoparasites, we can find many examples such as the 'human' flea, *Pulex irritans*, which have a narrow preferred range (badgers and foxes) on which most insects are found, but which can be occasionally discovered on a wide range of other hosts (in this case man, pigs and other large mammals) (Marshall 1981). When considering the largely free-living, temporary ectoparasites we find that many have catholic tastes; *Culex salinarius*, for example, has a wide range of potential hosts which includes birds (45% of the blood meals tested), equines (17%) and canines (15%) (Cupp & Stokes 1976). This mosquito will demonstrate its cosmopolitan tastes further if it is disturbed during the meal by moving willingly from one host species to another to complete this single, full blood meal. This is a regular and natural occurrence for this insect. In one investigation 13% of the blood meals tested were found to be from a mixture of hosts (Cupp & Stokes 1976).

The above rule of thumb works well in a broad sense but often breaks down when particular insects are considered, as can be seen with the help of two examples. The amblyceran louse, *Menacanthus eurysternus*, has poor locomotory abilities off the host and according to our rule of thumb it should be limited to a small number of hosts; in fact it has been recorded from 123 different species of bird (Price 1975)! In contrast, each of four closely related species of winged streblid were found to be specific to a different genus of bats and this relationship held true even in caves containing three of the bat genera in crowded conditions (Maa & Marshall 1981).

Before looking at some of the factors which underpin host choice let us make some general comments of particular relevance to temporary ectoparasites. The commonest hosts are probably large herbivores as large, social herbivores present an abundant and easily visible food source for many temporary ectoparasites. They are also a reliable food source because normally they only move slowly from one pasture to another. Carnivores are less abundant than their prey, often solitary, less visible and less predictable, occupying a large home range. For these reasons carnivores are less likely to be a primary element in the host choice of a blood-sucking insect. Also large vertebrates are likely to be preferred over smaller ones. This is because small vertebrates, which suffer greater losses from the attentions of a blood-sucking insect, use their agility to develop efficient defences.

Moving from these generalizations to consider the specific choices made by particular insects is a more complex issue. Sometimes a reasoned case for the evolution of a particular pattern of host choice can be made, but on most occasions we are in the dark. The choice may be determined by a large number of factors (usually acting in combination) including behavioural, physiological, morphological, ecological, geographical, temporal and genetic considerations. To give an idea of the complexity of the issue let us look at a small number of examples, showing how each of these factors may affect host choice.

The behaviour of the host can have a considerable effect on the attention it receives from blood-sucking insects. This has been demonstrated in the anti-mosquito behaviour displayed by a range of ciconiiform birds. At roost some of these birds perform up to 3000 defensive movements per hour! But the green heron and the crowned night heron are far less active, performing as few as 650 defensive movements per hour; as a consequence they receive far more bites than their more active relatives (Weber & Edman 1972). If no other factors were in operation it is easy to see how natural selection would bear on such a case and would lead a blood-sucking insect to select and eventually specialize in feeding from the easiest available target. The fact that under natural conditions mosquitoes still feed from a wide range of ciconiiform birds, including those displaying the highest levels of defensive behaviours, is a testament to the complexity of the factors which determine host choice. None the less, grooming and host defensive behaviour in general are likely to be highly efficient agents of natural selection. For man the concept of defensive behaviour needs to be extended to include such practices as sleeping under bednets, screening houses and the use of repellents (modern preparations containing active ingredients such as di-ethyl toluamide or traditional concoctions such as mixtures of cow dung, urine and ash) (MacCormack 1984).

The intensity of grooming behaviour correlates with the number of blood-sucking insects attacking the host. Different host species are likely to differ in their relative tolerance to insect attack. Thus it is possible that the seasonal changes in the numbers of blood-sucking insects may play a part in the seasonal changes in host choice which have been reported due to alterations in the defensive behaviour of hosts to the different levels of attack (Edman & Spielman 1988). Such seasonal changes in host choice may be important in the transmission of some zoonoses to man. For example, in the USA the arbovirus eastern equine encephalitis is enzootic in birds and the mosquito *Culiseta melanura* is its vector. During the spring and early summer this insect feeds almost exclusively on passerine birds, but later in the season its choice of hosts becomes more catholic and includes other bird groups and mammals. It is also in this later part of the season that epizootics of the virus occur in birds and horses and epidemics occur in humans.

I suggest that physiological factors are mainly important in determining host choice in the sense that once the insect has become associated with a narrow range of hosts then specializations in its physiology will occur which may limit the range of other hosts it can exploit. This is because natural selection will ensure that all of the insect's systems will

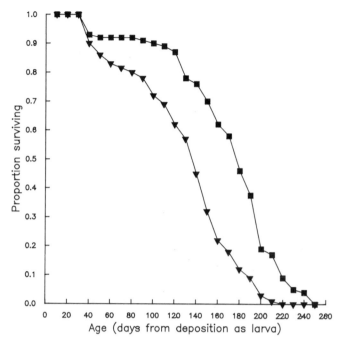

Figure 3.1 The tsetse fly, *Glossina m. morsitans*, shows increased survivorship when fed on rabbits (■) rather than on goats (▼). (Redrawn from Jordan & Curtis 1972.)

become tuned to the exploitation of the resources of its major hosts which may restrict the insect's ability to deal with unusual situations. In some cases choice of the wrong host can, for physiological reasons, lead to the death of the insect. For example, a blood meal from the guinea-pig may form oxyhaemoglobin crystals which will rupture the intestine of several blood-sucking insects (Krynski *et al.* 1952). The physiological consequences of moving on to an unusual host can have a less obvious, but a nevertheless damaging, effect on the success of the insect. Several experimental studies have shown that the fecundity of a blood-sucking insect depends on the host on which the insect is fed. Reduced fecundity can be generated by a reduced rate of development of the insect, reduced longevity (Fig. 3.1), a skewed sex ratio, or reduced food intake or rate of digestion (Nelson *et al.* 1975, Rothschild 1975). We can look at the tsetse fly as an example. *Glossina austeni* fed on rabbit blood showed a consistently higher fecundity than flies fed on goat blood (Jordan & Curtis 1968). In a further series of experiments it was shown that both *G. austeni* and *G. morsitans morsitans* fed on pig blood produced the heaviest puparia, closely followed by flies fed on goat blood, while the puparia produced from flies fed on cow blood were considerably lighter. Clearly under natural conditions it would be a selective advantage to the insect to

17

develop mechanisms which avoided such a restriction on its reproductive success by careful choice of hosts.

The source of the blood meal or its quality may also influence the course of an infection in the insect. If tsetse flies are fed on goats or cattle they will develop a much higher infection rate with *Trypanosoma vivax* than if they are fed on mice (Maudlin *et al*. 1984). Whether this is due to a difference in the physiological or nutritional status of the flies or to a direct effect on the trypanosomes is unknown.

As with the physiological restrictions on host choice just discussed, morphological factors are probably important as a limitation on the range of available hosts once a certain degree of specialization for a particular host has taken place. With permanent and periodic ectoparasites especially, restriction in host range usually involves morphological specialization of the insect's mouthparts, ovipositor, or locomotory or attachment apparatus. Specializations of any of these can easily restrict the host range of the insect concerned; indeed, in many cases specialization may even limit the area of the body of the preferred host which the insect can utilize (see Section 7.1). Morphological characteristics of the host are also important in determining host choice as can be seen in the frequently reported preference of *Anopheles gambiae* for human adults rather than children. In this case the choice is a direct consequence of the difference in size of these two potential hosts, the number of bites received by an individual being in direct proportion to the surface area that individual contributes to the total immediately available to the mosquito (Port *et al*. 1980).

A clear example of ecological influence on host choice is furnished by the fleas. The immature stages of most fleas live on the detritus in the homes of the host animal. In other words, host choice for fleas is largely determined by the ecology of the host, most fleas being restricted to hosts with long-term homes, or at least seasonal lairs which are used each year. Because they so commonly use long-term homes mammals are the main hosts for fleas, and rodents – the home builders *par excellence* – are the most afflicted, harbouring 74% of all known flea species (Marshall 1981). Man is also an habitual home dweller and because of this is the only primate which is flea-ridden.

A second example of ecological influence on host choice, of considerable importance in the health of Europeans, can be seen in the changing feeding patterns of the anopheline mosquitoes of Europe. Until the middle of the 19th century benign tertian malaria (caused by *Plasmodium vivax*) was endemic in Europe as far north as Scandinavia, with *P. malariae* as far north as Holland and the UK, and *P. falciparum* extensive in southern Europe. At this time malaria began to retreat from Europe and transmission of the disease now rarely occurs. Why did this happen? Ronald Ross described the full cycle of malaria in 1898 and as soon after this as 1901 it was realized that there were areas of Europe which abounded with anopheline mosquitoes, and into which malaria carriers routinely passed, but in which the disease was not transmitted (Falleroni 1927). Several theories were put forward to account for this apparent anomaly, but it was not until after the introduction of the precipitin test for the identification of mosquito blood meals (Box 3.1) that useful hard

Box 3.1 The precipitin ring test

The widespread use of immunological techniques for the identification of the source of an insect's blood meal has followed work begun in the 1920s. The precipitin ring test has been the basis of most of these investigations and the procedure is outlined in the diagram. A variety of modifications of the basic technique are now available. The usefulness of the most widely used of these modifications have been evaluated by the World Health Organization in an interlaboratory trial (Pant, Houba & Engers 1987, WHO/VBC/87, 947) and a summary of their results is given.

1 *Precipitation ring test* (PRT)

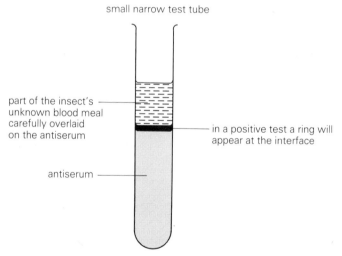

small narrow test tube

part of the insect's unknown blood meal carefully overlaid on the antiserum

in a positive test a ring will appear at the interface

antiserum

A range of tubes are prepared containing antiserum to the blood of likely host animals.

Probably the most widely used test. It is simple, but expensive (up to US $4 per sample) as a lot of antisera is required. It is relatively insensitive.

2 *Double gel diffusion test* (DGD)

Easy to perform, very economical (US $0.07 per sample). This gave the best results of all the methods tested in the WHO interlaboratory trial of 1986–7.

3 *Counter-current immunoelectrophoresis*
A relatively difficult test to perform but is good for the identification of small blood meals and mixed blood meals.

4 *Haemagglutination*
A difficult but very sensitive assay. The inhibition assay based on this technique is good for distinguishing between blood meals from closely related species.

5 *Latex agglutination*
Potentially this is a very simple, cheap and rapid technique, but it is currently not widely used because of problems with the stability of the antibody-coated polystyrene particles.

6 *Complement fixation*
A difficult and relatively insensitive assay.

7 *Enzyme-linked immunosorbent assay* (ELISA)
This is a simple and sensitive technique. There are good possibilities for the production of dip-stick and/or dot blot modifications which can be used directly in the field.

evidence began to accumulate. The key factor was that anopheline mosquitoes in the malaria-free areas were 400 times as likely to feed on animals other than man while in malarious areas one in eight feeds were from man (Hackett & Missiroli 1931). The importance of these man-biting rates for the transmission of malaria is explained in Box 3.2.

Why should the mosquitoes of one area bite man so regularly and mosquitoes of the same species in another area bite him so infrequently? Although it was suggested that the mosquitoes were actually changing in their relative attractions for man and domesticated animals (particularly pigs, horses and cattle) no such change could be demonstrated experimentally. It is now widely accepted that the switch from man to animals was a response to the changing proportions of human and non-human hosts available. At this time animal husbandry practices in Europe were changing so that far more animals were being kept and were available to the mosquitoes as food sources. A contributory factor was the increasing standard of living for the population which led to larger and brighter housing which was much less attractive as a resting site for the mosquito. The mosquitoes were diverted to the darker animal sheds around the house, reducing considerably the likelihood of them biting man (Harrison 1978). Also of importance was the decreasing birth rate among the population, reducing the numbers of human hosts available, a factor accentuated by the increased mechanization of agriculture which also reduced the populations in rural communities. So we can see that host choice is not only a behavioural, decision-making process on the part of the insect, but can also be strongly influenced by the relative availability of different host species.

Box 3.2 The importance of man-biting rates for the transmission of malaria. (From Burkot 1988)

The vectorial capacity (C) of a population of mosquitoes transmitting malaria can be described by the equation

$$C = m \times a^2 \times \frac{P^n}{-\ln P}$$

where m is the number of mosquitoes per person and a is the proportion of these mosquitoes that bite man (biting once to acquire the infection and transmitting it on the occasion of a subsequent meal, hence a^2). a is usually estimated using the immunological techniques described in Box 3.1.

$$\frac{P^n}{-\ln P}$$

describes the probability of the mosquito surviving through the development period of the malaria parasite.

If we take the example described in the text we can see that in those parts of Europe where malaria was being transmitted $a = \frac{1}{8}$, whereas in those areas where mosquitoes were present but no malaria was being transmitted $a = \frac{1}{400}$. If we assume that m and $\frac{P^n}{-\ln P}$

are constants then the equation tells us that humans in the malarious areas are 2500 times as likely to contract malaria as those in non-malarious areas. The difference is entirely due to the different proportion of mosquitoes biting man in the two areas.

Geographical considerations are also important in host choice because if there is no overlap in the range of insect and potential host then the host is not available as a food source. This factor can work on a variety of scales. The effect is obvious from a global viewpoint, but not so obvious when more localized issues are considered. The tsetse fly can be used as an example which has considerable economic importance. On the largest geographical scale we can see that tsetse flies are only found in Africa where they are widely distributed from the southern borders of the Sahara to Mozambique. Within this region they transmit trypanosomes causing sleeping sickness in humans and nagana in animals. The game animals of Africa have some degree of immunity to nagana, but normally it quickly kills introduced domesticated animals. This disease and the tsetse fly transmitting it have had a tremendous influence on the history of tropical Africa. The combination largely prevented the invasion from the north of armies dependent on horses. They also impeded Africa's development by limiting the use of draft animals and prevented (and indeed still prevent) the use of large tracts of land for ranching. In other words large herbivores entering the tsetse's geographical range, including horses and cattle, become available as hosts and are often chosen as food sources. But geographical overlap is also important in tsetse fly biology on a much smaller scale. Long before Bruce (1895) demonstrated that the tsetse fly transmits the trypanosomes which cause illness and death, the herdsmen of southern Africa had learned by bitter experience that the fly was deadly. This fact became obvious to them because the tsetse flies of this region are not evenly spread throughout the land, but are restricted to certain areas known as fly belts. The herdsmen learned that if they avoided these belts, and the flies that lived there, then their stock did not acquire nagana (McKelvey 1973).

Geographical effects on host choice can be seen on an even smaller scale. The treehole mosquitoes, *Aedes triseriatus* and *Ae. hendersoni* are sibling species living sympatrically in woodlands in the eastern and midwestern USA. Intuitively it might be expected that two such closely related species living in the same woodland would be feeding on similar hosts, but geographical separation again affects host choice. *Aedes triseriatus* feeds mainly at ground level and consequently feeds on ground-dwelling animals such as deer and chipmunk. In contrast, *Ae. hendersoni* feeds mainly in the canopy of these woodland trees and so feeds on tree squirrels and raccoons (Nasci 1982). Clearly geographical overlap is very important in determining host choice; indeed, host abundance and proximity may often be the ultimate arbiter of host choice. As pointed out above, mosquitoes in cattle sheds bite cattle, mosquitoes in houses bite man.

As we have just seen, availability of hosts is often a prime factor in determining host choice. To be available there must not only be geographic overlap between the insect and host but also a temporal overlap because most temporary ectoparasites only feed during a well-defined period of the day. Let us look at two contrasting ways in which temporal overlap could occur. Some hosts show such efficient defensive behaviour that temporary ectoparasites can only feed when the host is at rest. The degree of overlap between the activity period of the insect and the resting period of such a host will then be important in determining host choice. In contrast, animals living in deep burrows are unlikely to be hosts when at rest, but may be hosts if their activity period coincides with that of the insect.

Seasonal variation in the choice of host has been recognized in some blood-sucking insects, and it can have serious consequences for disease transmission to man. In North America the arbovirus, St Louis encephalitis, is transmitted during the summer months by the mosquitoes *Culex nigripalpus* and *C. tarsalis*. Both these mosquitoes show a marked seasonal change in their feeding patterns, switching from bird feeding in the winter and spring to mammal feeding in the summer when arbovirus transmission occurs (Tempelis & Washino 1967, Edman & Taylor 1968, Edman 1974).

It has often been suggested that there are genetically determined behavioural traits which give different feeding patterns in insect populations of a single species (rather than the well-recognized phenomenon of different feeding patterns seen among members of species complexes – see below). Most of the evidence for this is anecdotal, but convincing evidence has been reported for *Aedes simpsoni* and *Ae. aegypti* (Mukwaya 1977). These species showed direct evidence of zoophilic and anthropophilic populations both in the laboratory and the field. It was established that single species were involved when no crossing or hybrid sterility appeared during crossing and backcrossing experiments.

3.2 HOST CHOICE AND SPECIES COMPLEXES

For most blood-sucking insects the spectrum of host choice differs with changing place and season and given different proportions of available hosts. Selective adaptation to such changes in local circumstance is the factor driving speciation. This process is continual and the different species emerging are not always obvious from their morphological characteristics. Indeed, several blood-sucking insect 'species', including *Anopheles gambiae sensu lato* and *Simulium damnosum s.l.*, are known to be species complexes made up of reproductively isolated sibling species, having differing biological characteristics. These differences commonly include host choice. Let us consider the *An. gambiae* complex in more detail.

Anopheles gambiae s.l. is an enormously important malaria vector throughout Africa south of the Sahara. A major filariasis vector, it probably also transmits some arboviral diseases; as such, it has received considerable attention from entomologists. Before the realization that *An. gambiae s.l.* was a complex of species, the conflicting nature of the biological data gathered was a source of puzzlement and controversy. It is now known that the *An. gambiae* complex is made up of six very similar sibling species. Three of these breed in fresh water: *An. gambiae sensu stricto* which is distributed throughout sub-Sahelian Africa, particularly the more humid regions, *An. arabiensis* which is also found widely distributed in Africa but shows some preference for rather drier areas, and *An. quadriannulatus* which shows a more restricted range being found only in Ethiopia, Zanzibar and parts of southern Africa. A fourth species, *An. bwambae*, is only known from mineral springs in the Semliki forest of Uganda; it is a malaria vector, but because of its narrow geographical range is of minor importance. Finally there are two species which breed in brackish water: *An. melas* which is found along the West African coast, and its East African equivalent, *An. merus*. All three freshwater species and *An. bwambae* are morphologically indistinguishable and require cytotaxonomic or biochemical methods for identification. The salt-water species can be identified from the careful use of morphological criteria. Because of their differing biological characteristics, the six sibling species have very different vectorial capacities. The principal vector of human malaria and one of the two major vectors of filariasis in the complex is *An. gambiae s.s.*, which is primarily an endophilic species biting man. In many parts of this mosquito's range man is the major available host and, in these circumstances, blood meal identifications often reveal that virtually 100% of the meals of this species are of human origin (Davidson & Draper 1953). Interestingly, when cattle or other animals are housed next to, or in, the dwelling place, the proportion of feeds from man can fall to 50% or less (White & Rosen 1973, see White 1974), which brings to mind the European malaria story told above. In complete contrast, *Anopheles quadriannulatus* is a strongly zoophilic species which is not a vector of human disease. Over most of its range it is exophilic, but in the highlands of Ethiopia, probably as an adaptation to the cold nights, it is endophilic. The endophilic form may rarely feed on man but does not transmit human disease. *Anopheles arabiensis* falls between the two: ecologically and behaviourally it is an extremely plastic species; across its geographic range, exophilic and endophilic, anthropophilic and zoophilic forms are to be found. Generally, *An. arabiensis* is exophilic and zoophilic, but in the absence of other hosts it can live quite happily on man and may rest inside his dwellings. It is a malaria vector, but is less efficient than *An. gambiae s.s.* This can be seen in the sporozoite rates of *An. arabiensis* (the proportion of the mosquito population carrying sporozoites) which are commonly about 1/15th of those of *An. gambiae s.s.* (White *et al.* 1972). The higher sporozoite rates in *An. gambiae s.s.* are not a reflection of different susceptibilities to malaria parasites – the two species are equally susceptible – but arise because *An. gambiae s.s.* lives longer and feeds more often on man (White 1974). Despite being an important vector of malaria, *An. arabiensis* is not a major filariasis vector as it tends to occur seasonally which precludes it from maintaining the high transmission rates required for the establishment of endemic foci of this disease (White 1974). The two salt-water species feed primarily on non-human hosts but in the absence of these they can survive quite happily on man. Of the two species, *An. melas* feeds more readily and regularly on man and is therefore the more important vector of both malaria and filariasis, particularly in coastal areas where alternative hosts to man are scarce. So, even with six very closely related insects innate host preference can vary greatly, and, as mentioned previously, despite these preferences host choice is ultimately dependent upon local circumstance and in particular the proportions of available hosts.

CHAPTER FOUR

Location of the host

The difficulty hungry blood-sucking insects have in locating their next blood meal depends upon the closeness of their association with the host. At one extreme we have the permanent ectoparasites which are in the happy position of having food continually 'on tap'. Only by accident will they find themselves more than a few millimetres from the skin of the host and the blood that it holds. At the other extreme are those temporary ectoparasites, like blackflies and tabanids, which do not remain permanently in the vicinity of the host. When these insects are hungry their first problem is to locate the host, often a difficult and complex behavioural task. These differences in life-style are reflected in the number of antennal receptors which different types of blood-sucking insect possess (Chapman 1982); not surprisingly the more independent, host-seeking insects possess the most receptors. Thus, lice have only 10–20 antennal receptors and fleas about 50, but the stablefly, which spends most of its time at some distance from the host, has nearly 5000 antennal receptors. Considering two bugs, we see that *Cimex lectularius* has only 56 antennal receptors compared to 2900 on the more adventurous *Triatoma infestans*. Most of the detailed information on host finding is restricted to a small number of temporary ectoparasites and the discussion which follows will concentrate largely on these.

4.1 THE BEHAVIOURAL FRAMEWORK OF HOST LOCATION

The location of the host is an integrated, but flexible, behavioural package which gathers momentum as the host is tracked down. The behaviour patterns involved are not arranged in a strict hierarchy, that is they do not occur in a strict sequence with behaviour one always being followed by behaviour two, followed by three, etc. This versatility allows a flexible response on the part of the insect to the differing circumstances in which it will encounter hosts: however, it is probable that insects

25

mostly encounter host-derived stimuli in a particular sequence. The insect often makes use of this predictability by permitting the current behavioural pattern in the host location sequence to lower the response threshold for subsequent host-related stimuli. For example, an insect which would not normally respond to a certain visual stimulus may respond strongly if it has just been exposed to an increase in carbon dioxide levels. In this way a behavioural momentum is built up during host finding. This behavioural momentum is further enhanced by the wide range of increasingly strong host stimuli which the insect encounters as host location proceeds (Sutcliffe 1987).

From observations of blood-sucking insects both in the laboratory and the field, and from the clear evidence on the discrimination and selectivity which they can show, we can predict that a variety of host signals are used in host finding. Information on what signals are used and the processes involved is still fragmentary. In general, visual and olfactory stimuli, aided by anemotactic and optomotor responses, are the most important signals when the insect is still at some distance from the host. Nearer to the host different stimuli become important, particularly heat.

For the purposes of explanation (in reality the whole process is a continuum with one behaviour pattern dovetailing into the next) the various behaviour patterns involved in host location can be conveniently divided into three phases (Sutcliffe 1987):

(a) *Appetitive searching* – Driven by hunger the insect indulges in non-oriented behaviour likely to bring it into contact with stimuli derived from a potential host. (This usually occurs at specific times of the day regulated by the insect's internally programmed activity cycle.)

(b) *Activation and orientation* – Upon receipt of host stimuli (activation) the insect switches from behaviour patterns driven from within (appetitive searching) to oriented host location behaviour. This is driven by host stimuli which are of increasing variety and strength as the insect and host come closer together.

(c) *Attraction* – The final phase, in which host stimuli are used to bring the insect into the host's immediate vicinity, and in which the decision of whether or not to contact the potential host is made.

This categorization of host location has another benefit. It clearly indicates that host location is a series of behavioural events, rather than just one. This may help explain the conflicting reports on insect behavioural responses to certain host stimuli, and at the same time give us a clear warning of the need for great care in the interpretation of results from behavioural experiments. The effect of reflected ultraviolet (UV) light on the tsetse fly can be used to illustrate this point. Reflected

UV light is a deterrent to the fly during the orientation phase of host finding, but will increase the number of flies landing during the attraction phase. It would be easy to confuse the two effects in a poorly designed experiment.

Throughout the rest of the chapter, to avoid confusion over which phase of host location is being discussed I will only use the words appetitive, searching, activation, orientation and attraction to refer to the different phases of host finding outlined above.

4.2   APPETITIVE SEARCHING

Blood-sucking insects usually have a delay period between their emergence from the egg, or previous developmental stage, and their first blood meal. The reasons for this are not clear especially as other activities, such as mating and dispersal, commonly occur during this period. One reason for the delay may be that, after adult emergence, the female reproductive system of many blood-sucking insects undergoes a maturation period lasting several days. Blood meals taken before maturation has passed a certain point do not add to the reproductive output of the insect, but visiting the host to get them will greatly increase the insect's chances of being damaged or killed. For example, in the cat flea, *Ctenocephalides felis felis*, the reproductive system takes about four days from emergence to mature and a further two days after the blood meal to produce eggs. It will be selectively advantageous for the insect to remain off the host (and unfed) for the first four days after emergence, as this will minimize the danger from the host's grooming activity without affecting the insect's reproductive success (Osbrink & Rust 1985). Another reason for the delay in blood feeding may lie in the progressive thickening and hardening of the cuticle which takes place during the teneral period. This thickening may mean, as it does in mosquitoes (Bates 1949), that for approximately the first 24 h after emergence the mouthparts are insufficiently hard to permit efficient skin penetration.

As the post-emergence delay period progresses, or as the time since the last blood meal lengthens, the insect becomes increasingly hungry for blood and more likely to begin host seeking. Bouts of activity will be mainly restricted to particular times of the day (Dethier 1976) because activity in blood-sucking insects, as in other animals, occurs in set patterns during each 24 h (circadian) cycle. The timing of these activity bouts is internally programmed and the time of day at which they occur is characteristic for each species (Fig. 4.1). The patterns are not inviolable; species commonly show variations in periodicity when collected from different habitats or at different times of the year. As hunger increases, these periods of activity intensify (Fig. 4.2) and also occupy longer

27

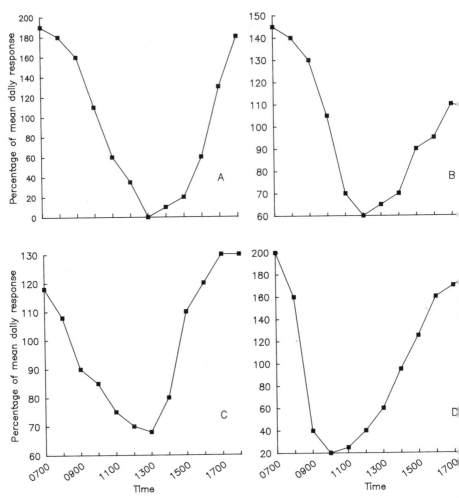

Figure 4.1 Insects will commonly show just a single peak of activity in a day (Lewis & Taylor 1965), but there are plenty of exceptions. The tsetse fly, for instance, shows two endogenously controlled peaks in activity: one at dawn and one at dusk (Brady 1975). The key behaviours in the life of the fly most frequently take place at these times. Some of these are shown in the graph where they are expressed as percentages of mean daily response: (A) field biting activity, (B) spontaneous flight in actographs, (C) optokinetic responsiveness, (D) olfactory responsiveness. (Redrawn from Brady 1975.)

periods of time (Brady 1972). Much of the activity seen in the hungry insect is appetitive behaviour, that is behaviour which maximizes the chances of the insect contacting a signal derived from a host animal. The simplest appetitive behaviour pattern is to sit still and wait for a host stimulus to arrive. This may sound a very chancy business, but providing the insect chooses the resting site carefully, it could be a sound strategy, combining maximum energy conservation with a strong chance of encountering a host. This strategy is likely to be employed to a greater or lesser degree by virtually all blood-sucking insects, but it is likely to be used more by forest-dwelling species than by those living in more open terrains (Sutcliffe 1986). Although there is little direct evidence, insects almost certainly also engage in active appetitive behaviour. Non-activating, non-orientating, non-attracting sampling devices, such as suction traps sunk below ground or electric nets, will often catch large numbers of hungry female insects which are impregnated but not ready for egg laying. As these insects are often trapped at peak biting times it is quite possible that they are engaged in non-oriented appetitive activity.

We know virtually nothing of the search patterns used by insects undertaking active appetitive behaviour. Based on the reasonable assumption that the insect will attempt to optimize its chances of encountering a host while minimizing its energy expenditure, theoretical work has been carried out to determine such patterns for flying insects.

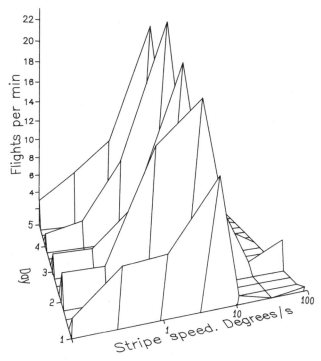

Figure 4.2 Hunger will lead to an increase in the overall activity of the insect. This can be demonstrated by monitoring various behavioural responses. Here the changing kinetic responsiveness of male tsetse flies to a moving 'target' (stripe speed) are recorded over a five-day period. As the fly becomes hungrier it shows increased flight activity in response to a given stimulus. It is also interesting to note that the pattern of response stays the same throughout the period, showing flies are 'tuned' to particular patterns of movement. (Redrawn from Brady 1972.)

In a wind blowing consistently from one direction the optimum strategy is for the insect to fly across the wind, allowing the maximum number of air streams to be monitored for a particular energy expenditure (Linsenmair 1973). In the field, winds often veer rapidly from one direction to another. If they veer by more than 30° from the mean, then downwind flight becomes the most energy efficient method of sampling the maximum number of airstreams (Sabelis & Schippers 1984). Whether these theoretical conclusions relate to the appetitive behaviours displayed by blood-sucking insects in the field remains to be seen.

4.3   ACTIVATION AND ORIENTATION

Activation occurs when the insect comes into contact with a suitable signal from a potential host animal. Such a signal may simply change the behavioural awareness of the insect without causing any observable activity on the insect's part. In such a case, orienting behaviour would be released by a subsequent stimulus, for which the insect has now been primed. Alternatively, the stimulus may directly cause the insect to switch over from endogenously driven appetitive searching to oriented host location behaviour. The insect then uses the information contained in host-derived signals to orientate towards the host. Despite entomologists' convictions that a range of stimuli are likely to be used by insects in activation and orientation, conclusive and unambiguous evidence is available for the involvement of relatively few.

Olfaction

There seems to be an olfactory component to host finding in virtually all blood-sucking insects. It might be assumed that it is of most importance to forest-dwelling insects, because direct visual contact with the host is most restricted for them. But the evidence from tsetse flies suggests that the species showing the clearest response to odours are those (*Glossina morsitans*, *G. pallidipes* and *G. longipennis*) living in fairly open situations. The forest-dwelling *palpalis* group show little response at a distance. This is probably explained by the fate of the odour plume in the two situations. The plume is likely to be a much more continuous, relatively linear guide for insects in an open, more breezy situation than in a forest (Elkinton *et al.* 1987). Odour is probably most important in orientation for night-feeding forms living in open situations. Olfactory stimuli implicated in host location to date include carbon dioxide, lactic acid, acetone, butanone, octenol and phenolic components of urine.

It is generally accepted that carbon dioxide is involved in both the activation and orientation of virtually all blood-sucking insects. It is

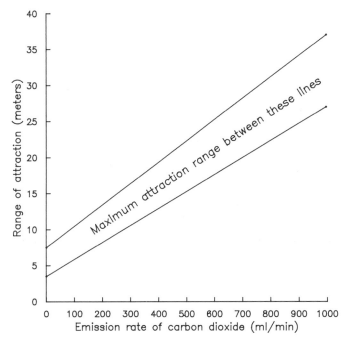

Figure 4.3 The effect of varying the emission rate of carbon dioxide on its drawing power for mosquitoes has been measured by various authors under field conditions. The lower edge of the 'attraction range' shown in the figure is the furthest trapping point at which an effect of the carbon dioxide was noted. The upper edge of the zone is the nearest trap at which no effect of the carbon dioxide was seen. (Redrawn from Gillies 1980.)

normally present in the atmosphere at between about 0.03 and 0.05%, occasionally rising to 0.1% in dense vegetation at night. It is secreted by the skin of hosts, but the major emissions occur in exhaled breath which, in humans, contains about 4.5% carbon dioxide. So a plume produced from a solitary human would remain above background until the exhaled breath had been diluted by a factor of about 100 (Gillies 1980). Predicting how dilution of the plume will occur is no easy task. Packets of relatively undiluted carbon dioxide are likely to travel for quite some distance downwind and the distance from the host at which activation and orientation are still likely to occur will vary according to local meteorological conditions. Perhaps the easiest way of dealing with the problem is to look directly at the effects of carbon dioxide in the field (Fig. 4.3). The range over which a host animal can activate and orientate an insect, on the basis of odour alone, has been calculated for some blood-sucking insects. For the tsetse flies *Glossina morsitans* and *G. pallidipes* the maximum range for an ox is estimated to be about 90 m (Vale 1977). Oxen draw the tabanid *Philoliche zonata* at 80 m, but this reduces to only

31

30 m for species of *Tabanus* (Phelps & Vale 1976). Calves draw a number of mosquito species at distances between 15 and 80 m (Gillies & Wilkes 1969, 1970, 1972).

Carbon dioxide on its own has been shown to cause activation in several blood-sucking insects (Omer & Gillies 1971, Bursell 1984, 1987, Warnes & Finlayson 1985) and its involvement has been suggested from circumstantial evidence in others (Nelson 1965, Compton-Knox & Hayes 1972, Roberts 1972). In mosquitoes it is the change in concentration of carbon dioxide rather than the level of carbon dioxide encountered which is the important factor eliciting behavioural responses (Wright & Kellogg 1962). This is probably true for all blood-sucking insects. To

Box 4.1 Hormonally controlled sensitivity changes in mosquito lactic acid receptors

It is generally held that behaviour patterns are set in train after a decision-making process in the central nervous system. Interestingly, it has been argued that some behavioural responses in mosquitoes are not dependent on the decision-making process in the central nervous system. In mosquitoes the lactic acid sensitive receptors are in the antennal grooved pegs (Davis & Sokolove 1976) (Fig. 4.4). The receptors are of two types; one is excited by lactic acid, the other is inhibited by it. The responsiveness of the lactic acid excitable receptor is variable, being about 10 times less sensitive in the fed, non-actively host-seeking mosquito compared to a hungry one. This change in responsiveness is induced by a hormone released from the insect's fat body during oogenesis (Klowden & Lea 1979, Davis 1984, Klowden et al. 1987). It has been argued that the variable responsiveness of the lactic acid receptors is sufficient to account for changes in the behaviour patterns of mosquitoes without the need to invoke central nervous system involvement (Davis et al. 1987). The theory is that in the unfed mosquito the summed output of the excited and inhibited receptors in the presence of lactic acid is positive (i.e. above the spontaneous level produced by the 'resting' receptors) and drives host-seeking behaviour. Conversely, in the fed fly, because of the lowered responsiveness of the lactic acid excited receptor (brought about by the fat body hormone), the summed output of the receptors in the presence of lactic acid is negative, leading to inhibition of host-seeking behaviour. This idea is illustrated in the graph, which shows the stimulus intensity–response functions for the lactic-acid-excited (LA-exc) and lactic-acid-inhibited (LA-inh) receptors of host-seeking (a) and non-host-seeking (b) female *Aedes aegypti* mosquitoes. The dashed lines represent the summed response for the two receptors (net). The central horizontal line ($\Delta F = 0$) represents the spontaneous activity level of the receptors. In the hungry state the summed response of the receptors to the presence of lactic acid is above the spontaneous activity level, while in the fed state it is below. So this system has a binary switching characteristic capable of activating or de-activating a particular behaviour pattern without the need to invoke a central nervous system decision-taking mechanism. (Redrawn from Davis et al. 1987.)

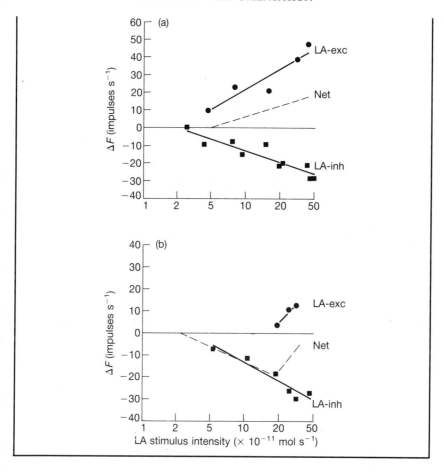

facilitate this, mosquitoes are sensitive to very small changes in carbon dioxide levels; changes as small as 0.05% will elicit behavioural responses in a wind tunnel (Mayer & James 1969) and electrophysiological recordings from the carbon dioxide receptors on mosquito palps show responses to changes as small as $+0.01\%$ (Kellogg 1970). The antennal receptors of *Stomoxys calcitrans* are slightly less sensitive, responding to an increase in carbon dioxide levels of 0.023%, but starvation lowers the response threshold of these receptors (Warnes & Finlayson 1986). It has been suggested that this change in sensitivity is hormonally controlled as has been shown for mosquito lactic acid receptors (Klowden & Lea 1979, Davis 1984, Klowden *et al.* 1987) (see Box 4.1, and Figure 4.4). It has also been reported that the activation response to carbon dioxide in *S. calcitrans* rapidly habituates. The stimulus had to be increased from 1.04 to 2.04% carbon dioxide to induce the same activation response in habituated flies (Warnes & Finlayson 1985). It would be interesting to

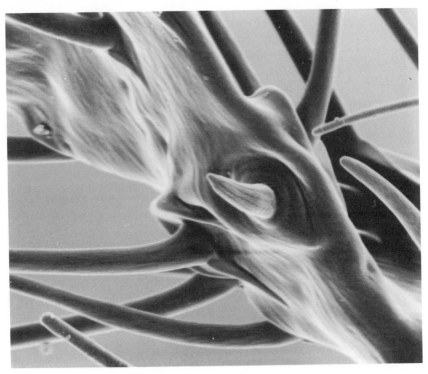

Figure 4.4   Antennal grooved pegs of the mosquito, *Aedes aegypti*. Lactic acid excites some of these receptors but inhibits others (Davis & Sokolove 1976).

know if habituation occurs when the odour source is pulsed because it has been shown in mosquitoes that carbon dioxide only causes continued upwind flight when received at continually varying concentrations (Omer & Gillies 1971). It is also worth noting that in moths the responses to pulsed pheromone are considerably different to the responses obtained after exposure to an homogeneous pheromone cloud (Baker 1986). Under field conditions odours are likely to be received in the form of pulses. Mammals release breath in pulses, and turbulence will ensure the odour plume is continually mixed and diluted in an irregular fashion creating further pulses. If pulsing is important it could explain those reports, produced on the basis of wind tunnel experiments using an uninterrupted stimulus, where carbon dioxide has been dismissed as an activating and orientating stimulus for mosquitoes (Kellogg & Wright 1962, Daykin *et al.* 1965, Khan & Maibach 1966).

   As well as acting as an activating agent, carbon dioxide also acts as a stimulus, orienting blood-sucking insects to hosts (Omer & Gillies 1971, Warnes & Finlayson 1985). This can be neatly shown by filtering carbon dioxide from the breath of a host (Laarman 1958) or by using two sets of inanimate traps, only one of which is baited with carbon dioxide (Fallis & Raybould 1975). In both cases fewer insects arrive at the source minus carbon dioxide.

In the natural situation, carbon dioxide is only one of several host stimuli which the insect receives. Carbon dioxide can act in concert with other stimuli giving a response which is different from that of either stimulus given alone (Gillies 1980). There is a spectrum of responses which can be seen to these dual stimuli, ranging from synergism (where the two stimuli give an overall reaction which is greater than the sum of the two stimuli given separately) to an interaction where one stimulus primes the insect to respond to the second which, if given alone, has no effect (Laarman 1958, Bos & Laarman 1975, Bar-Zeev et al. 1977). For example, lactic acid is an activating and orientating stimulus for mosquitoes, but only if carbon dioxide is also present in the air stream (Smith et al. 1970, Price et al. 1979). Other components of host breath can also act in concert with carbon dioxide. Octenol, which is a component of ox breath, enhances the catches of tsetse flies in traps, especially when carbon dioxide is also released (Hall et al. 1984, Vale & Hall 1985a, b). This enhancement is possibly explained by the fact that octenol, which clearly can be used by tsetse for orientation, does not cause activation of tsetse flies. Acetone, another constituent of ox breath, further enhances the effectiveness of these traps.

An interesting aspect of the interaction of host stimuli is that even closely related species may respond differently to combinations of stimuli. Looking at the levels of efficiency of a synthetic odour consisting of 1.2 l of carbon dioxide per hour, 5 mg of acetone per hour and 0.05 mg of octenol per hour for two tsetse species, we find that it is almost as effective in drawing *Glossina morsitans morsitans* to a field trap as natural ox odour, but is only half as effective as the ox for *G. pallidipes*. Similarly, if we look at the drawing power of these bait components for muscoid biting flies, we see that they are lured strongly by carbon dioxide alone, or carbon dioxide–acetone mixtures, but that octenol has little, if any, effect on catches when released at the levels given above (Vale & Hall 1985b). Perhaps the clearest example of the species-specific nature of the response to odour mixtures is seen in the drawing power of the bovine urine components, 4–methylphenol and 3-*n*-propylphenol, for different tsetse species. Used singly, 3-*n*-propylphenol drew roughly equal numbers of *G. pallidipes* and *G. m. morsitans*. When used in combination, trap catches increased by up to 400% for *G. pallidipes* (Vale et al. 1988), in contrast to trap catches for *G. m. morsitans* which decreased.

Why have insects developed distinctive responses to mixtures of different host-derived stimuli? Multiple stimuli are likely to be a far surer guide to the presence of a host than one stimulus received alone. This is because while one stimulus alone may be of non-host origin, this is very unlikely for combinations of stimuli, especially when they are received in particular proportions. In other words, responding to stimuli received in combination is likely to maximize the chances of host encounter while minimizing energy consumption. Also, insects show preferences for particular hosts (see Section 3.1). Responding to particular combinations of host signals may permit a degree of selection for a particular host which is still some distance away. Carbon dioxide is released by all potential hosts and, although it may be good at alerting the insect to the possibility of a meal, it does not allow the insect to discriminate between hosts. Body odours, however, may be characteristic for particular groups of host animal or even for particular species. An extreme example of selection at a distance is seen in the blackfly, *Simulium euryadminiculum*, which is drawn to a unique, non-polar product of the uropygial gland of a particular water fowl, the loon (Fallis & Smith 1964, Lowther & Wood 1964). This is probably an exceptional case, based as it is on a very specific chemical signal. In most instances, it is more likely that the insect discriminates between hosts on the basis of the varying proportions of a number of less specific stimuli. It seems that the response to combinations of stimuli can also be used to detect the presence of hosts the insect wishes specifically to avoid. This is seen in the reduced catches of tsetse flies using stationary bait animals if man remains in the vicinity of the bait (Vale 1974a). In this case man can be recognized by tsetse flies not only by his characteristic odour, but also visually.

How are synergistic responses to host odours generated? Two possibilities present themselves. First, each odour may be treated separately by the peripheral nervous system with integration and amplification of the signals occurring in the central nervous system. Or second, the response of peripheral receptors to a substance may be significantly higher in the presence of the second chemical. Electrophysiological recordings from mosquito receptors strongly suggest that the former theory is correct (Mayer & James 1970, Davis & Sokolove 1975) (see Box 4.1).

Although it is clear that blood-sucking insects use olfactory clues to trace the host animal, it is still uncertain how they achieve this feat. The odour plume is not a smooth homogeneous cone of molecules fanning out from the source. In field conditions the plume is a series of loosely gathered lamellae and filaments of odour which are all continually mixed and dispersed by the vagaries of the wind (Wright 1958, Murlis 1986). As a consequence, the plume is a heterogeneous patchwork of odours of differing concentrations. In addition, the wind often veers from its mean direction with the effect that the odour plume meanders as it moves away from the source, possibly breaking into discrete packets. Wind direction within each meander or packet of odour remains remarkably consistent over considerable distances (David *et al.* 1982). Tracing the source of such a plume is a major task.

Several hypotheses have been proposed to account for the ability of insects to accurately find the source of an airborne odour trail. Early theories tended to stress the importance of features of the odour plume, such as the increasing concentration of odour as the source is approached (Sutton 1947, 1953), the pulse frequency of the odour in the plume (Wright 1958), or the boundaries of the plume where odoriferous air meets odour-free air (Farkas & Shorey 1972). Currently, the most widely held theories suggest that the concentration of the odour, or other features of the plume, are largely irrelevant and that, above a threshold level, odour is merely used as a switch to release particular behavioural responses (Kennedy 1983, Payne *et al.* 1986). The best-worked examples of flying insects following an odour trail are male moths homing in on a pheromone source. Within an odour plume these insects continually fly in an internally programmed, upwind, zig-zagging pattern. This mainly serves to hold them within the plume and to move them closer to the odour source (Baker 1986). To be able to fly upwind (positive anemotaxis) the insect must know in which direction the wind is blowing. The flying insect is of low inertia and is blown along by the wind such that, to all intents and purposes, it seems to be in still air except for the movements it generates in flight. For this reason the flying insect is unable to sense wind direction by using air movements as a guide (David 1986). Instead it determines wind direction by observing the apparent movement of fixed objects in its environment (this is called optomotor anemotaxis and is explained further in Fig. 4.5) (Kennedy 1940, David 1986). When contact with the plume is lost, narrow, zig-zagging, upwind flight at a small angle to wind direction gives way to a much wider, casting flight pattern which is at right angles to the wind. This change in the angle of flight direction to wind direction, combined with the longer period between turns, optimizes the chances of the insect regaining contact with the plume (Linsenmair 1973, Kennedy 1983, Sabelis & Schippers 1984). On regaining contact with the odour plume narrow, zig-zagging, upwind flight is resumed. There is so little evidence available for blood-sucking insect species that it is not yet possible to say if this theory can account for their olfactory-based, host-finding behaviour. But one insect for which information is rapidly accumulating is the tsetse fly.

It seems likely that tsetse flies can use upwind, optomotor anemotaxis as an efficient method for locating hosts (Vale 1974a, Colvin *et al.* 1989). Understanding of the methods they employ would allow improvements in the design of traps which show promise as a principal tsetse control mechanism (Challier & Laveissiere 1973, Vale *et al.* 1986). It is known that tsetse flies are rapid fliers (groundspeeds of 6.5 m s^{-1} have been recorded; Gibson & Brady 1988), that they live in habitats characterized by low wind speeds, and that they are capable of locating hosts using upwind anemotaxis in air speeds as low as 0.1 m s^{-1} (Vale 1983). It has been suggested that, in such conditions, flying tsetse would have

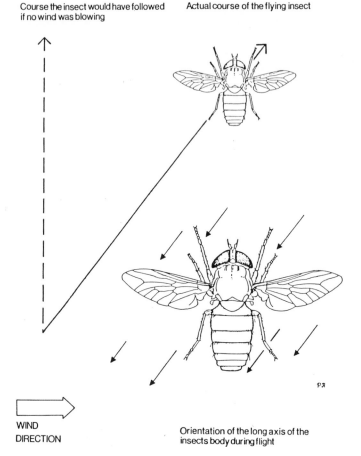

Course the insect would have followed
if no wind was blowing

Actual course of the flying insect

WIND
DIRECTION

Orientation of the long axis of the
insects body during flight

Figure 4.5 Fixed objects in the field of view of an insect flying in a straight line in still air will apparently move backwards and parallel to the insect's flight plane. If wind is blowing at an angle to the insect's flight plane the insect will be blown off course. The fixed objects in the insect's field of view will now apparently move at an angle to a plane through the long axis of the body (short arrows). Flying insects can use this information to determine wind direction (Kennedy 1940).

difficulty in using optomotor anemotaxis, and an alternative method for determining wind direction has been proposed. This suggests that tsetse flies reach the host in a series of short flights. According to the theory, the fly uses mechanoreceptors to determine wind direction while it is standing at each stopping point before setting off once more on a short upwind flight (Bursell 1984). This mechanism of locating an odour source in a series of short flights is reportedly used by both the onion fly, *Hylemya antigua*, and the cabbage root fly, *Delia brassica* (Hawkes & Coaker 1979, Dindonis & Miller 1980). In support of this theory it has been noted that some of the tsetse which fly upwind in an odour plume turn into the wind before take-off, suggesting that the standing fly can indeed determine wind direction (Bursell 1987). Further supporting evidence comes from laboratory work suggesting that tsetse flies engage in short bursts of activity separated by longer periods of rest (Brady 1970). Each individual flight has been estimated to be about 15 – 60 m (Bursell & Taylor 1980, Gibson & Brady 1988). However, these short flights have not yet been recorded in the field. Against this theory is evidence gathered in field experiments in which the behaviour of tsetse flies encountering odour plumes was recorded on video (Gibson & Brady 1985, 1988). These recordings give strong evidence that tsetse flies can determine wind direction in flight, in conditions of low wind speed (<0.5 m s^{-1}), which suggests that intermittent landing to determine wind direction is unnecessary. These recordings also showed no evidence of flies landing on leaving the odour plume. Instead flies entering and leaving a plume engaged in turning behaviour most likely, in the former case, to keep them in the plume and to bring them nearer to the odour source and, in the latter case, to bring them back to the plume. This is the sort of behaviour which would be predicted if flying tsetse engaged in host location were using upwind anemotaxis in the presence of an odour plume. No evidence of the sort of zig-zagging flight used by moths (Kennedy 1983) has been found in tsetse flies and therefore the means by which they maintain contact with the odour plume, while assessing wind direction, is still uncertain. The video recordings do show that flying tsetse travel almost exclusively in wide turns rather than straight lines and in some unknown way these may help the fly to locate the host. At the low wind speeds characteristic of the environments in which tsetse are found, the trajectory of the plume is highly non-linear (Elkinton *et al.* 1987), and so is not a particularly good guide to the host. Perhaps under such conditions these turns shown by tsetse flies can fulfil the function of zig-zagging by moths, but an explanation of their use awaits further investigation.

It is interesting to note that many flying temporary ectoparasites, including tsetse, tabanids, mosquitoes and simuliids, all approach the host while flying close to the ground, or slightly above the top of the predominant vegetation. What advantage this is to the insect is unknown. Flying nearer to the ground increases the perceived angular rate of change of fixed objects on the ground. The insect is also able to resolve finer detail in the visual field, and both these visual inputs may be of importance. It has been pointed out that shear forces generated by the frictional drag of the moving air with the ground increase rapidly near to the ground. It has been suggested that the low approach to the host, used by many blood-sucking insects, may allow the insect to use these shear forces to determine wind direction, but no experimental evidence has been produced to test this hypothesis. The explanation of why an insect chooses to approach the host at a particular height is probably complex. The insect probably adopts a compromise between, on the one hand, the height giving the most favourable wind speed for host odour detection, optomotor anemotaxis and upwind flight and, on the other hand, a height permitting a stealthy approach to the host and one giving maximum protection from any predators.

Finally there have been reports that blood-feeding insects may release odours which attract other insects to the host (Alekseev *et al.* 1977, Schlein *et al.* 1984, Ahmadi & McClelland 1985). It is hard to see any selective advantage to the female in expending energy to help other females find a host. Drawing more insects to the host is likely to reduce the feeding success of the original feeder through density-dependent effects (see Section 7.5). Even if she escapes with a full meal, she may well have impaired her offspring's chances of prosperity due to the increased competition for larval resources which will take place. These arguments do not exclude the possibility that insects are drawn by substances being released by successfully feeding insects, it merely suggests that such substances are not being deliberately released by the insect for the purposes of bringing others to the host. More work is needed on this interesting subject.

Vision

Vision is also important in the activation and orientation of many blood-sucking insects. Not surprisingly, it is most widely used by diurnal insects which live in open habitats. This can be seen in blackflies where savannah and open country species, unlike their forest-dwelling relations, are drawn strongly to silhouette traps even in the absence of carbon dioxide or other odour sources (Fredeen 1961, Peschken & Thorsteinson 1965, Thompson 1976). It is likely that in most instances vision is not used alone in the search for the host, but is integrated with information from the other senses. A common pattern of events is that the insect is initially activated by host odour, and then uses smell to track the host from a distance (to give an impression of scale, let us say from about 100 m – see above). As it closes on the host, visual contact is made and visual information is then used in the final stages of orientation. This explains results such as the five- to seven-fold increase in efficiency of odour-baited traps for tsetse flies over the same traps relying on visual appeal alone (Politzar & Merot 1984).

At what distance does visual orientation become important? Clearly this will vary with the visual acuity of the insect involved and the size and visibility of the object concerned. The more contrast a bait has with the background, and the larger and more mobile it is, the more visible it will be over greater distances. Estimates for the range of visible effectiveness are available for some insect–object combinations. For a range of mosquito species visual orientation to stationary, unpainted, thin plywood traps, 2.4 m long by 1.5 m high, begins between about 5 and 20 m (Bidlingmayer & Hem 1980). Tsetse flies respond to a moving screen, 1.3 m long by 1.0 m high, at distances up to about 50 m (Chapman 1961) and to a herd of moving cattle at up to 183 m (Napier Bax 1937).

The main visual organs in insects are the compound eyes. They provide the sensory input used in the discrimination of pattern and form, movement, light intensity, contrast and colour. The initial detection of an object depends on differences in colour contrast, relative brightness (intensity contrast) or motion between the object and the background. Once detected, it is possible that the shape of the object may be of importance in deciding whether it is worth pursuing. However, shape discrimination at a distance seems to be poor – for example, tsetse quite happily pursue Landrovers or a variety of other inanimate objects dragged through the bush. The relative importance of each visual capacity is likely to vary with the habits of the insect. For example, utilization of colour information is unlikely to be well developed in insects active at night, but these insects may be particularly sensitive to intensity contrast. In savannah-dwelling, day-biting flies, utilization of colour information and the ability to detect movement may be particularly highly developed. In addition to their compound eyes, insects possess other light receptors called ocelli. There is still speculation about their function; the most likely suggestion is that they are rapidly responding horizon detectors which play an important role in maintaining a level attitude in flight (Wilson 1978, Stange 1981).

Although insects have eyes and are sensitive to much of the light which we can see, it is important to appreciate that they do not see the world as we do. Behavioural work shows us that blood-sucking insects utilize colour information (Table 4.1) (Bradbury & Bennett 1974a, Allan & Stoffolano 1986b, Green 1989). It also shows us that insects are sensitive to UV light which is invisible to us and, conversely, that most insects are probably insensitive to parts of the spectrum, particularly the red end above 650 nm, to which we are sensitive. In other words, the coloured patterning of objects seen by insects is not necessarily that seen by humans. Electroretinogram responses in the compound eyes of those blood-sucking insects studied show peaks in two or more areas of the spectrum (Table 4.1). In common with most other insects, blood feeders show peaks of sensitivity in the near UV at about 355 nm and in the blue/green part of the spectrum between 450 and 550 nm. In addition, in the higher Diptera, another shoulder of responsiveness appears in the red/orange part of the spectrum around 620 nm. But crude sensitivity measured in electroretinograms is not always a good guide to

41

Table 4.1 Different blood-sucking insects respond in dissimilar ways to spectral information, but as we can see in this table, a range of flies are generally attracted to blue/black targets while being repelled by yellow ones.

	Spectral sensitivity				Most attraction	Least attraction or repulsion
	UV	Blue	Green	Orange/red		
mosquitoes	+	+	+		blue, black, red	white, yellow
blackflies	+	+	+		blue, black, red	white, yellow, UV
tabanids	+ (350 nm)	+ (477 nm)	+ (520 nm)		dark colours, blue, red, black some species only to white	green, silver, yellow white for most species
stablefly	+ (360 nm)	+ (450–550 nm)	+	+ (625 nm)	low-intensity colours, blue, UV	high-intensity colours
hornfly	+ (360 nm)	+ (460 nm)			UV	
tsetse flies	+ (350–365 nm)	+ (450–550 nm)	+	+ (600–625 nm)	UV, blue, black, white	yellow, green

behavioural responses. Let us look at one example of the complexity of behavioural responses to a visual signal. Tsetse flies are drawn to UV light sources (Green & Cosens 1983), but are repelled by UV-reflecting surfaces as shown by the reduced efficiency of traps covered with them (Green 1989). So whether the light is presented as a point source or a surface can be important. Things are even more complicated because the response to the same stimulus can vary under different behavioural conditions. So while UV-reflecting surfaces are repellent during orientation of the tsetse fly to a trap, the same surfaces promote landing if the fly arrives in the trap's immediate vicinity (Green 1989). Underpinning all of these differing behavioural responses, but giving no clue to the complexity of their nature, would be a strong electroretinogram reaction to UV light.

Theoretically, there are two different types of visual information available to blood-sucking insects. The first set of information comes from the intensity contrast between the target and the background. If the insect is not using colour information, the maximum contrast will be between black and white or between grey and either black or white depending on the shade of grey used. The ease with which the target–background combination is seen by the insect also depends on the intensity ('brightness') of the objects. The second set of information comes from colour. Sensitivity of the insect to colour can be shown in experiments demonstrating either an increased or decreased response to a colour compared to an achromatic series (the spectrum of greys running from black to white) of equal intensity and contrast with the background. In a similar way to the intensity contrast information discussed above, the information to be gained from colour depends on both its intensity and contrast with the background. The colour combination giving maximum contrast will depend on the visual system of the insect, which may be 'tuned' to detect red/green divisions or blue/green divisions or possibly both as in humans. In the real world the two sets of information (from intensity contrast and colour) are effectively intermingled. The experimental separation of the responses to the two different sets of information is, at best, extremely difficult and requires careful measurement of the visual characteristics of the targets involved.

The importance of colour and intensity contrast has been illustrated in work on the tabanid, *Tabanus nigrovittatus*. Large numbers of these flies are drawn to low intensity blue panels presented against grey backgrounds irrespective of the intensity of those backgrounds. This strongly suggests that the flies can respond to colour alone, but the most efficient targets are those showing not only colour contrast but also maximum intensity contrast with the background. In the case of the high intensity blue panels used in these experiments, this was achieved by presenting them against low intensity, grey background panels (Allan & Stoffolano

1986a). High intensity contrast is also an efficient draw for stableflies (Pospisil & Zdarek 1965, LeBreque et al. 1972), simuliids (Bradbury & Bennett 1974a, Browne & Bennett 1980) and mosquitoes (Browne & Bennett 1981). The responsiveness of blood-sucking insects to targets showing strong intensity contrast with the background is not surprising as large homeotherms are low intensity objects which will appear dark in contrast to high intensity vegetation (Allan et al. 1987). Work such as this has important implications for the design of traps which are economically as well as biologically efficient.

For man, the colour giving the greatest contrast against a green background is blue (Hailman 1979). We know that blue targets are highly efficient for tsetse flies and tabanids (Challier et al. 1977, Allan & Stoffolano 1986a, b) but we do not know why. Only when we have determined the spectral discrimination functions and colour matching functions of the eyes of these insects will we know if the high efficiency of these traps is because for the insect, as for man, blue gives maximum colour contrast with the green background.

Patterning of the target normally decreases its power as an activating and orienting agent as can be seen from the work on Diptera. If the pattern is a simple contrasting one then considerable numbers of tabanids still arrive, but as the patterning becomes more complex, progressively less insects are caught (Bracken & Thorsteinson 1965; Hansens et al. 1971, Browne & Bennett 1980). Similarly, the most effective tsetse targets are of a uniform colour; increasing the complexity with chequered patterns, stripes or complex edges lowers their appeal (Turner & Invest 1973). It is likely that patterned targets are less effective because of reduced visibility of the target from a distance. While considering patterning, it is interesting that not all zebras are striped; only those sympatric with tsetse flies bear the characteristic stripes, suggesting that striping may be a means of reducing the attention zebras receive from tsetse flies (Waage 1981).

The insect compound eye can be very sensitive to movement. One measure which relates to movement detection is the flicker fusion frequency of the eye (the rate of flicker which the eye is just unable to distinguish). Flicker fusion frequency is a function of the recovery time of the photoreceptors. In man the flicker fusion frequency lies somewhere between 20 and 30 flashes per second (20–30 Hz). Fast flying, diurnal insects typically have a flicker fusion frequency of 200–300 flashes per second while in slow flying or nocturnal insects this drops to between 10 and 40 flashes per second (Mazokhin-Porshnyakov 1969). The ommatidia of the tsetse fly G. m. morsitans are sensitive to frequencies of over 200 flashes per second (Miall 1978) enabling the insect to detect rapid movements in its visual field. Tsetse flies are also sensitive down to 1 flash per 25 s (0.04 Hz) (Turner & Invest 1973) suggesting that the insect may also be responsive to very slow movements across its visual field. Another gauge of movement detection is angular velocity sensitivity which is a measure of responsiveness to the rate of movement of an object across the insect's visual field. For tsetse flies, activation is optimal at angular velocities of about $3-7°\,s^{-1}$. This is the equivalent of an antelope trotting at $10\,km\,h^{-1}$ at $20-60\,m$ from the resting fly or a vehicle travelling at $30\,km\,h^{-1}$ at 60–180 m from the fly (Brady 1972).

How widespread the use of movement detection is in the finding of hosts is unclear. A complication, when designing experiments to look at the effect of movement, is that a moving animal breathes faster and sweats more, increasing the output of odour components which themselves draw blood-sucking insects. In consequence targets are commonly substituted for hosts in these experiments. Tsetse flies, males in particular, are drawn by large, dark, moving objects (Brady 1972, Gatehouse 1972, Vale 1974b), but the situation here is complicated. Under field conditions, it seems that most tsetse drawn to moving objects mate rather than feed, while those drawn to stationary objects are almost exclusively interested in feeding, not mating (Vale 1974b, Owaga & Challier 1985). Superficially, this seems to be a useful strategy acting to minimize wasted effort in tsetse flies. Inseminated female flies moving to stationary targets do not have to fend off males, while the males (grouped in the following swarm around moving animals) spend less time chasing previously inseminated females. Movement is also important for host finding in some mosquito species (Sippel & Brown 1953), but the evidence for its use in host finding by tabanids (Bracken et al. 1962, Bracken & Thorsteinson 1965, Browne & Bennett 1980) and simuliids (Underhill 1940, Thompson 1976) is equivocal.

For many species of blood-sucking insect, the evidence on shape preference is also conflicting. One reason for the confusion is that orientation and attraction are often confused in these studies. Flies may well show a preference for a particular shape to orientate towards but, on close approach to the target, they are not attracted to it and sheer off. Unless specific measures are taken to collect or count these insects, no useful information on orientation will come from the experiment (Vale 1974b). Another reason for confusion is that much of the work on shape preference has been carried out using two-dimensional rather than three-dimensional targets. There are reasons for this. Often the work has had the aim of trying to produce cheap targets which could be used for the control of the insects and the work has often shown that insects have preferences. For example, tsetse are drawn to two-dimensional targets in the ranked order: circle > square > horizontal oblong = vertical oblong (Torr 1989). For work aimed at investigating the natural host-finding mechanisms of insects, insights are more likely to be gained from three-dimensional targets. Tabanids, for example, show no preference

45

for two-dimensional targets of different shape (Roberts 1977, Browne & Bennett 1980), but when three-dimensional targets are employed, they show considerable preference for spheres over cubes or vertical cylinders (Bracken et al. 1962). Clearly, the three-dimensional target comes closest to resembling the host and it is reasonable to conclude that shape is important in drawing tabanids to hosts. Mosquito species can also distinguish between three-dimensional targets (a pyramid and a cube), with different species showing a preference for each (Browne & Bennett 1981). But some blood-sucking insects, most simuliids studied, for example, appear to show no preference for visual targets of a particular shape (Fredeen 1961, Bradbury and Bennett 1974a, Browne & Bennett 1980).

The size of traps is important; large traps draw or catch more tabanids (Bracken & Thorsteinson 1965, Thorsteinson & Bracken 1965, Thorsteinson et al. 1966), simuliids (Anderson & Hoy 1972) and tsetse flies (Hargrove 1980). In addition three-dimensional traps catch more tabanids than two-dimensional screens (Bracken et al. 1962, Thorsteinson et al. 1966). The increased drawing power of large traps is probably explained by the greater area over which large and/or three-dimensional traps are visible. This is likely to be only part of the explanation of their increased catching efficiency as alighting responses of tsetse flies may also increase with increasing trap size (Hargrove 1980).

Odours and visual clues are probably the most important factors in the orientation of most insects to the host, but they are not the only mechanisms. It has often been speculated that parts of the electromagnetic spectrum other than the visible (infrared radiation in particular) may be involved in host location. Sound has also been reported to be important as a factor guiding some insects to their hosts (McKeever 1977).

4.4   ATTRACTION

Vision, as well as being important in activating and orientating insects to the host, is also important in the attraction of blood-sucking insects and in their decision of whether and where to land. In general, the colour preferences displayed by alighting insects are the same as those involved in their activation and orientation, but not always (see above). To generalize, host-seeking insects prefer to land on dark, low intensity colours similar to those of many host animals.

Host animals are contoured and shaped and blood-sucking insects can recognize different parts of the animal. For example different tabanid species select markedly different landing sites on cattle (Fig. 7.3) (Mullens & Gerhardt 1979). This is probably visual recognition of the

46

preferred site because flying insects show definite preferences for particular parts of complex inanimate targets and also for targets of particular shapes. For example, many simuliids and mosquitoes show a preference for the extremities of a target (Fallis *et al.* 1967, Browne & Bennett 1981). Simuliids, attacking sticky animal silhouettes, are mainly caught on the projections (for example 'ears') (Wenk & Schlorer 1963), with the inner ('body') portions of the target catching few insects (Browne & Bennett 1980). To give another example, the blackfly, *Simulium euryadminiculum*, is activated by the odour of the uropygial gland of the loon which it also uses to orientate towards the host, but it only lands if the target has a prominent extension mimicking the neck of the bird, which is the normal feeding site of the fly (Fallis & Smith 1964, Bennett *et al.* 1972). A neat example of the effect of shape on the alighting response is given in the five-fold increase in tsetse flies landing on horizontally elongated targets over the same target presented vertically (Vale 1974b, Torr 1987). Given a target of the right shape, blood-sucking insects will often choose to alight at a colour or intensity border, or at the edge caused by the confluence of two angled planes (Fallis *et al.* 1967, Turner & Invest 1973, Bradbury & Bennett 1974b, Browne & Bennett 1980, Allan & Stoffolano 1986a, Brady & Shereni 1988). The size of the target may also be important in determining landing rates for tsetse flies (Hargrove 1980).

In many cases visual stimuli alone are not sufficient to lead to alightment. Mosquitoes are activated and orientate towards large visible objects, but on closing with an inanimate target (*c.* 30 cm.), and in the absence of other host associated signals, they sheer away from it and do not land (Bidlingmayer & Hem 1979). Similar responses can be seen in many tsetse and stableflies reacting to unbaited traps (Hargrove 1980, Vale 1982). Clearly there are other host-associated stimuli which are important to the insect in the attraction phase of host location. Odour is important once again. Generally, those odours which were involved in activation and orientation are also of significance in the attraction phase of host location, but new stimuli are also becoming available to the insect. One of the most important of these is heat.

Blood-sucking insects can be very sensitive to heat. The mosquito *Aedes aegypti* shows maximal spike frequency changes, in the cold and hot receptors of their antennal sensilla coeloconica, in response to a temperature change of $+0.2°C$ (Davis & Sokolove 1975). Little information is available on the distances over which heat is effective. It has been demonstrated that there are marked convection currents, with local thermal differences of $1°C$ or more, at up to and beyond 40 cm from a human arm (Wright 1968). Clearly *Ae. aegypti* could detect such a target thermally from a considerable distance. The more sedentary bedbug appears to be less sensitive to heat, requiring a temperature differential of

1–2°C (Marx 1955, Overal & Wingate 1976) and only being drawn to warm (37°C) objects when they are presented at a range of about 5 cm.

Heat may act in various ways as a clue to the presence of the host. The insect may respond to the radiant heat emitted by the host, to temperature gradients between the host and insect (convective heat), or directly to body heat of the host once it has been contacted (conducted heat). Unfortunately it is often difficult to establish which heat stimulus is being utilized by the insect. Both the hemipteran, *Rhodnius prolixus*, (Wigglesworth & Gillett 1934) and the yellow fever mosquito, *Aedes aegypti*, (Peterson & Brown 1951), use convective heat in attraction. The cat flea, *Ctenocephalides felis felis*, is drawn to warm targets with a maximum response to objects at 40°C, but only if the target is moving or otherwise creates air currents (Osbrink & Rust 1985). Whether this is a response to radiant or convective heat is unknown.

Heat is not a short-range orientation factor in all blood-sucking insects. The drawing power of sticky targets for the blackfly *Simulium venustum*, (Fallis *et al.* 1967) and horseflies (Bracken & Thorsteinson 1965) is not enhanced by heating the target.

Heat, as well as acting as an attracting agent, can also be used to gain information about the host animal. The louse, *Pediculus humanus*, rarely leaves the host, but it will do so if the host dies or suffers from a serious febrile illness. The louse is responding to temperature changes, leaving the host if its temperature falls or rises substantially from normal (Buxton 1947). The louse can then move as far as 2 m in quest of a new host, using heat as one of the locating signals (Wigglesworth 1941). It has occasionally been suggested that hosts with fever, due perhaps to a parasite transmitted by a vector insect, may show increased drawing power for a blood-sucking insect, compared to an uninfected host (Gillett & Connor 1976, Mahon & Gibbs 1982, Turell *et al.* 1984). This suggestion, if correct, would have considerable epidemiological consequences. It was not found to be the case for mosquitoes feeding on malaria-infected small mammals, where hyperthermia had no significant impact on the numbers of feeding insects (Day & Edman 1984).

As with other host location stimuli (Fig. 4.2), the heat response of the insect can be influenced by its physiological state. This is shown well by the streblid, *Trichobius major*, which lives in association with the bat, *Myotis velifer*. In summer, the bat's temperature rises to about 33°C before it takes flight. Fed flies are repelled by these temperatures and migrate off the bat onto the wall of the roost. In contrast, hungry flies are drawn by the heat and migrate onto the bat (Overal 1980). In winter, when the bats are not active, body temperature remains below 10°C and the flies are found permanently on the bat.

Most of the studies on water vapour perception by blood-sucking insects have been physiological or structural (Altner & Loftus 1985) rather than behavioural, and therefore the degree of involvement of water vapour in short-range orientation–attraction is unclear. When it has been reported as important it is usually as a synergistic agent with another stimulus. Thus the upwind movement of the mosquito *Ae. aegypti* in a stream of warmed air is enhanced if the relative humidity of the air is between 40 and 60% (Bar-Zeev *et al.* 1977).

4.5 MOVEMENT BETWEEN HOSTS

There is considerable variation in the extent of the association between the blood-sucking insect and the host. At one extreme the permanent ectoparasites are continuously associated with the host, and at the other extreme some temporary ectoparasites, including many of the blood-sucking Diptera, only visit the host for long enough to obtain a blood meal. Clearly insects in the latter category, which actively seek the host, must be capable of a considerable degree of movement, but even permanent ectoparasites must disperse from one host to another and so must also have an efficient means of locomotion.

For those insects which live within the pelage or feathers of the host, having wings is an encumbrance. For this reason permanent ecto-parasites, and most periodic ectoparasites, have secondarily lost their wings, relying on their legs to enable them to move around on the host and to transfer between hosts. When hosts are in close bodily contact, such as during copulation or suckling, these ectoparasites walk from the current host onto a new one. Permanent and periodic ectoparasites commonly have other adaptations which make movement within the covering of the body easier. These include the modification of the tarsal claws to enable them to grip the hairs or feathers of the host efficiently. A common modification is in the effective diameter of the claw and a loose correlation is found between the size of the claw and the diameter of the host's hairs, etc., when a range of species is considered (Hocking 1957). As well as tarsal claws, the legs of ectoparasitic insects are often well endowed with backwardly pointing setae. These may help the insect to move through and/or maintain its position in the host's covering; for example, polyctenid bugs, such as *Eoctenes* spp., swim rather than walk in the fur of their hosts, and extensions of the limbs act as flippers to help this type of movement (Marshall 1981). A modification seen in the limbs of several highly active bat ectoparasites are pseudojoints (unsclerotized rings of cuticle) which increase the flexibility of the limb (Marshall 1981) and presumably aid the movement of the insects on or between host animals.

To return to the question of wings, some periodic ectoparasites, such as keds of *Lipoptena* spp., have the best of both worlds. They emerge with wings which enable them to rapidly and effectively move on to a host, but once the insect successfully reaches the host the wings are shed. Presumably the loss of flight capability is more than offset by the advantage gained by being able to move freely in the host's pelage. Increased mobility in the pelage, enabled by the loss of the wings, may be a crucial factor in permitting the insect to escape from the host's grooming activities. Interestingly, keds can still transfer between hosts even after shedding their wings, using the classic permanent ectoparasite's method of walking from one to the other when the hosts are in close bodily contact (Samuel & Trainer 1972). Not all periodic ectoparasites find wings a significant encumbrance in their movements around the host animal. For example, many adult hippoboscids and streblids retain their wings throughout their life, but in these cases the wings are considerably modified, being either toughened or constructed so that the insect can fold them in a way that avoids abrasion by the host's covering (Bequaert 1953, Marshall 1981).

Temporary ectoparasites tend to take very large blood meals, thus limiting the danger from the host by minimizing the number of visits paid to it, and acting as an insurance policy in case hosts are difficult to find in future. Natural selection has ensured that these insects have optimized their locomotory systems to maximize the size of the meal they can take, while at the same time minimizing the risks attendant in carrying it off. In tsetse flies the meal is often two to three times the fly's unfed body weight (Langley 1970). Such a meal seriously impairs the insect's manoeuvrability, and, after feeding, flight speed falls from 15 to only 3–4 miles per hour (Glasgow 1961). Clearly this is a dangerous time for the fly. Maximum generation of lift by the tsetse fly is attained at 32°C (Hargrove 1980). As tsetse flies normally feed around dawn and dusk ambient temperature will usually be below this optimum. To compensate for this the fed fly generates heat endogenously. Immediately following the blood meal the fly raises its thoracic temperature towards the optimum of 32°C by 'buzzing' (i.e. producing the characteristic sound after which the tsetse is named). Buzzing is caused by the rapid contraction of the flight musculature while the wings are uncoupled from the flight motor mechanism. By raising the thoracic temperature in this way, the tsetse maximizes its lift and flight speed, allowing it to take an enormous blood meal while retaining the maximum chance of avoiding host defensive responses and escaping predators (Howe & Lehane 1986).

The fleas have lost their ancestral wings but many have found a way of putting the thoracic musculature to good use – it helps them to jump. Jumping fleas represent a halfway house between insects which rely on walking to transfer from host to host and those which fly. For their size fleas can jump amazing distances. For example the cat flea, *Ctenocephalides felis*, which is only about 3 mm long, can jump a height of 330 mm –

110 times its own length (Rothschild *et al.* 1972). These feats are achieved by using their thoracic musculature which progressively compresses an elastic protein called resilin held in the pleural arches. Simultaneously, cuticular catches are engaged which ensure that the resilin remains compressed when the thoracic musculature relaxes. Jumping occurs when the flea releases the catches and the energy stored in the compressed resilin is released with explosive force. The energy is transmitted to the substrate through the flea's hind legs (Rothschild *et al.* 1972). At room temperature a flea can jump about every 5 s and can do so repeatedly almost indefinitely.

Resilin is a remarkable rubber-like substance. It is the most efficient material known, natural or man-made, for the storage and sudden release of energy. It is little affected by temperature, and therefore the flea can manage to jump at temperatures which would certainly have grounded a flying insect. How is this achieved? Although the efficiency of the muscles is reduced at lower temperatures, as long as they can compress the resilin, no matter how long it takes, the fleas can still jump. This is unlike flying insects, where decreased muscular efficiency at lower temperatures rapidly limits the ability to fly. Indeed flight becomes impossible below a certain threshold temperature.

Fleas use their prodigious jumping feats to contact hosts passing in their vicinity. A flea in the correct physiological state is triggered to jump by host-related stimuli. The jump is aimed towards the source of the stimuli, and the legs, especially the middle pair, are held out ready to grasp on to a host if the jump is successful. The success of fleas is a testimony to the efficiency of this means of moving on to a host. Jumping seems to combine the ability to move rapidly, normally restricted to flying insects, with the capacity to do away with wings which are such a hindrance to movement within the host's covering. Jumping is also compatible with the production of a flattened body which further enhances the insect's mobility on the host, but the mechanism is clearly not as efficient as flight in host location as we can see from the distribution of fleas. Fleas are mainly limited to hosts which rear their young in some sort of nest and are therefore generally not found on grazing and browsing animals (Traub 1985). Why should this be so? Fleas, in common with the majority of temporary ectoparasitic Diptera, have larval stages which live independently of the host. When the adult flea emerges it must locate a host and jumping is clearly going to be efficient over a much smaller range than flight. So although the newly emerged dipteran can fly considerable distances to find a host, the flea must be confident that it will find a host in its immediate vicinity.

CHAPTER FIVE

Ingestion of the blood meal

5.1 VERTEBRATE HAEMOSTASIS

Bleeding occurs at sites of tissue damage in the vertebrate host. The bleeding is stopped by a series of interrelated mechanisms involving the blood vessels themselves, coagulation and platelet activity. This process is collectively known as haemostasis. Haemostasis is clearly of great importance to the blood-feeding insect which has developed a series of mechanisms to overcome the problems posed by it. Before discussing these adaptive features of blood-sucking insects, a brief outline will be given of the major mechanisms involved in vertebrate haemostasis. I emphasize that this is only a brief outline because haemostasis, with its central importance in human health, has been studied in considerable depth, and complete books are produced on this subject alone (Ogston 1983).

Vasoconstriction and platelet plugs

The amount of blood lost from a damaged vessel is directly proportional to the quantity of blood which flows through it. As a consequence venules and arterioles contract after injury, reducing the local flow of blood and in turn blood loss. The capillaries lack the necessary muscle layer to contract in this way, but it is possible that following injury blood flow in the affected capillary bed is reduced by contraction of the precapillary sphincter.

The blood vascular system is lined by endothelium. The cells of the endothelium have a regulatory role, inhibiting inappropriate episodes of coagulation or platelet aggregation, which could lead to thrombus formation in the closed vascular system. The rupture of the wall of a blood vessel leads to the exposure of blood to non-endothelial tissues. Many of the components of these tissues, collagen in particular, carry negatively charged groups on their surface. Platelets adhere strongly to these negatively charged surfaces and are induced to synthesize and secrete a number of factors. One of the most powerful of these substances is thromboxane A_2 which, along with serotonin (5HT) (which is also secreted by platelets), causes local vasoconstriction. The platelets also secrete ADP, which supplements the ADP released during cell damage at the site of injury. ADP in combination with thromboxane A_2 and thrombin induces aggregation of platelets at the site of injury, and the forming platelet plug begins to block the hole in the vascular system (Fig. 5.1). The platelet plug is sufficient on its own to rapidly (a few seconds) block small injuries to capillaries. To block more extensive injuries the platelet plug needs to be stabilized and strengthened. This is achieved by the introduction of a network of fibrin into the plug during blood coagulation.

Coagulation

Damage to the blood vessels, and the subsequent exposure of the sub-endothelial tissues, not only induces platelet aggregation at the site of injury, but also blood coagulation. The coagulation system is organized as a biological amplifier. A small initial stimulus is turned into a major response by a chain of enzyme reactions (the coagulation cascade) interlinked by a series of positive feedback loops. The coagulation cascade can be initiated in two ways. First, damage to the tissues leads to the release of tissue thromboplastin; this begins coagulation by the extrinsic pathway. Alternatively, the exposure of factor XII (which is present in blood plasma) to negatively charged surfaces such as collagen (as happens when the endothelial lining of the vascular system is broken) causes its conversion into activated factor XII (i.e. XIIa). This begins coagulation by the second route, the intrinsic pathway. The extrinsic and intrinsic pathways are connected by feedback loops and both lead into a common final portion of the coagulation cascade (Fig. 5.1).

Activated factor XII activates prekallikrein to kallikrein which, in the presence of high molecular weight (HMW) kininogen, causes the conversion of more XII to XIIa in a positive feedback loop. Factor XIIa in the presence of HMW kininogen causes the conversion of factor XI to XIa. Factor XIa in the presence of calcium ions in turn converts factor IX to IXa. This step is the point of convergence of the extrinsic and intrinsic pathways. Conversion of IX to IXa can also be caused by tissue thrombo-plastin in the presence of activated factor VII. Factor IXa can then cleave factor X to Xa in the presence of calcium ions, phospholipid (which may be supplied by the platelet membranes) and thrombin-altered factor VIII. This conversion (X to Xa) is probably achieved on the surface of the forming platelet plug (phospholipid co-factor) whereas achieving the correct symmetry for the conversion is probably the function of the other co-factors. Factor VIIa can also stimulate the conversion of X to Xa, and

53

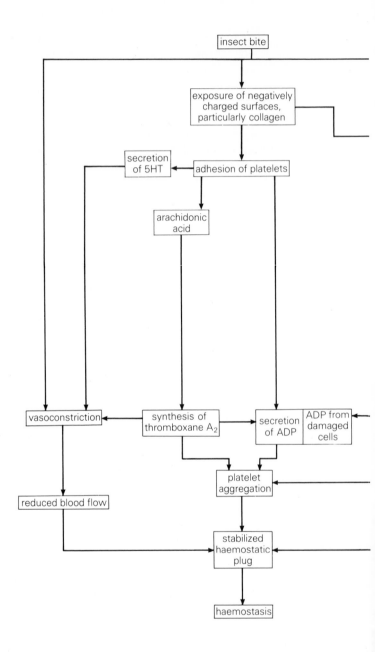

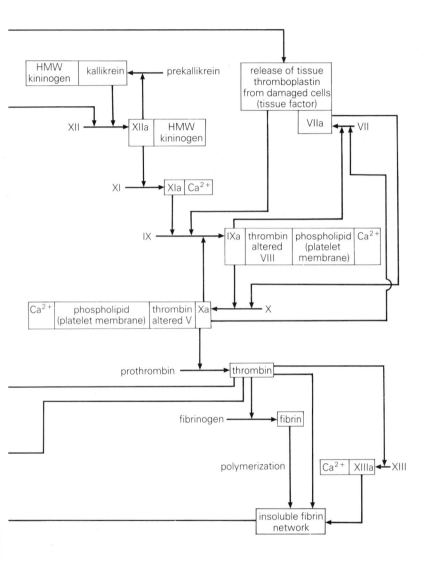

Figure 5.1 An outline of haemostasis. Mechanisms leading to vasoconstriction and plugging of the puncture by platelet activity are shown on the left hand page. The main elements of the coagulation system are shown on the right hand page. (Based on information taken from Ogston 1983.)

factor Xa stimulates the conversion of IX to IXa in a positive feedback loop which acts to amplify the system. Both factor IXa and Xa can stimulate the further conversion of VII to VIIa in two positive feedback loops, but the main function of Xa (in the presence of thrombin-altered factor V, phospholipid and calcium ions) is the conversion of prothrombin to thrombin. Thrombin stimulates the further aggregation of platelets and their secretion of ADP. It also cleaves fibrinogen to fibrin which polymerizes into an insoluble fibrin network. Thrombin also stimulates the conversion of factor XIII to XIIIa which, in the presence of calcium ions, acts to stabilize the forming fibrin network. The network of fibrin strengthens the platelet plug which blocks the hole in the blood vascular system and stops further bleeding (Fig. 5.1).

5.2   INSECT ANTI-HAEMOSTATIC FACTORS

The introduction of saliva into the host stimulates an immune response which alerts the host to the presence of the insect. In the long term an immune response may deny the insect access to a meal altogether (see Ch. 7). Given these negative aspects of saliva injection into the wounds, it seems clear that it must also carry considerable benefits.

Saliva is produced by most, possibly all, terrestrial animals; its original function was probably as a lubricant for the mouthparts as they worked against one another. It is proposed that the specific components of the saliva, which vary according to the specific diet of the animal, appeared later in evolution to provide secondary functions to the lubricatory role. Perhaps the most obvious advantage that saliva can bring is the presence of enzymes which are used in the digestion of the meal, but in blood-sucking insects this has not happened. Few digestive enzymes have been found in these insects, and these are at concentrations which make it unlikely that they are important factors in blood meal digestion (Gooding 1972). Clearly the introduction of hydrolytic enzymes into the host is likely to stimulate a rapid and pronounced response in which the host may well kill or damage the insect, so their absence in blood-sucking insects may well be an adaptation designed to eliminate this response.

Anticoagulants

Blood-sucking insects take their blood meal either from a pool which forms on the surface of the skin, from a haematoma formed beneath the surface of the skin by the actions of the mouthparts, or sometimes directly from a blood vessel. It is always vitally important for the insect that the blood remains in a liquid form until feeding is complete. Should the blood coagulate, not only will the insect be unable to complete the

Table 5.1   Anticoagulants are found in the saliva of many (+) but not all (−) blood-sucking species. (A more extensive list can be found in Gooding 1972.)

Insect	Salivary anticoagulants	Insect	Salivary anticoagulants
Bugs		Mosquitoes	
Cimex lectularius	+	*Aedes aegypti*	−
Rhodnius prolixus	+	*Anopheles quadrimaculatus*	+
Triatoma infestans	+	*Culex pipiens*	+
		Culex salinarius	−
Lice		Biting flies	
Pediculus humanus	+	*Glossina morsitans*	+
		Stomoxys calcitrans	−
Sandflies			
Phlebotomus papatasi	+		

blood meal, but its mouthparts will be blocked by the forming clot. Given the possibility of this unpleasant and probably fatal event, it is not surprising that the saliva of most blood-sucking insects contains anticoagulants (Table 5.1).

Consistent with the polyphyletic origins of blood-feeding in insects, it has been shown that different insects produce different anticoagulins which act at various points in the coagulation cascade. For example, the saliva of the tsetse fly, *Glossina morsitans*, contains an antithrombin (Hawkins 1966). This molecule is similar to the antithrombin hirudin produced by blood-sucking leeches (Parker & Mant 1979). In contrast, the salivary anticoagulant of *Rhodnius prolixus*, which has been named Prolixin S, disrupts the coagulation cascade by acting against factor VIII (Hellman & Hawkins 1964). On the face of it this might not seem a particularly good means of disrupting coagulation as factor VIII is a co-factor in the conversion of factor X to Xa, which can also be achieved along other pathways, but factor VIII is also necessary for the correct interaction of platelets with collagen. The importance of factor VIII in haemostasis is clearly illustrated by the profuse bleeding which occurs in individuals suffering from the genetic disorder haemophilia, in which factor VIII is lacking.

From the widespread occurrence of anticoagulants in the saliva, it seems clear that they play a useful role in feeding by many blood-sucking insects, but it is not clear if the presence of anticoagulins is necessary in order to be a successful blood-feeding insect. The doubt arises from two lines of evidence. First, the salivary glands of some blood-sucking insects (including such influential examples as the yellow fever mosquito, *Aedes aegypti*) have been reported to lack anticoagulins (Gooding 1972). Second, experiments in which the salivary glands are surgically removed from insects known to possess anticoagulins, have shown that the operated insects are still capable of taking blood meals for a limited time. Operated *R. prolixus* can successfully feed from a live rabbit (Ribeiro & Garcia 1981a), and tsetse flies with their salivary glands removed can still successfully complete a number of feeds on a host (although eventually lethal clots did form in the mouthparts of these flies) (Lester & Lloyd 1929).

So we have deduced that the saliva must have considerable benefits for the insect to overcome the problems that its injection causes. But, from the evidence available on anticoagulants in the saliva, it seems unlikely that they convey sufficient benefit alone to merit all the problems associated with their injection. Indeed, it could be argued that the prevention of clot formation in the mouthparts and foregut could be achieved by local action inside the insect without the need to undertake the dangerous step of injecting anything into the host.

Other anti-haemostatic factors

So what is the vital function of saliva in blood feeding by insects? In the last ten years a convincing story has begun to emerge (Ribeiro 1987), centring around the ability of the insect to minimize haemostasis by platelet action in the host animal. First, the haematologists showed that mechanical injuries to small blood vessels, such as those which occur during the probing of insect mouthparts, are plugged within seconds by aggregating platelets (Vargaftig *et al.* 1981) (the process is outlined above and in Fig. 5.1). This method of controlling blood loss from small blood vessels is very important for the animal. If platelet activity is inhibited even minor damage to small blood vessels can lead to severe bleeding. Several different blood-sucking insect species possess powerful anti-haemostatic factors which act on this platelet mechanism for blocking small blood vessels. The saliva of mosquitoes contains an enzyme, apyrase, which acts on the platelet aggregating factor ADP, converting it to non-active AMP and orthophosphate (Ribeiro *et al.* 1985). This inhibits platelet aggregation (Fig. 5.2). The time taken by a mosquito to feed is directly dependent upon the amount of apyrase which it can inject into the wound (Fig. 5.3). An apyrase which inhibits platelet aggregation is also contained in the saliva of *Rhodnius prolixus* (Sarkis *et al.* 1986) together with anti-thromboxane activity which further reduces platelet aggregation (and in addition decreases vasoconstriction) (Ribeiro & Sarkis 1982). *R. prolixus* saliva also contains anti-serotonin (5HT) and anti-histamine which inhibit vasoconstriction, and an anti-factor VIII which slows down thrombin formation. This inhibits thrombin-induced platelet aggregation (as well as inhibiting coagulation) (Hellman & Hawkins 1964). Apyrase activity has also been demonstrated in *Glossina* (Mant &

Parker 1981) and fleas (Ribeiro *et al.* 1989c). Sandflies also possess apyrase (Ribeiro *et al.* 1989a), and in addition they have the most powerful vasodilator known (Ribeiro *et al.* 1989b).

The key function of apyrase and the other antagonists of host haemostasis, appears to be in minimizing contact time between the host and the feeding insect. This has a strong selective advantage, particularly for the temporary ectoparasite, helping it survive this most dangerous period in its life. Strong evidence that salivary agents permit the minimization of probing and feeding times was obtained from experiments in which *R. prolixus* had their salivary glands surgically removed. While these insects could feed equally rapidly from blood presented in an artificial feeding apparatus, the operated insects fed more slowly from a host than the unoperated controls (Ribeiro & Garcia 1981b). Similarly, operated *Ae. aegypti* fed more slowly from a host animal than did controls with intact salivary glands (Mellink & van den Bovenkamp 1981). Recent work has established and quantified the inverse relationship between the quantity of anti-haemostatic factors in the saliva and the time it takes the insect to feed from a host (Fig. 5.3). How does apyrase minimize contact time?

The mechanisms which allow rapid feeding to occur are obvious for

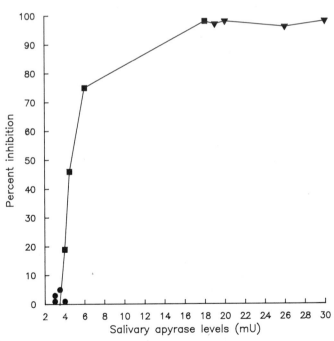

Figure 5.2   Platelet aggregation is inhibited by apyrase from the salivary glands of anopheline mosquitoes. ●, *Anopheles* sp. nr. *salbaii*; ■, *An. stephensi*; ▼, *An. freeborni*. (Redrawn from Ribeiro *et al.* 1985.)

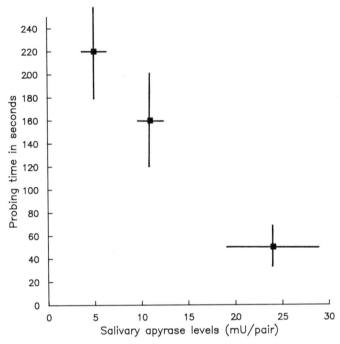

Figure 5.3 Feeding time in mosquitoes is directly related to the quantity of apyrase in their salivary glands. This is illustrated here, where median probing time for *Anopheles freeborni* (lowest probing time), *An. stephensi* and *An.* sp. nr. *salbaii* (highest probing time) are plotted against the mean quantity of salivary apyrase present in the pair of salivary glands of each species. Bars = standard errors. (Redrawn from Ribeiro *et al.* 1985.)

pool-feeding insects, because anti-haemostatic factors allow a haematoma to form more quickly once a blood vessel has been located in the probed skin. Although anti-haemostatic factors also enable rapid feeding to occur in insects believed to be vessel feeders, the mechanisms involved are not so clear. It has been suggested that the presence of these factors enables rapid location of the blood vessels. How this occurs has not been convincingly explained. One possibility suggested is that vessel feeders initially feed from the haematoma they cause in the probed skin, and as this is sucked dry of blood their mouthparts are drawn toward the lacerated vessel (Ribeiro 1987). Another suggestion is that the vasodilatory substances in the saliva may increase the blood flow in the area of skin being probed. This would increase the chances of encountering a vessel with a good supply of blood, which in turn would decrease feeding time (Pappas *et al.* 1986). Yet another possibility is that the apyrase system of saliva may prevent clumping of platelets at the tip or inside the feeding canal of the feeding insect, which would also expedite feeding.

60

So it seems likely that minimization of insect–host contact time (due to minimization of feeding time), in combination with the introduction of an anticoagulant to prevent clotting in the mouthparts, are the benefits which outweigh the problems associated with antigen introduction into the host during feeding.

Other salivary functions

Secondary roles for saliva have been suggested, including playing a part in skin penetration. It is common for small parasites (the infective third stage larva of hookworms and the cercaria of schistosomes for example), to use softening agents and enzymes to facilitate their passage across the skin, but blood-sucking insects seem to rely almost entirely on mechanical means of penetration. One possible exception is the cat flea, *Ctenocephalides felis*, the saliva of which is reported to have a skin softening agent which makes it easier for the mouthparts to penetrate (Feingold & Benjamini 1961). A characteristic of chemical-based systems to soften the skin is their slow action. So temporary ectoparasites, at least, are likely to rely on mechanical means of penetration because of the selective advantage to be gained from feeding quickly.

The saliva of some blood-sucking insects, (*R. prolixus*, for example) is known to contain antihistamine (Ribeiro 1982) which acts as an antagonist to vasoconstriction (see Fig. 5.1), thereby increasing the amount of blood in the wound. Another possible function of antihistamine is as an anti-inflammatory agent. Such an agent may extend the 'safe feeding period', i.e. the period before the inflammatory reaction and the itchiness associated with it, draw the attention of the host to the biting site. Even relatively modest increases in this period could have considerable advantages for the biting insect and particularly for temporary ectoparasites (see Ch. 7).

It has been suggested that the saliva of *Ae. aegypti* contains an anaesthetic which serves to deaden the response of local nerve endings when the insect is probing. The benefits of such a component in the saliva are clear, but its presence has not been substantiated (Mellink & van den Bovenkamp 1981). Another suggested role for saliva in mosquitoes is in holding the stylet bundle together by surface tension (Lee 1974).

5.3 PROBING STIMULANTS

Probing and engorgement occur in response to the quality and quantity of host-related stimuli (Friend & Smith 1977), but as in the other phases of insect response to host stimuli, the response is not performed in a completely stereotyped way. Anyone who has slept under a mosquito net is likely to have had first hand experience of this flexibility of responsiveness; if the skin becomes pressed against the net, mosquitoes are quite happy to probe and feed through it. The set of stimuli received in these circumstances must be very different to the range of host-related stimuli which they would normally receive. Post-landing responses can also vary with internal changes of circumstance. Responsiveness may be modulated by the insect's degree of hunger (Brady 1972, 1973, Friend & Smith 1975) or water deprivation (Khan & Maibach 1970, 1971), feeding experience (Mitchell & Reinouts van Haga-Kelker 1976) or reproductive state (Tobe & Davey 1972). Even if internal and external factors are carefully controlled, different individual insects still show a considerable degree of innate variation in their response to a host (Gatehouse 1970). This flexibility of close range responsiveness allows the insect to make the most of the differing circumstances in which it contacts hosts.

External factors affecting the readiness of insects to probe include vibration, surface texture, skin, hair and feather thickness, carbon dioxide levels, odours, visual stimuli, chemical stimuli, heat and moisture levels. Of these, heat is an important probing stimulant in many insects (Friend & Smith 1977). It can be sufficient on its own to stimulate probing in hungry R. prolixus, which will attempt to probe the inside of glass containers warmed on the outside by a hand. In these insects the heat receptors are restricted to the antennae (Wigglesworth & Gillett 1934). In the tsetse fly the heat receptors are found on both the antennae and the prothoracic leg tarsi (Dethier 1954, Reinouts van Haga & Mitchell 1975). Using these receptors the tsetse fly can monitor substrate temperature and, providing there is a temperature differential of 14°C between the substrate and the air, probing may be initiated (Dethier 1954). Similarly, a rapid increase in substrate temperature induces probing in the stablefly, Stomoxys calcitrans (Gatehouse 1970). In contrast the bedbug, Cimex lectularius, may be induced to probe by a temperature differential of only 1–2°C (Aboul-Nasr 1967). Heat is also required to stimulate feeding in simuliids (Sutcliffe & McIver 1975) and tabanids (Lall 1969). The situation is not so clear-cut in mosquitoes where heat may be an important factor inducing probing (Davis & Sokolove 1975), but is not always a necessary stimulus (Jones & Pillitt 1973).

Thickness of the substrate being probed can also be an important part of the series of events leading up to blood feeding. Tsetse flies feed far more readily when presented with blood under thicker, agar membranes than under thinner, parafilm membranes (Langley & Maly 1969, Margalit et al. 1972). It has been suggested that the thicker membrane may be more effective because of the stimulation of increased numbers of external receptors on the inserted mouthparts (Rice et al. 1973). The inclusion of ATP in the thicker membranes further increases the probing

62

response. ATP would be encountered naturally when skin is probed and may well have a role in promoting the probing response.

5.4 PHAGOSTIMULANTS

Having successfully probed the skin, the insect is now in contact with the blood meal. It makes no assumptions, but carefully looks for a set of blood-associated cues before it begins ingestion. In line with their polyphyletic origin, there are several cues which different species of blood-sucking insect may look for, but each particular species tends to have a very small range of clues which it uses. The fact that relatively few signals are acceptable to a particular insect is reflected by the relatively small numbers of receptors found on the mouthparts of most blood-sucking insects (Chapman 1982).

Blood-sucking insects can be classified into three groups dependent on the blood-associated clues they use as phagostimulants (Galun 1986). This classification (outlined below) is interesting in that it shows that, even within a closely related evolutionary group, different clues have emerged for the recognition of blood. The mosquitoes are such an example; culicines respond maximally to ADP while aedines prefer ATP (Galun 1987, Galun et al. 1988) and anophelines will feed in the absence of nucleotides.

(a)   Insects using clues which are associated with the cellular fraction of the blood meal.

Extrapolating from the few species studied to date, it seems probable that most blood-sucking insect species fall into this group. Tsetse flies (Galun 1975, Galun & Kabayo 1988), tabanids (Friend & Stoffolano 1984), blackflies (Smith & Friend 1982), culicine mosquitoes (Hosoi 1958, 1959, Galun et al. 1963, Friend 1978) and reduviid bugs (Friend 1965) all feed readily on whole blood, but are very reluctant to feed on plasma presented alone. It is clear that the convergent evolution that has occurred among the different members of this group, has produced a range of slightly different mechanisms allowing them to discriminate between plasma and whole blood. So we can see that while ATP is the major stimulant for most of the insects in this group, *Simulium venustum* is maximally responsive to ADP (Smith & Friend 1982), and that while the dipteran members of the group only respond to adenine nucleotides, *R. prolixus* also responds to a range of other nucleotides and phosphate derivatives. Work with ATP analogues suggests that the range of responsiveness seen reflects differences in the binding of the stimulants to the membrane receptors of the gustatory sensillae in the insect's foregut. Stimulation is not dependent on ATP acting as an energy source for the receptor. Rather, it has been suggested that stimulation occurs in response to conformational changes in the membrane brought about by nucleotide binding, which in turn opens membrane channels permitting an influx of sodium ions to the receptor (Galun 1986). Like so many of the responses occurring during host location, it is notable that sensitivity to ATP is not a fixed phenomenon. As the insect becomes hungrier so it becomes more sensitive to ATP (Friend & Smith 1975).

(b)   Insects using clues present in the blood plasma.

The sandfly, *Lutzomyia longipalpis*, and the anopheline mosquitoes, *Anopheles freeborni*, *An. stephensi*, *An. gambiae* and *An. dirus*, all feed happily on cell-free plasma (Ready 1978, Galun et al. 1985). In fact the sandfly feeds readily on isotonic NaCl, while *An. freeborni*, *An. stephensi* and *An. gambiae* feed on isotonic saline when 10^{-3} M NaHCO$_3$ is added. *Anopheles dirus* feeds on this solution if albumin is added. It seems that the major consideration in this group is the tonicity of the solution, and that ATP has no impact on feeding.

(c)   Insects intermediate between (a) and (b) above.

These insects rely on more than one stimulus for full engorgement. The flea, *Xenopsylla cheopis*, is a straightforward example. It responds to the tonicity of the meal and feeds on solutions isotonic with plasma; a full feeding response can be elicited to such solutions if they are supplemented with ATP (Galun 1966).

Whether they use nucleotides as a stimulant or not, all blood-feeding insects require isotonic saline (NaCl) before they will display optimal feeding in an artificial system. Tonicity and the presence of adenine nucleotides are not the only engorgement stimuli used by blood-sucking insects. This can be seen in the body lice which will feed, but not fully engorge, on plasma, but they will not feed on isotonic saline. Nor can they be stimulated to take a full meal by the addition of ATP to plasma. It seems that other, small molecular weight components of the cellular fraction are important here (Mumcuoglu & Galun 1987).

5.5   MOUTHPARTS

Vertebrate skin is divided into an outer epidermis and an inner dermis. Only the dermis is vascularized and any insect in search of a blood meal must first find a means of getting to the blood. Some insects are opportunists and will feed on blood when it leaks to the surface through a wound. This is fairly common in the Muscidae; for example, the non-biting flies *Fannia benjamini* and *Hydrotaea armipes*, take blood from wounds caused by tabanids (Garcia & Radovsky 1962). *Hydrotaea* spp. may even become impatient and crowd around the mouthparts of the biting fly, feeding concurrently with it (Tashiro & Schwardt 1953), sometimes interrupting the feeding of the biting fly and forcing it away from the wound in the process.

Other insects are more 'professional' in their blood-feeding activities and do not depend on chance for their blood meal. Species such as *Philaematomyia lineata* have prestomal teeth which may be sufficiently well developed to break through the partially dried and clotted surface of a wound (Patton & Cragg 1913). Other species have gone a step further and have mouthparts which can penetrate intact skin. Mouthparts designed for this purpose have arisen independently in several different insect groups. They can be crudely divided into two categories. Piercing and sucking mouthparts are seen in the bugs, lice, fleas and mosquitoes. In all of these the mandibles and/or maxillae have been modified to form long, thin, piercing stylets which are also interconnected to form a long tube through which blood can be sucked. Commonly, insects with these mouthparts take their food directly from a blood vessel which has been lanced by the mouthparts. The second category includes mouthparts which are used to rip, tear or otherwise cut the skin, and then to lap or suck blood from the haemorrhagic pool which forms. These mouthparts are seen in the tabanids, blackflies and biting flies. A guide to the different mouthpart components of the different insect groups, and an outline of the ways in which they have been modified for each feeding task, is given in Table 5.2. The way that these mouthparts are used to cut or pierce the skin is still not clear for many species. Some examples, of the better studied and understood cases, are described below.

Lice

The labrum forms a short snout at the front of the louse. This is everted during feeding to expose a series of teeth which grip the host's skin. Within the head is the trophic sac which contains the three stylets which are the penetrative elements of the mouthparts (together these are known as the syntrophium or fascicle) (Jobling 1976). The fascicle is formed of a dorsal (hypopharyngeal) stylet, a median stylet (the homologies of which are uncertain but it carries the salivary canal within it), and a ventral (labial) stylet (Fig. 5.4). The ventral stylet is armed with teeth and is used for piercing the skin (Stojanovich 1945).

Bugs

The structure of the mouthparts, and the way in which they are used in feeding, has been extensively studied in the reduviid bugs *Triatoma* and *Rhodnius* (Snodgrass 1944, Lavoipierre *et al.* 1959, Friend & Smith 1971).

Table 5.2 Adaptations of mouthpart components for different purposes in various haematophagous insect groups. (Information largely from Marshall 1981 and Smith 1984.)

	Sheath	Salivary canal	Food canal	Anchorage	Puncture	Penetration	Site of feeding
Hemiptera	labium	maxillae	maxillae	mandibular teeth	mandibular stylets	maxillae (+ mandibles in *Cimex*)	vessel
Anoplura	invagination of the labium	median stylet (labial?)	hypopharynx	haustellar teeth (labral?)	haustellum	dorsal, median and ventral stylets	vessel
Siphonaptera	labium	maxillae	epipharynx	?	maxillae (laciniae)	maxillae	vessel or pool
Diptera Culicidae	labium	hypopharynx	labrum (and mandibles?)	maxillary teeth	maxillae	maxillae	vessel or pool
Simuliidae	labium	hypopharynx	labrum and hypopharynx	maxillary teeth	mandibles	mandibles and maxillae	pool
Tabanidae	labium	hypopharynx	labrum	maxillary teeth	mandibles (and maxillae?)	mandibles and maxillae	pool
Muscidae	maxillary palps	hypopharynx	labrum	—	labellar teeth	labellar teeth	pool
Hippoboscidae	labium	hypopharynx	labrum	—	labellar teeth	labellar teeth	pool or vessel

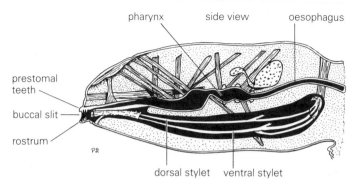

pharynx   side view   oesophagus

prestomal teeth

buccal slit

rostrum

dorsal stylet   ventral stylet

Figure 5.4 A longitudinal section of the head showing the arrangements of the mouth-parts of a blood-sucking louse. (Redrawn from James & Harwood 1969.)

Figure 5.5 The left maxilla of *Rhodnius* will slide backwards on the right maxilla when a blood vessel is entered. This allows the left maxilla to fold outwards away from the food canal, increasing the diameter of the feeding channel and so decreasing the time taken to obtain a blood meal. (Micrograph kindly provided by J.J.B. Smith.)

The mouthparts are in the form of a long thin beak which, when the insect is not feeding, is folded back under the head. The food canal and salivary canal are both formed by the maxillae, which are joined together by hooked structures which have been named coaptations. The maxillae are flanked by the mandibles which have backward pointing teeth on their lateral edge. The maxillae and mandibles are in the form of long thin stylets. The labium is the largest structure in the mouthparts of these bugs. The dorsal surface of the labium is deeply indented to form a sunken groove in which the stylet-like maxillae and mandibles lie. When the bug is about to feed the mouthparts are swung forwards in front of the rest of the head and are placed in contact with the skin of the host. The mandibles are then alternately retracted and protracted, when it is presumed that the mandibular teeth cut the skin in a saw-like manner. The mandibles do not penetrate far into the skin, but act as an anchorage point through which the maxillae are projected. The maxillae are flexible and capable of changing direction within the skin which they penetrate deeply. Once a blood vessel is penetrated the left maxilla slides backwards over the right for some distance, thus disconnecting a catch mechanism which has been holding the two together, and permits the tip of the left maxilla to fold outwards from the food canal (Fig. 5.5). Whether this mechanism is to give blood cells easier access to the food canal or to hold open a punctured capillary is a matter of conjecture.

Blackflies

These are pool-feeding flies and their mouthparts and feeding method are thought to represent the primitive condition in the Nematocera. The food canal is a chamber lying between the labrum and the hypopharynx. The salivary canal is contained in the hypopharynx. The skin is cut by the mandibles and laciniae of the maxillae. Both elements have backward pointing teeth at their distal ends. The prominent snout of blackflies is formed by the labrum which has a deep groove on its ventral surface which holds the other elements of the mouthparts when they are not in use. The cutting of the skin is a serial process (Wenk 1962). The mouthparts are opened and applied to the surface of the skin. The mandibles are repeatedly protracted and retracted, making the initial cut in the skin on the retracting stroke. The wound is opened by the insertion of the labrum and the laciniae. The laciniae appear to be used as anchorage points while all the mouthparts except the labium are thrust into the wound. This series of events is repeated and the wound is deepened until blood vessels are lacerated and a pool of blood forms. The labella and ligula of the labium form a seal around the entrance to the wound, which aids the fly in efficiently sucking blood from the forming pool (Sutcliffe 1978).

Figure 5.6   As the mosquito fascicle penetrates the tissues of the host it springs out of the channel in the labium which is progressively bent as the fascicle passes increasingly deeper into the host.

Mosquitoes

Mosquitoes have piercing mouthparts which they use for feeding directly from blood vessels or from a pool of blood in the skin. These mouthparts, and this feeding method, are thought to be the most highly evolved for blood feeding in the Nematocera. During feeding the mouthparts divide into the fascicle (or stylet bundle) and the labium which encloses it. When feeding, only the fascicle passes into the skin, while the labium becomes progressively bent as the mosquito probes increasingly deeper into the tissues (Fig. 5.6). The fascicle is formed of the labrum, maxillae, hypopharynx and mandibles which are all drawn out into long thin stylets. These individual components are usually held tightly together. It has been suggested that adhesion of the stylets is aided by the surface tension of the saliva (Lee 1974). While this may be the case in the non-feeding mosquito, it is not true during feeding as the mouthparts are immersed in liquid. The maxillae have a pointed tip and recurved teeth at their distal ends; they are believed to be the main penetrative elements of the mouthparts. The maxillae are thought to thrust alternately and, by using the teeth to anchor themselves in the tissues, to pull the labrum (food channel) and hypopharynx (salivary channel) with them as they penetrate the tissues (Robinson 1939). The function of the mandibular stylets is unclear, but may involve protection of the bevelled end of the labral food canal during penetration, or the separation of the salivary canal from the food canal in the feeding process (Robinson 1939, Lee 1974). Retraction of the mouthparts from the wound is also thought to involve

69

the active participation of the maxillae. When the mouthparts are not in use the fascicle is enclosed in a dorsal or ventral channel in the labium.

Tabanids

Close personal contact with some of the larger tabanids can be an unforgettable experience. They are pool feeders, drawing blood from a considerable wound. The maxillae and mandibles bear teeth which are used to cut the skin in a scissor-like fashion on the inward sweep (Dickerson & Lavoipierre 1959). The toothed maxillae are probably used for anchoring the mouthparts, giving the mandibles the purchase necessary to work on the skin. The labellar lobes, at the tip of the labium, are muscoid in type and bear the characteristic pseudotracheal (sponging) channels of these flies. The labium, however, does not enter the wound and is not used in blood feeding; instead the food channel is formed by a deep gutter in the labrum.

Tsetse, stableflies, etc.

These pool-feeding flies pierce the skin in an essentially different manner to other blood-sucking Diptera. Instead of using the mandibles and/or maxillae as cutting or penetrating stylets, the biting flies use the labium. This forms the major part of a semi-rigid set of mouthparts normally called the proboscis, or less often the haustellum. The labium bears two labellar lobes at its distal end. These lobes are everted during feeding, exposing an array of highly sclerotized teeth which are hidden from view in the non-feeding fly (Fig. 5.7). To cut the skin the lobes are repeatedly and rapidly moved outwards and upwards (which is the cutting stroke) (Gordon et al. 1956). It is unclear how the labium is forced into the skin, but it has been suggested that the labellar teeth anchor the mouthparts in the tissues so that the upward stroke of the labellar lobes forces the tip of the labium increasingly further into the tissues. Once the skin is pierced the proboscis is often partly withdrawn before being thrust in again at a slightly different angle, possibly to locate suitable blood vessels or to increase the size, and rate of formation of the blood pool. The whole tip of the proboscis is also capable of considerable rotation about the long axis of the proboscis; in the stablefly this rotation occurs when the proboscis is fully inserted into the skin (Lehane 1975). This may also be a means of increasing the damage to the blood vessels and the rate at which the blood pool will form. From personal experience, it also seems to be the most painful stage in the bite of the stablefly. The dorsal surface of the labium has a deep gutter which forms the floor of the food canal. It also holds the hypopharynx (through which the salivary canal runs) and the labrum (which forms the roof of the food canal).

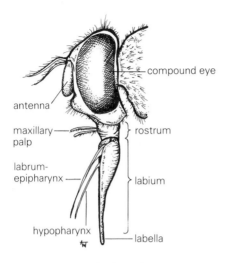

- compound eye
- antenna
- maxillary palp
- rostrum
- labrum-epipharynx
- labium
- hypopharynx
- labella

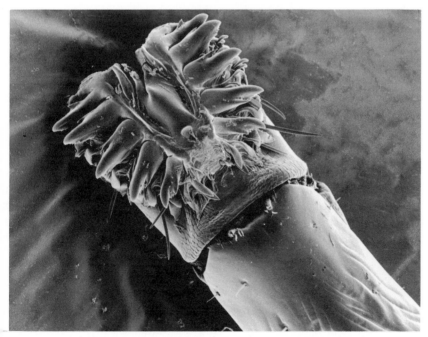

Figure 5.7 Top, the mouthparts of the stablefly, *S. calcitrans*, are shown with the labrum and hypopharynx artificially sprung out of the deep gutter in the labium. Bottom, the labellar lobes at the tip of the proboscis are everted during feeding to expose these highly sclerotized teeth. (Top redrawn from James & Harwood 1969.)

71

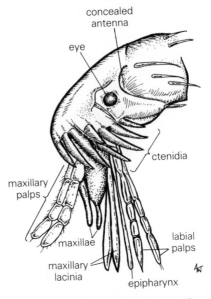

Figure 5.8 The mouthparts of the flea.

Fleas

The mouthparts are shown in Figure 5.8. The food canal is formed by the apposition of the epipharyngeal stylet dorsally and laterally, with the laciniae of the maxillae ventrally. Both the epipharyngeal stylet and the maxillae have an extended stylet-like form, and the sides of the maxillae have backward pointing teeth at their distal end (Snodgrass 1946, Wenk 1953). The teeth on the maxillae are used to cut the skin of the host. When a blood vessel has been located and feeding is taking place, only the epipharyngeal stylet is within the blood vessel (Lavoipierre & Hamachi 1961).

5.6 BLOOD INTAKE

Having located the blood the insect has to transfer it from the host to its gut. To do this it uses pumps located in the head capsule. There may be only one, the cibarial pump, as in the bugs and the muscoid flies, or this may be supplemented by a second, pharyngeal pump, as in the mosquitoes, blackflies and tabanids. Essentially the pumps work by creating a negative pressure difference between the tip of the mouthparts and the pump by muscular cavitation of the pump's lumen. The action of the cibarial pump of *R. prolixus* has been studied in some detail (Bennet

72

Clark 1963, Smith & Friend 1970, Smith 1979); let us look at its structure and action as an example.

The head of a fifth instar nymph of *R. prolixus* is *c*. 5 mm long and 0.8 mm in diameter. The pump and its associated musculature nearly fill the head capsule. The pump is formed from a rigid, ventrally positioned U-shaped girder 3.5 mm long and 0.28 mm wide. Attached to this, by 'rubbery' ligaments on either side, is a more flexible overlying element which acts as a piston. The piston is attached to the well-developed muscles of the pump (Fig. 5.9). The fifth instar nymph is capable of ingesting a meal of 300 mg in 15 min. The terminal diameter of the mouthparts is *c*. 8 μm (or 10 μm if the left maxilla is fully retracted (Fig. 5.4)). This gives an astonishing flow rate at the opening of the mouthparts of between approximately 4.4. and 6.6 m s^{-1}. The cibarial pump supplies the force to move the blood. Given that the viscosity of blood at 37°C is probably about 3 centipoises (cP)(Altman & Dittmer 1971), the pressure differential the cibarial pump must generate to move the blood can be calculated from the Hagen-Poiseuille Equation to be about 576 kPa (5.76 atm) at a terminal diameter of 10 μm, and a massive 1405 kPa (14.05 atm) at a terminal diameter of 8 μm. But a vacuum pump can generate a negative pressure of only 1 atmosphere! The blood in the capillaries may be under some positive pressure, but not 13 atmospheres. Clearly there is some fundamental misunderstanding in our view of the mechanics of the feeding process.

Many of the calculations carried out on the movement of blood in insects have used the Hagen-Poiseuille Equation which describes the mechanics of Newtonian fluids. There is some uncertainty as to the validity of these calculations because blood is a non-Newtonian fluid, that is, its apparent viscosity will depend upon the shear rate to which it is subjected. However, at the tube radii and feeding pressures seen in piercing and sucking insects blood behaves essentially as a Newtonian

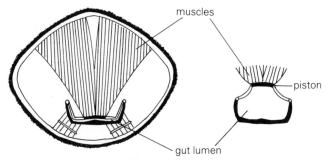

Figure 5.9 The cibarial pump of *Rhodnius prolixus*, showing the relaxed pump with a closed lumen (left) and the cavitated lumen (right), caused by the contraction of the pump's well-developed musculature. (Redrawn from Bennet-Clark 1963.)

fluid and so the Hagen-Poiseuille Equation is still used in calculations on blood movements. Another problem encountered when performing calculations on blood flow in insect mouthparts is that there is still considerable doubt about the viscosity of blood at the tube diameters and pressures which are seen in insect mouthparts. Estimates for the viscosity of blood *in vivo* range from c. 1.5 cP, which is the measured value for plasma alone, to c. 3 cP, which has been measured for whole blood (but in tubes with much larger diameters than insect mouthparts).

Despite these uncertainties models describing the flow of blood in narrow tubes have been developed and used to describe the mechanics of feeding in blood-sucking insects (Tawfik 1968, Daniel & Kingsolver 1983). Analysis of these models suggests that the biomechanics of feeding is reasonably accurately described by the Hagen–Poiseuille Equation:

$$Q = \frac{\pi r^4 p t}{8 \eta l}$$

where Q is the meal size, p is the pressure difference between the tip of the mouthparts and the lumen of the pump, r is the radius of the feeding tube, l is the length of the feeding tube, t is the feeding time and η is the dynamic viscosity of the blood.

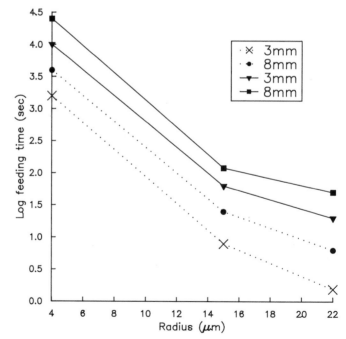

Figure 5.10 The time needed to complete a blood meal of 4.2 mm³ is plotted against the radius of the food canal for two lengths of feeding canal and for two pressure differences, 100 kPa (····) and 10 kPa (——). (Redrawn from Daniel & Kingsolver 1983.)

The feeding models developed have demonstrated several interesting relationships. First, the time required to complete a blood meal is most sensitive to the radius of the food canal. As a rule of thumb, this means that for a food canal of fixed length, and for a fixed pressure drop applied by the cibarial pump, feeding time will increase 16-fold for a halving in the radius of the canal. To a lesser extent feeding time is also directly proportional to the length of the food canal, and is in inverse proportion to the pressure difference between the cibarial pump and the blood source. These relationships are illustrated in Figure 5.10.

Because of the dangers inherent in obtaining a blood meal, short feeding time is likely to be a character which is strongly selected for in temporary ectoparasites. As we have seen, small increases in the radius of the feeding canal can bring large decreases in the time taken for the insect to achieve a full blood meal. Therefore we might expect that natural selection is acting to maximize the diameter of the feeding canal. The fact that insects such as the triatomine bugs retain a food canal with a terminal diameter (8–10 μm) little larger than that of blood cells is strong evidence that other equally important selection factors must be in operation to prevent a larger terminal aperture from emerging. The most obvious of these selection pressures is that a small terminal diameter is required in order to efficiently (and painlessly?) penetrate the tissues of the host. If we look at the terminal diameter of the food canal of other species of piercing insects we see a convergence in size which supports this argument. For example, the terminal diameters of the food canal of *Cimex*, *Aedes* and *Pediculus* are 8 μm, 11 μm and 10 μm, respectively (Tawfik 1968). However, the factors defining the size of the terminal aperture in *R. prolixus* must be in very fine balance because even an increase in terminal diameter from 8 to 11 μm (which is the diameter successfully used by *Aedes*) has not arisen, although this would theoretically decrease the feeding time by a factor of approximately 3.6, i.e. from *c.* 15 min to under 5 min! This would seem to be an enormous potential advantage to the insect, yet it has not been adopted. It is possible that the morphological adaptation seen in the triatomine bugs, where the left maxilla is redrawn on the right to increase the effective opening of the terminal aperture from 8 to 10 μm, is an evolutionary attempt to have the best of both worlds, a fine diameter stylet during probing and a wider food canal during feeding. In many host species the diameter of red blood corpuscles is considerably greater than 10 μm (Table 5.3) and this may have interesting consequences in terms of feeding time and/or host choice.

Another notable relationship arising from these models is that between the haematocrit of the blood (i.e. the proportion of red cells present) and its viscosity. A decrease in the haematocrit leads to a decrease in the apparent viscosity of blood. In the original interpretation of the model it

Table 5.3   The size of red blood corpuscles varies widely in different animals. Given that many blood-sucking insects have mouthparts with a terminal diameter of around 10 μm in diameter, this may be a factor affecting the feeding efficiency of blood-sucking insects feeding on different host species. (Data from Altman & Dittmer 1971.)

Host	Red blood corpuscle dimensions (μm) estimated from dry films	Host	Red blood corpuscle dimensions (μm) estimated from dry films
mammals		reptiles	
man	7.5	alligator	23.2
horse	5.5	tortoise	18.0
cow	5.9		
sheep	4.8	amphibians	
		congo 'snake'	62.5
birds		(*Amphiuma means*)	
turkey	15.5	frog	24.8
chicken	11.2	(*Rana catesbeiana*)	

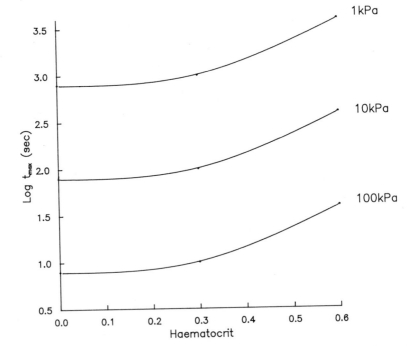

Figure 5.11   The feeding time (expressed as log t_{max} s) for *Aedes* decreases with decreasing haematocrit, irrespective of the pressure differential exerted by the insect. Because many parasitic infections cause a fall in the haematocrit of the host, this would favour those insects feeding on infected hosts (but other factors must also be taken into consideration). (Redrawn from Daniel & Kingsolver 1983.)

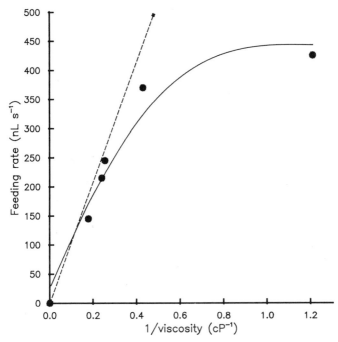

Figure 5.12 The measured feeding rate for R. *prolixus* is plotted on the reciprocal of the viscosity of the fluid ingested (●, solid line). The dashed line shows the feeding time predicted from the Hagen–Poiseuille Equation. Blood viscosity at 37 °C is estimated to be about 3 cP. Increasing viscosity will lead to a decrease in feeding rate which can be predicted by the Hagen–Poiseuille Equation. Decreasing viscosity, due to, for example, decreased haematocrit, will cause an increase in feeding rate, but this will not necessarily be predicted by the Hagen–Poiseuille Equation. (Redrawn from Smith 1979.)

was suggested that a decrease of 30% in the haematocrit would lead to about a 30% decrease in the time needed to complete the meal (Fig. 5.11) (Daniel & Kingsolver 1983). This is predicted from the Hagen–Poiseuille Equation, but experimental work on R. *prolixus* suggests this is an overestimate and that a 15% decrease in ingestion time would be expected (Fig. 5.12) (Smith 1979) – still a significant decrease. The insect is quite likely to encounter lowered haematocrits in the field as it is known that blood-feeding insects may directly cause anaemia in their hosts; the parasites that they transmit may also decrease the hosts' haematocrit. It would be advantageous for the insects to feed on these hosts, in the sense that this would decrease the time necessary to complete feeding.

One fact which seems to have been ignored in the exploration of these models is that blood viscosity is inversely related to temperature. The measured absolute viscosity of blood from female humans rises from c. 3 cP at 37°C to c. 4.46 cP at 20°C (Altman & Dittmer 1971). The models have been interpreted with the assumption that the blood being ingested is always at 37°C. This is not necessarily the case. The temperature of the

77

skin at many points on the host's body will be significantly lower than core body temperature. The Hagen–Poiseuille Equation predicts that because of the higher viscosity of blood, at 20°C it will take *c.* 1.49 times longer to complete a full blood meal taken at this temperature. The effect of viscosity of the meal on feeding time has been verified experimentally (Fig. 5.12) (Smith 1979). Because of this effect on the rate of ingestion, increased blood viscosity at lower temperatures may well be a factor which influences the choice of insect feeding sites on the host.

CHAPTER SIX

Managing the blood meal

6.1 MIDGUT ANATOMY

Blood-sucking insects can be divided into two groups depending on the design of the alimentary canal for the storage of the blood meal. In one group, typified by Hemiptera and fleas, the alimentary canal is a simple tube and the blood is stored in the midgut. In the second group, typified by Diptera, the gut has between one and three diverticulae which may be used, in addition to the midgut, for the storage of the blood meal (Fig. 6.1). In this second group some insects, such as tsetse flies, always store blood in the diverticulae. In others, such as the stablefly, passage of blood to the diverticulum is more variable among feeds (Gooding 1972).

The midgut is the site of blood-meal digestion and absorption. Two basic patterns of digestion are seen in blood-sucking insects: a batch system and a continuous system (Fig. 6.1). In the batch system, which is well illustrated by mosquitoes, sandflies and fleas, digestion proceeds almost simultaneously over the entire surface of the food bolus. The continuous system is typical of higher Diptera and Hemiptera, the blood meal being held in a specialized portion of the anterior midgut where no digestion takes place. Portions of the blood meal are then gradually passed down through the digestive and absorptive mid and posterior regions of the midgut. In this continuous system much of the meal will have been completely processed and defecated before some has even entered the digestive section of the midgut.

The blood meal is normally separated from the midgut epithelium by an extracellular layer known as the peritrophic membrane. The major constituents of this layer appear to be mucopolysaccharides (including chitin), with some protein and possibly lipid components. Two types of peritrophic membrane are recognized based on their method of production (Waterhouse 1953). The most common production method in insects is by secretion from cells along the complete length of the midgut. This type of peritrophic membrane (Type I) is absent in the unfed insect, and is only produced when the blood meal has been taken. These Type I

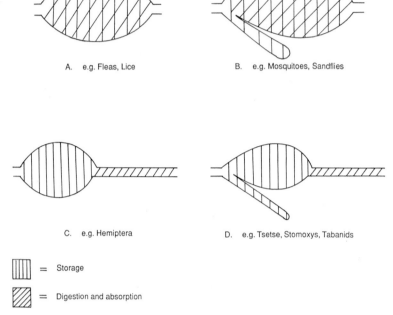

A. e.g. Fleas, Lice

B. e.g. Mosquitoes, Sandflies

C. e.g. Hemiptera

D. e.g. Tsetse, Stomoxys, Tabanids

⫿⫿⫿ = Storage

▨ = Digestion and absorption

Figure 6.1 Divided on the basis of blood storage, there are two basic designs of alimentary canal. Either the straight tube (A) and (C) or an alimentary canal with diverticulae (B) and (D). Blood-sucking insects can also be divided into two categories based on their method of digesting and absorbing the blood meal. Batch processors (A) and (B) commence digestion over the entire surface of the food bolus, continuous processors (C) and (D) segregate the intestine into a 'production line', passing packets of food down the 'line' from the storage region for digestion followed by absorption.

peritrophic membranes are found in adult mosquitoes (Fig. 6.2), black-flies, sandflies and tabanids. Type II peritrophic membranes are found in adult muscids, tsetse flies and the hippoboscids; they are produced by a special organ, the proventriculus, which is sited at the junction of the fore- and midgut (Lehane 1976b) (Fig. 6.3). The membranes are secreted continuously and form a complete cylinder lining the midgut in both fed and unfed flies. The peritrophic membrane is a semi-permeable filter with a pore size of about 9 nm in tsetse flies (Miller & Lehane 1990). This pore size permits the passage of globular proteins of about 140 kDa and so presents no barrier to either trypsins (about 25 kDa; Moffatt & Lehane 1990) or to the passage of the peptides produced during digestion. A peritrophic membrane is not produced in hemipteran bugs, but they do produce an extracellular coating in response to the blood meal (Lane & Harrison 1979, Billingsley & Downe 1983, 1986). This surface coating is

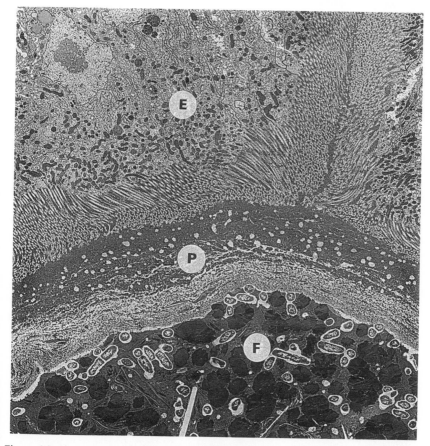

Figure 6.2 An example of a Type I peritrophic membrane from the posterior midgut of *Anopheles stephensi* at about 60 h after the blood meal. The peritrophic membrane (P) can be seen lying between the midgut epithelium (E) and the food bolus (F). (Micrograph kindly provided by W. Rudin.)

known as the extracellular membrane layer, and possibly fulfils the same function as a peritrophic membrane.

Peritrophic membranes are still an enigma to the physiologist, although several suggestions have been made for their function. The peritrophic membrane is the first line of antimicrobial protection for the cells of the midgut, and the presence of lectins capable of binding bacteria has been shown in blowflies and tsetse flies (Peters *et al.* 1983). Other suggested functions are based on the division of the midgut lumen into an endoperitrophic and ectoperitrophic space by the peritrophic membrane. It is suggested that this division increases the efficiency of digestion by spatially separating the various digestive processes, or that it permits the conservation of digestive enzymes which can be recirculated back up the midgut in the ectoperitrophic space (Terra 1988), or that the production of an unstirred layer next to the midgut cells may increase absorption efficiency. These various suggestions are at least partly true, but there is still a nagging feeling that there is some fundamental and still unrecognized role for the peritrophic membrane in digestive physiology. (See Section 8.7 for further discussion of peritrophic membranes.)

6.2 THE BLOOD MEAL

The size of the blood meal is affected by a range of factors including ambient temperature, insect age, mating status, stage of the gonotrophic cycle, previous feeding history, and source of the blood meal. Nevertheless, it is fair to say that temporary ectoparasites often take very large blood meals. Adult insects will commonly take meals which are twice their unfed body weight, while nymphal stages of the blood-sucking Hemiptera may take meals of ten times their unfed weight (Table 6.1)! These large blood meals certainly impair the mobility of the insect, increasing the short-term chances of it being swatted by the host or eaten by a predator. For example, the flight speed of *Glossina swynnertoni* is reduced from 15 to 3–4 miles per hour after feeding (Glasgow 1961), with

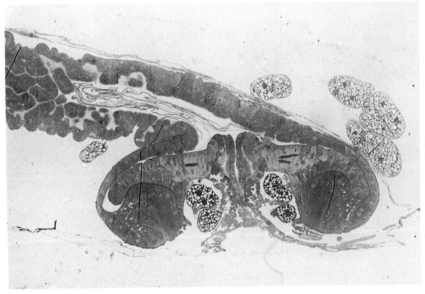

Figure 6.3 Type II peritrophic membranes are produced continuously from the proventriculus. Above, transverse section through the proventriculus of *Stomoxys calcitrans*. Right, transverse section through the fully formed, five-layered, peritrophic membrane of *S. calcitrans*.

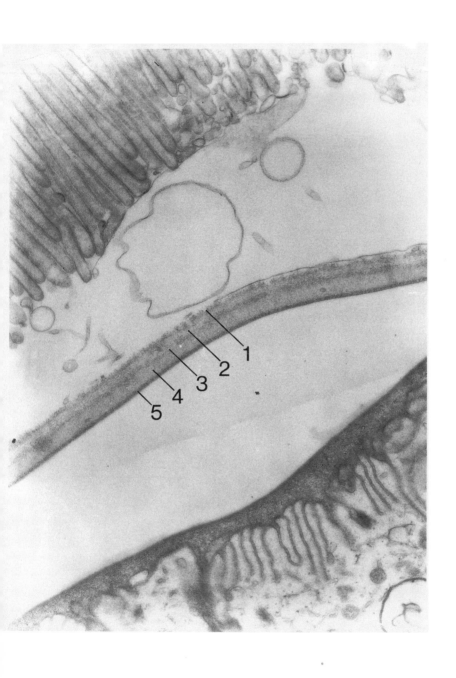

Table 6.1 The size of the red blood meal and the time taken in its digestion are affected by a range of factors including ambient temperature, age of the insect, mating status, stage of the gonotrophic cycle, previous feeding history, source of the blood meal, etc. The figures given here are a rough guideline to the 'average' meal size and time for digestion in a variety of haematophagous insects.

Species	'Average' time for the digestion of a blood meal by the adult female insect (h)	'Average' meal size for an adult female insect (in an early phase of the reproductive cycle) (% unfed body weight)
Pediculus humanus	4	30
Cimex lectularius	168	130
Triatoma infestans	336	210
Aedes aegypti	60	109
Anopheles quadrimaculatus	60	122
Culex quinquefasciatus	70	140
Culicoides impunctatus	130	—
Stomoxys calcitrans	48	110
Glossina m. morsitans	48	170
Pulex irritans	36	—

the fly often only capable of a downward 'glide' away from the host. However, as can be seen by its widespread adoption by so many different groups of blood-sucking insect, the disadvantages of taking such large blood meals must clearly be outweighed by the advantages.

The main benefit of taking very large blood meals is probably the minimization of the number of visits that the temporary ectoparasite must pay to the host; this could be advantageous in several ways. If hosts are difficult to find, then the insect must make the most of each encounter. It would also be selectively advantageous to take the largest meal possible if the extra time and energy required for more frequent host visits could be better spent in reproductive activity. But the main advantage to the insect is probably the reduced risk of being swatted which goes with visiting the host as few times as possible. Taking such large and discomforting meals is likely to be a disadvantage to permanent ectoparasites because it would restrict their mobility in the host's covering and none of the benefits outlined above would be gained. In consequence many of the permanent ectoparasites take small, frequent meals. For example, the anopluran lice feed every few hours and take meals which are only 20–30% of their unfed weights (Buxton 1947, Murray & Nicholls 1965).

Temporary ectoparasites have adapted their morphology and physiology to minimize the risks involved in taking such large meals. When a blood meal is taken it imposes considerable mechanical stresses on the storage zone of the gut and the abdominal wall. The crop of tsetse flies and the midgut storage regions of other blood feeders are capable of considerable stretching to accommodate the blood meal. Undoubtedly the nature of the intercellular junctions (Billingsley 1990) and underlying muscular coat of the midgut cells helps blood feeders to withstand these mechanical stresses. The abdominal wall is also very elastic or has folds permitting the considerable distension during feeding which is so characteristic of these blood-sucking insects. There is some evidence that the abdominal wall of *Rhodnius prolixus* may be plasticized in response to feeding (Bennet-Clark 1962), the elasticity of the abdominal wall being switched on and off in response to the blood meal. The abdominal wall is provided with stretch receptors to prevent overdistension (Fig. 6.4) (Maddrell 1963, Gwadz 1969). This is neatly shown in female tsetse flies which retain the developing larva inside the abdomen until the larva is fully mature. As the larva grows, the size of each blood meal diminishes so that the abdomen never exceeds a certain volume (Tobe & Davey 1972).

The disadvantages of taking very large blood meals are also overcome

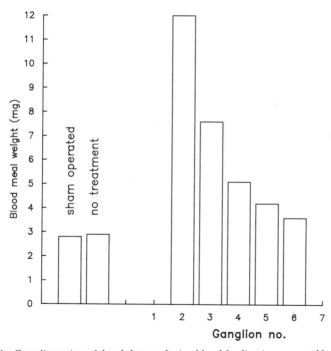

Figure 6.4 Overdistension of the abdomen during blood feeding is prevented by abdominal stretch receptors. This can be illustrated by the overdistension of the abdomen which occurs when the abdominal nerve cord of *Aedes aegypti* is cut at various points along its length. (Redrawn from Gwadz 1969.)

Table 6.2   The major constituents of the blood are reasonably uniform in most host animals. The exception is in the high levels of nucleic acids in the birds and reptiles because of their nucleated red blood cells. Proteins are far and away the most abundant nutrients in blood and nutrients are unevenly distributed between whole blood (B), red blood cells alone (E), or plasma alone (P). (Data drawn from Albritton 1952, and Altman & Dittmer 1971.)

		Water g/100 ml	Lipid g/100 ml	Carbohydrate g/100 ml	Protein g/100 ml	Nucleic acid g/100 ml
man	B 80	0.652	0.088	20.5	—
	E 72	0.596	0.074	36.8	0.136
	P 94	0.60	0.097	7.41	0.057
cow	B 81	—	0.046	—	—
	E 62	—	—	29	—
	P 91	0.348	—	8.32	—
chicken	B 87	—	0.170	—	—
	E 72	—	—	29	4.216
	P 94	0.520	—	3.6	—

by making use of the fact that about 80% of the blood meal is water (Table 6.2). Most of this excess water is not required by the insects, and they possess very efficient physiological systems for its rapid excretion, thereby reducing their weight and restoring their manoeuvrability. To achieve this the meal is held in a distinct region of the midgut where the epithelium is adapted for rapid water transfer (Fig. 6.5). In tsetse flies and triatomine bugs water movement across this epithelium is linked to a ouabain sensitive, Na^+K^+ ATPase located in the basal membranes of the epithelium with chloride as the counter-ion (Gooding 1975, Peacock 1981, 1982, Farmer et al. 1981). This very efficient pump works by generating an osmotic gradient across the epithelium which drags water passively after it. It is possible that the pump in *Rhodnius* is switched on by the same diuretic hormone, released from the mesothoracic ganglion in response to the blood meal, which stimulates a 1000-fold increase in fluid secretion from the Malpighian tubules. These systems are so efficient that tsetse flies can shed about 40% of the weight of the meal in the first 30 min following feeding (Moloo & Kutuza 1970, Gee 1975), and most of the fluid in the very large blood meals of triatomine bugs is discarded within four hours of ingestion (Wigglesworth 1931). A possible danger in this massive flux of water through the haemolymph of the insect is that the haemolymph balance will be disturbed. This does not occur probably because any flux in the haemolymph volume alters the concentration of the diuretic hormone. This in turn causes a feedback response in the rate of absorption by the midgut and/or secretion by the Malpighian tubule epithelium to correct the situation (Maddrell 1980).

The rapid and energy efficient removal of water from the blood meal would be assisted by the precipitation of the soluble materials in the plasma, because this would reduce the osmotic pressure of the gut contents. Given the rubbery consistency of the blood meal soon after its ingestion in many insects, it is possible that they have mechanisms to do precisely this, but the removal of water from the meal is also likely to effect this rubbery consistency. The 'coagulins' secreted into the ingested meal by many blood-sucking Diptera (Gooding 1972) precipitate at least a part of the soluble components of the plasma, thereby reducing its osmotic pressure. However, coagulation cannot explain the consistency of the blood meal in all blood feeders because blood-sucking Hemiptera do not produce coagulant, but anti-coagulins, in the midgut (Gooding 1972). Whether the jelly-like consistency of the blood is brought about by water removal, haemaglutination, the introduction of precipitating agents, or a combination of factors, requires further investigation.

Type II peritrophic membranes, by virtue of their filtration properties, may make water removal from blood meals more energy efficient. The large meal size in tsetse flies produces hydrostatic pressure within the

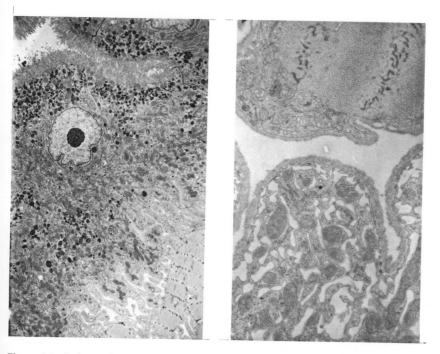

Figure 6.5   Left, an electron micrograph of the epithelium of the reservoir region of th anterior midgut of *Stomoxys calcitrans* in which the blood meal is stored. Right, th extensive basal infoldings of this region and the accumulations of mitochondria aroun them which are typical of tissues designed for the rapid transport of water.

peritrophic membrane leading to bulk filtration of the blood meal such that plasma, and those blood solutes having a diameter of less than 9 nm, are exuded into the ectoperitrophic space (Miller & Lehane 1990). Once in the ectoperitrophic space this filtrate can be efficiently absorbed by active transport through cells in the anterior part of the midgut. Bulk filtration saves the fly energy by reducing the osmotic pressure of the fluid it is absorbing.

There are other ways in which insects can minimize the dangers of taking such large blood meals. Lift in tsetse flies is proportional to the square of wing beat frequency, and wing beat frequency increases with temperature up to 32°C (Hargrove 1980). Fed tsetse flies maximize their manoeuvrability by generating heat in the thorax (Howe & Lehane 1986). They do this by rapidly vibrating the thoracic flight-box after uncoupling the wings, producing not only heat but also the characteristic buzzing sound after which the tsetse flies are named. This endogenously generated heat can have an important impact on the fly. A fly at an ambient temperature of 20°C can increase its lift potential by about 17% by raising thoracic temperature to the optimum of 32°C. Buzzing also increases abdominal temperature, which allows more rapid excretion of water from the meal, further improving the manoeuvrability of the fed fly.

The time taken for digestion of the blood meal varies widely both intra- and interspecifically, and is strongly influenced by ambient temperature, blood source, meal size, and several other factors (guideline figures are given in Table 6.1). This can be of importance when using serology in epidemiological studies to determine host choice rates (see Section 3.1) because it will determine the time period over which the blood meal is still identifiable. Determining digestion time can be complicated in some insects by their habit of jettisoning the semi-digested remains of one meal if another is offered. This potentially wasted resource is put to good use in some fleas where the larval forms feed on this supply of semi-digested blood from the adults.

One of the most important events in blood digestion is the lysis of the red blood cells because these contain much of the protein in the meal (Table 6.2). Blood meal haemolysis is achieved by a variety of physical or chemical means. Some mechanical haemolysis occurs in a few blood-sucking insects. In fleas this is achieved by repeatedly pushing the food bolus forwards against a series of backwardly projecting proventricular spines, and some mosquitoes possess a cibarial armature which ruptures as many as 50% of the blood cells as they flow past (Coluzzi et al. 1982). Cibarial armatures are found in other blood-sucking Diptera as well as mosquitoes, although it is not known if they cause haemolysis of their blood meals. It is known that cibarial armatures damage ingested filarial worms, and it has been suggested that they have developed as a defence against these parasites. It seems unlikely that their primary role is in haemolysis given their relative inefficiency.

Chemical haemolysins may be produced in the salivary gland, as in *Cimex lectularius* (Sangiorgi & Frosini 1940), but are more normally produced in the midgut. In *Rhodnius prolixus* haemolysin is produced in the anterior, storage region of the midgut where no proteolytic digestion of the blood takes place (De Azambuja *et al.* 1983); the molecule involved is a small basic peptide. In the tsetse fly and stablefly the haemolytic agent is secreted in the posterior, digestive region of the midgut, in response to some component of the cellular fraction of the blood meal (Gooding 1977, Spates 1981, Spates *et al.* 1982). In the tsetse fly the haemolysin is proteinaceous, but in the stablefly particular species of free fatty acid may bring about haemolysis.

The salivary glands of blood-sucking insects are unusual in producing virtually no digestive enzymes (Gooding 1972). The loss of hydrolytic enzymes in the saliva is probably related to the need for the insect to cause the least possible disturbance to the host during feeding. Most digestive enzymes in blood-sucking insects arise from the midgut cells. Because the blood meal is predominantly protein (c. 95%) it is not surprising that the major digestive enzymes are proteinases. Trypsins (proteinases active at alkaline pH) are predominant with carboxypepti-dases and chymotrypsins playing a subsidiary role in blood meal digest-ion in most blood-sucking insects. Hemipteran bugs are exceptional in using cathepsin-like proteinases (active at acid pH). This is consistent with a proposed evolutionary path for blood-sucking hemipteran bugs from sap-sucking ancestors. It is argued that Hemiptera with a sap-sucking life-style did not need trypsins and lost the ability to produce them. When they became blood feeders they required digestive protei-nases once more. All cells produce cathepsins for use in lysosomes and hemipterans may possibly have re-routed these enzymes for extracellular digestion (Houseman *et al.* 1985, Billingsley & Downe 1988, Terra 1988).

Some blood-sucking insects, such as *Stomoxys calcitrans*, secrete digest-ive enzymes by the regulated route; that is, digestive enzymes are stored in secretory granules in midgut cells and are secreted in response to the meal. In these insects digestion starts immediately the meal is taken with more enzymes being rapidly synthesized and secreted (Lehane 1976a, 1987, 1988, 1989, Moffat & Lehane 1990). In other insects, such as the mosquitoes, no significant store of enzymes is held in the unfed midgut; enzymes are only produced in significant quantities some hours after the meal is taken (Akov 1965, Hecker & Rudin 1981, Graf *et al.* 1986). In all blood-sucking insects, however, the levels of digestive enzymes increase after the blood meal, reaching a peak before declining to resting levels as digestion is completed. Secretion of digestive enzymes is regulated by the quantity of protein found in the midgut (Fig. 6.6)

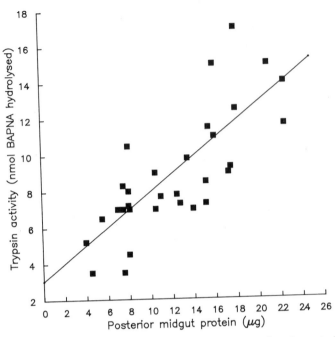

Figure 6.6 The level of trypsin secreted into the intestine is directly controlled by the quantity of protein present in the digestive region of the midgut. (Redrawn from Houseman *et al*. 1985.)

(Gooding 1974, Lehane 1977a, Houseman *et al*. 1985). It is not known if control is exercised directly on the enzyme-secreting cells or if it involves hormones released from the large number of endocrine cells which are interspersed among the digestive cells of the midgut. Some of these cells have been shown to contain FMRFamide and pancreatic polypeptide-like immunoreactivity indicating a role for these hormones in the regulation of digestive events (Billingsley 1990). It seems that different ratios of the various digestive enzymes can be induced by blood meals from different hosts. For example, blood from bird hosts, which contains nucleated red cells, may induce high levels of digestive DNAase. Variation in the digestive response to blood of different origins may have an important bearing on the susceptibility of the insect to parasites; *Phlebotomus papatasi*, for example, is readily infected with *Leishmania tropica*, but not if it is fed on turkey blood (Schlein *et al*. 1983).

Vertebrate blood contains proteinase inhibitors. These are designed to control tissue destruction by endogenously produced serine proteinases such as those produced by granulocytes. Insects must overcome the effect of these inhibitors if they are to digest the blood meal. These inhibitors slow the rate of blood meal proteolysis, especially in the early phases of digestion when relatively small amounts of insect trypsin are present (Huang 1971). Blood from different vertebrates displays different levels of inhibition of insect trypsins. Pig and human serum possess only about a third of the inhibitory effect of cow plasma on tsetse fly trypsins (Gooding 1974). It is possible that these inhibitors may play a part in causing the different rates of digestion seen in an insect dealing with blood from different sources. But this is certainly not always the case, as we can see if we look at the tsetse again. Despite a 300% difference in trypsin inhibition levels in the sera of the two blood meals, tsetse flies can digest a cow meal and a pig meal at the same rate. There is some evidence that insects compensate for the effects of these inhibitors. For example, blood possessing high levels of proteinase inhibitor may stimulate tsetse flies to secrete increased levels of trypsin (Gooding 1977). The tabanid *Chrysops excitans* appears to deal with the problem by having two populations of trypsin-like molecules: one which is serum-inhibited and another which is serum-activated.

Following extracellular digestion, further digestion of the meal takes place on and in the midgut epithelial cells. We know that aminopeptidase is bound to the microvillar membranes of midgut epithelial cells, and it is possible that intracellular digestion of absorbed peptides occurs in lysosomes (Billingsley & Downe 1985). Other digestive enzymes such as invertase, amylase, and esterases can also be found in the gut (Gooding 1972); these probably play a subsidiary role in blood digestion while some may be necessary for dealing with the insect's non-blood meals.

Absorption of the products of digestion also occurs in the midgut, but relatively little is known of the processes involved. In those insects showing continuous digestion, absorption of the digestive products occurs in a specialized region of the posterior midgut. The region involved is often characterized by the gradual appearance and accumulation of large lipid globules in the apices of its cells (Fig. 6.7), followed by their gradual disappearance as the insect is starved. This reflects the metabolic fate of the blood meal in adult insects where the absorptive products are largely converted to and stored as fats, even though the blood meal consists of about 95% protein and only about 4% lipid. These absorptive cells are probably involved in the conversion of the protein products of digestion to lipids (Lehane 1977b). Most of these lipids will eventually be transferred to other organs such as the fat body or ovaries. It is also possible, but less likely, that all conversions occur in the fat body and that the midgut wall is merely a storage organ as seems to be the case in one part of the midgut of *Rhodnius prolixus* (Billingsley & Downe 1989). In insects where the immature stages are also blood feeders, there may be less emphasis on conversion of digestive products to lipid because a considerable amount of protein is used in growth and development.

In insects where digestion is a batch process there is less evidence for

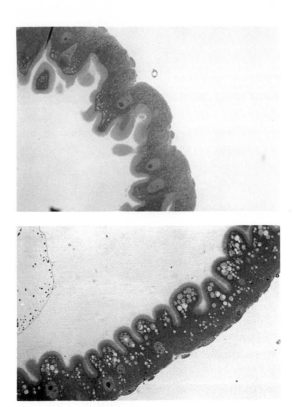

Figure 6.7 In those insects using the continuous rather than the batch method of blood meal digestion, the posterior region of the midgut is specialized for absorption. These cells may accumulate large amounts of lipid during the course of digestion and lose them again after digestion is complete. The figures show the absorptive region of the midgut of *Stomoxys calcitrans* in the (top) very early stages of, and (bottom) at the height of, blood meal digestion.

cell specialization in the midgut; most cells in these forms are probably responsible for secretion as well as absorption (Rudin & Hecker 1979, 1982). Lipid accumulations still occur in the midgut cells and the blood is still largely converted to lipid for storage.

6.3 GONOTROPHIC CONCORDANCE

In adult insects the blood meal is largely used to provide resources for the reproductive effort; thus in females digestion and ovarian development are physiologically integrated. The coordination of these events has been well studied in anautogenous mosquitoes where the blood meal triggers egg development. In these mosquitoes egg development is arrested at the 'resting stage' until a blood meal is taken which stimulates the release of egg development neurosecretory hormone (EDNH) from the corpus cardiacum. EDNH release may be induced by abdominal distension (Spielman & Wong 1974), or directly by a component of the blood meal (Chang & Judson 1977), indirectly by blood meal components stimulating the production of a releasing factor from the ovaries (Lea & Van Handel 1982), or by a combination of these mechanisms (Klowden 1986).

Many female blood-sucking insects will develop and lay a batch of eggs each time a blood meal is taken, providing the quantity of blood ingested exceeds a minimum threshold level. This process is called gonotrophic concordance (Swellengrebel 1929, Detinova 1962) and is exhibited by many mosquitoes, blackflies, sandflies, tabanids and, in a rather poorly defined way, triatomine bugs. The number of eggs produced by these insects is directly proportional to the size of the blood meal (Roy 1936, Goodchild 1955) up to a maximum which is determined by the number of ovarioles in the ovary. This in turn can be influenced by the nutritional history of the insect in its immature stages (Colless & Chellapah 1960). In insects showing gonotrophic concordance it is often possible, by the careful dissection and observation of the ovaries, to tell how many egg-laying (gonotrophic) cycles the insect has undergone. This in turn indicates how many blood meals have been taken by the insect and also allows us to estimate the insect's physiological age and possibly its calendar age (Detinova 1962). Such information can be of great value to the epidemiologist.

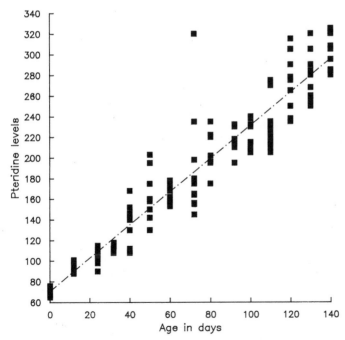

Figure 6.8 The graph shows the regular accumulation of fluorescent pigments in the head of an adult female tsetse fly (*Glossina m. morsitans*) as it grows older. This can be used to estimate the age of insects caught in the field. (Redrawn from Lehane & Mail 1985.)

Not all blood-sucking insects show gonotrophic concordance. Many, including biting flies, tsetse flies, hippoboscids and lice, require several meals for the production of each batch of eggs. Even some mosquitoes show a degree of gonotrophic dissociation, and some may need more than one meal to produce a single batch of eggs. It is often possible to estimate the number of gonotrophic cycles these insects have undergone from the structure of the ovaries. From this information estimates of physiological and calendar age can be made, but estimates of the number of blood meals taken are less reliable. Recently a new method for determining insect age has been introduced which relies on the regular accumulation of fluorescent pigments in the insect as it grows older (Lehane & Mail 1985, Lehane 1985) (Fig. 6.8).

6.4 NUTRITION

Blood-sucking insects can be divided into three groups on the basis of their feeding histories. The first group is a minor one, containing forms such as the man-feeding Congo floor maggot, *Auchmeromyia luteola*, in which only the larval stage is blood-feeding. The second group contains insects which feed exclusively on blood throughout their entire life cycle and includes tsetse flies, streblids, hippoboscids and nycteribiids, triatomine and cimicid bugs, and lice. The third group comprises insects in which the adults are blood feeders but the larval stages are not, and includes mosquitoes, blackflies, ceratopogonids, sandflies, biting muscoids, horseflies and fleas (although the young stages of fleas may feed upon semi-digested faeces derived from blood feeding adult fleas). The third group can be further subdivided into three categories: (i) species with adults who are obligatory haematophages, such as fleas, (ii) forms such as stableflies which are optional blood feeders, where both sexes will also take non-blood meals and (iii) forms in which only the adult female is an optional blood feeder, such as mosquitoes.

Insects relying solely on blood as a food source throughout their life harbour specialized symbiotic micro-organisms which are not found in insects using other food sources at some stage in the life cycle. This strongly suggests that blood is not a complete food source and that the symbionts supplement the insect's nutrition in some way. Various methods have been developed for the removal of symbionts from living insects so that their role in nutrition can be determined. Some of these methods depend on the physical location of the symbionts or their means of transfer from one insect to the next, both of which may differ among insect groups (Table 6.3). Symbiont-free reduviid bugs have been produced by physically preventing passage of the symbiont to the next generation (Brecher & Wigglesworth 1944). In the louse, *Pediculus*

94

Table 6.3 Symbionts are common in insects relying on blood as the sole food source throughout their lives. An outline is given of their anatomical locations and the means of transmission from one generation to the next in different insect groups. (Information from Buchner 1965.)

Insect group	Location of symbionts	Means of transmission between generations
Anoplura	intracellular in midgut epithelium, between the epithelial cells of the midgut or in a mycetome beneath the midgut epithelium	infection of the egg during its intra-uterine development in the female
	in mycetomes in the haemocoel	
Cimicidae	mycetomes in the haemocoel	the oocyte is infected and so the new individual contains symbionts at conception
Reduviidae	partly intracellular in the midgut epithelial cells and partly in the gut lumen	the first stage nymph acquires the infection from egg shells or the faeces of infected bugs
Hippoboscidae	intracellular in a specialized zone of cells in the gut epithelium or in a mycetome adjacent to the gut epithelium	transmitted to the intra-uterine larva in the secretion of the mother's milk gland
Streblidae	small numbers of symbionts are found in a variety of different cells of the gut, Malpighian tubules, fat body and milk glands	transmitted in the secretion of the mother's milk gland
Nycteribiidae	in mycetomes which may lie at various sites in the haemocoel	transmission is via the secretion of the mother's milk gland
	symbiont-containing cells may also occur in the fat body and in the female, intertwined with the tubules of the milk glands, or only cells of the milk gland may be infected	
Glossinidae	intracellular in a specialized zone of cells in the gut epithelium	transmitted to the intra-uterine larva in the secretion of the mother's milk gland

humanus, the organ containing the symbionts (the mycetome) has been physically removed (Aschner 1932, 1934, Baudisch 1958). In tsetse flies antibiotics have been used against the symbionts (Hill *et al.* 1973, Southwood *et al.* 1975, Pell & Southern 1976, Schlein 1977) or they have been attacked with lysozyme (Nogge 1976) or with specific antibodies (Nogge 1978).

Suppression of the symbionts in the juvenile stages impairs subsequent insect development. Loss of symbionts from the adult does not affect its general health measured, for example, in terms of longevity; symbionts are often infrequent or absent in adult males. For example, in the lice and the nycteribiid, *Eucampsipoda aegyptica*, despite being present in the juvenile stages, symbionts are absent from adult males (Aschner 1946). Symbionts do, however, affect the reproductive performance of the adult female. The general situation is well illustrated in tsetse flies where the number of symbionts halves in the emerging adult males while it doubles in emerging adult females to give a complement of four times that seen in the mature male (Nogge & Ritz 1982). This increase in symbiont numbers is necessary for female tsetse flies to reproduce normally as symbiont-suppressed females are sterile (Nogge 1976, 1978); sterility can be partially reversed by feeding the symbiont-free flies on blood supplemented with various vitamins (Nogge 1981). The essential supplements are the B group vitamins, thiamine and pyridoxine, along with biotin (vitamin H), folic acid and pantothenic acid (Fig. 6.9).

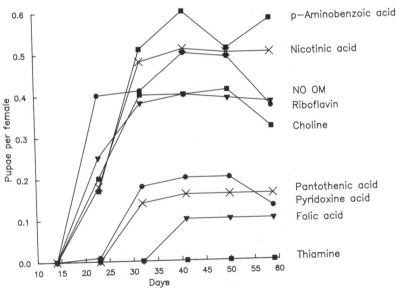

Figure 6.9   The diet of symbiont-free female tsetse flies was supplemented with a range of vitamins (no omission). In a series of experiments the named vitamin was then omitted from the full diet. The effect on fecundity (expressed as puparial output per surviving female per nine-day period) is shown. (Redrawn from Nogge 1981.)

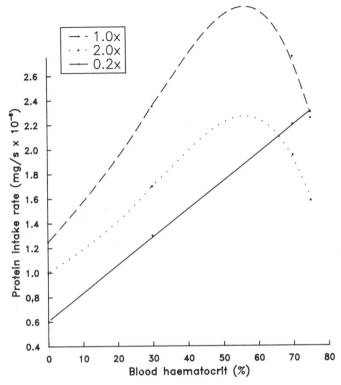

Figure 6.10 The rate of protein intake from a blood meal depends on the proportion of cells in the blood. The normal haematocrit for human blood is c. 40% but this is lowered by many parasitic infections. If no other factors are considered, lowering of the haematocrit will lower the rate of protein intake, a factor which will favour blood-sucking insects feeding from uninfected hosts. Natural selection has ensured the evolution of mouthparts showing efficient mouthpart design. We can see this if we look at pharyngeal height in *Aedes aegypti*. The broken line shows protein intake rate calculated for this mosquito. If pharyngeal height were either increased or decreased then protein intake rate would fall. This is illustrated here: protein intake rates are calculated for theoretical pharyngeal heights of 0.2 times and two times that in the mosquito. (Information kindly supplied by T.L. Daniel & J.G. Kingsolver.)

The antibacterial enzyme lysozyme is abundant in insect body tissues and is used to regulate the body areas which can be colonized by symbionts. In tsetse flies only the mycetome region of the midgut lacks this enzyme, so limiting colonization to this organ (Nogge & Ritz 1982). Because there are such clear-cut differences between the numbers of symbionts in male and female tsetse flies there obviously must be other factors at play within the mycetome cells themselves which regulate symbiont numbers (probably related to the lysosomal system of the cells) (Nogge & Ritz 1982).

The health of the host can influence the nutritional quality of its blood. Insects achieve theoretical maximum protein intake in a blood meal at high haematocrits (Fig. 6.10). The normal human blood haematocrit is about 40%, but many blood parasites cause a reduction in this value either by direct destruction of red blood cells or indirectly by interfering with the physiology of the host. So, in theory at least, blood-sucking insects might improve their nutritional status by selectively seeking out and feeding on uninfected hosts, but other factors also need to be considered (see Ch. 8).

Particular species of blood-sucking insect find some hosts are 'better' food sources than others; when they feed on a different host species they may show reduced fecundity generated by a reduced rate of development, a skewed sex ratio, reduced food intake or rate of digestion, or reduced longevity (Nelson *et al.* 1975, Rothschild 1975). This can be seen in the mosquito, *Culex pipiens*, which produced 82 eggs mg^{-1} of ingested canary blood compared to 40 eggs mg^{-1} of human blood (Woke 1937). Similarly the tsetse fly, *Glossina austeni*, despite showing similar survivorship when fed on rabbit or goat blood, showed a consistently higher fecundity when fed on rabbit blood (Fig. 6.11) (Jordan & Curtis 1968). Both the quantity and the quality of the offspring can be affected. *Glossina austeni* and *G. m. morsitans* produce the heaviest puparia when fed on pig

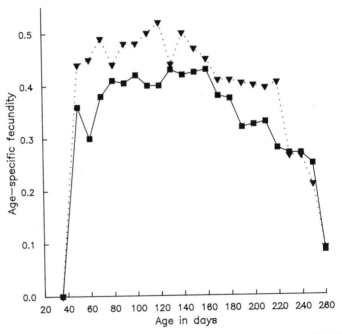

Figure 6.11 *Glossina austeni* shows increased fecundity when fed on rabbit blood (▼) rather than goat blood (■). (Redrawn from Jordan & Curtis 1968.)

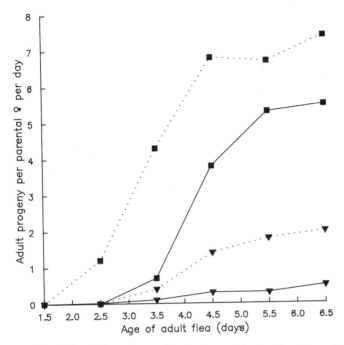

Figure 6.12   Fleas fed on baby mice produce far fewer offspring than those fed on adult mice. ···· ■ ····, Adult mouse, fleas held at 30 °C; ——■——, adult mouse, fleas held at 22 °C; ···· ▼ ····, young mouse, 30°C; ——▼——, young mouse, 22°C. (Redrawn from Buxton 1948.)

blood, followed closely by puparia produced by flies fed on goat blood with the puparia produced from flies fed on cow blood being considerably lighter. The differences in the suitability of hosts as food sources is not just interspecific; for example, *Xenopsylla cheopis* fed on baby mice produce far less eggs than fleas fed on adult mice (Fig. 6.12). From the studies performed to date it is often not clear what the particular differences between various blood sources are. The different effects observed may be a function of reduced food intake, due to immune or other defence responses of the different hosts, or to the ratio of specific elements in the meal such as the amino acids. It is known that the levels of host hormones in the blood meal can have a profound impact on the biology of some blood-sucking insects.

6.5   HOST HORMONES IN THE BLOOD MEAL

The best understood example of the interaction between a blood-sucking insect and host hormonal levels is that of the rabbit flea, *Spilopsyllus cuniculi*, feeding on the European rabbit, *Oryctolagus cuniculus* (Mead-Briggs 1964, Rothschild & Ford 1973). Throughout the rabbit's non-breeding season the distribution of fleas on its surface is related to skin temperature. The fleas are usually grouped on the rabbit's ears where they are attached by their mouthparts. In very cold weather the fleas move onto the body of the animal and in hot weather the fleas leave the rabbit completely if they cannot hold their temperature below about 27°C. This description of flea distribution on the body of the host is true only in a snapshot sense. If the association is continuously watched, fleas are constantly seen moving between rabbit and nest or directly from one rabbit to another.

The distribution described above, modulated by a range of environmental and physiological factors, persists throughout the rabbit's non-breeding season. With the onset of mating, in March and April, changes begin to occur. During mating, ear temperature of the rabbits rises by up to 7°C and this stimulates the movement of fleas from one host to another. Blood levels of various hormones such as progestin and oestrogen also change in the female rabbit (doe) when she copulates. The changed hormonal levels in the blood meal induce the fleas to attach firmly to the doe. The effect of the increased interchange of fleas between hosts, combined with the increased attachment to female rabbits, is that fleas tend to accumulate on the does during the mating season. About 10 days before the doe gives birth there is a large rise in the corticosteroid hormone levels in her blood. The raised levels of these hormones in the flea's food, combined with other hormonal influences from the rabbit, cause a series of dramatic changes in the fleas. Ovarian and testicular maturation take place accompanied by hypertrophy of the midgut, salivary glands and fat body. The fleas also increase their feeding rates from about once every 15–30 min to once every 4 min in males and to once every minute or so in females! More changes occur in the hormonal levels in the doe's blood when she gives birth causing the fleas to move from the ears to the face of the doe, from where 80% of the doe's fleas will eventually move onto the new litter of rabbits.

As the fleas feed on the young rabbits they encounter a new range of hormones in their food. They are also exposed to kairomones which are probably associated with the urine of the young rabbits. The combined effect of these factors is to stimulate the fleas into an orgy of mating and egg laying over the eight-day period following the birth of the litter. By this time the hormonal levels in the litter have changed and the fleas move back to the doe over the next 14 days. Under the influence of the new levels of luteinizing hormone and progestins in the doe's blood, the somatic changes which the flea underwent on the pregnant doe are reversed. The flea reverts to its non-reproductive phase until it encounters another pregnant rabbit.

While the events outlined above indicate the typical response of most fleas in a given population, not every flea responds to the changing physiological status of the rabbit in exactly the same way. This variation is to be expected as it will have survival value for the flea population, allowing it to withstand potential disasters such as the death of a litter or the abandonment of a burrow. Another such adaptation is seen in the development of the flea larvae in the burrow. One cohort develops rapidly and leaves the burrow with the does and the young rabbits. The second cohort are larger, better stocked with reserves than the first, and these develop more slowly, remaining in the burrow to await the reoccupation of the nest. In this way a disaster befalling one group of rabbits will not necessarily mean a catastrophe for that particular population of fleas.

Linking reproductive effort to that of the host is clearly advantageous to the insect as their offspring have the maximum chance of successfully finding a host which is not already saturated with parasites. This is a common strategy in many blood-sucking insects and also in other parasites. What is unusual here (and in a few other ectoparasitic forms) is the very complex physiological link between host and parasite ensuring the precise timing of reproduction in the invertebrate partner. It could also be argued that what is unusual here is our depth of understanding of the linkage and that more complex interactions of this sort will emerge with further detailed investigation. This complex reproductive control mechanism may have evolved in fleas because of the special feeding requirements of the larvae, which need the supply of dried blood found in adult flea faeces. Clearly this food resource is only likely to be generally available when rabbits habitually rest in a single place over a considerable period of time, the most reliable time being when a new litter is produced in the nest.

6.6 PARTITIONING OF RESOURCES FROM THE BLOOD MEAL

How are the resources gained from the blood meal partitioned into the various activities which the blood-sucking insect has to perform? Tsetse flies provide a well-studied example. The newly emerged adults use much of the resources of the first blood meals to build up their thoracic musculature (Bursell 1961). Subsequently resources are partitioned between basal metabolism, flight and reproductive effort; an energy budget is shown in Figure 6.13 (Bursell & Taylor 1980). Clearly, under each of the different conditions shown in Figure 6.13, the reproductive effort in the female consumes a considerable proportion of the available energy while it appears to be of negligible importance in the male.

101

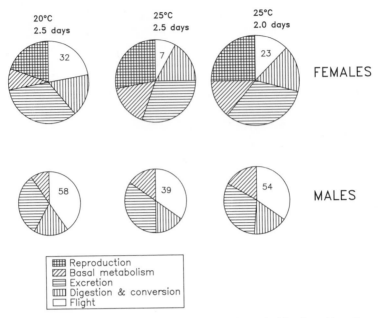

20°C	25°C
2.5 days	2.5 days

25°C
2.0 days

FEMALES

MALES

▦ Reproduction
▨ Basal metabolism
▤ Excretion
▥ Digestion & conversion
☐ Flight

Figure 6.13 Tsetse flies partition the energy derived from the blood meal into the various activities the insects must perform. The daily input of energy is represented by the area of the circle. Partitioning is affected by the sex of the fly and the conditions it is held under (the temperature and feeding intervals are given). The numbers in the flight sectors of the pie charts are the daily duration of flight in minutes which these reserves would permit. (Redrawn from Bursell & Taylor 1980.)

Again, regardless of the different conditions, nearly half the resources of the blood meal are devoted to excretion and the digestion of the meal. As basal metabolism is directly related to temperature, at lower temperatures more energy is left over each day for flight. The length of the inter-larval period is inversely related to temperature; therefore, at lower temperatures less resources are put into the female reproductive effort each day and so again more energy is available for flight. In reality it is likely that these theoretical effects are offset by the increasing intermeal period as the temperature falls, which will reduce the total energy available each day.

The females have considerably less energy available for flight than the males because of the major investment of their resources in reproduction (Fig. 6.13). Both sexes probably expend similar amounts of energy in host location and seeking out refuges in the intermeal period. As additional flights in the males are sexually appetitive ones (active seeking of receptive virgin females), they also invest any spare energy in reproductive effort (Bursell & Taylor 1980).

Tsetse flies are unusual in using an amino acid (proline) as the primary substrate for flight. This is possibly a biochemical adaptation to the blood-sucking habit (Bursell 1975, 1981), a view supported by the following analysis. A range of Diptera have been classified on the ability of their mitochondria to oxidize either the amino acid proline or pyruvate and α-glycerophosphate (both of which are markers of carbohydrate as a fuel source) (Bursell 1975). They were found to fall into three categories (Fig. 6.14). In the first category are flies which utilize carbohydrate as an

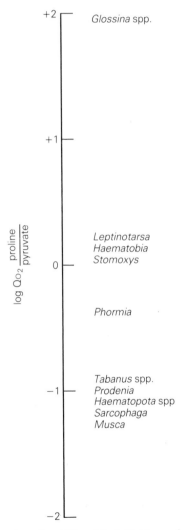

Figure 6.14 When compared on the ability of their flight muscle mitochondria to utilize either the amino acid proline or pyruvate (a marker for the use of carbohydrates as a fuel source), blood-sucking insects fall into three groups, as shown here. (Redrawn from information in Bursell 1975, 1981.)

energy source for flight; they are either non-blood feeders or only the female blood feeds. In the second category are facultative haematophages in which both sexes feed on blood as well as other foods; these insects can efficiently use both proline and carbohydrate for flight. In the third category are the tsetse flies, both sexes of which are obligate haematophages; proline is the only energy source used for flight. This analysis shows that as blood increases in prominence in the diet there is an increasing use of proline as an energy source; but proline is also used as an energy source in non-blood-sucking insects such as the colorado beetle. Clearly the biochemical capability for the use of proline is widespread in insects and the blood-sucking habit is only one life-style promoting its use over other energy sources. Why blood-sucking insects choose to utilize amino acids as an energy source instead of conventional carbohydrates or lipids has been considered by Bursell (1981), whose arguments are outlined below.

About 90% of the dry weight of the blood meal is protein and so the most abundantly available resources are amino acids (Table 6.2). These could be converted to carbohydrate through the gluconeogenic pathway to provide a good energy source, which is far more easily stored than amino acids and, because it is highly soluble, could be easily translocated to the flight muscles when required. Gluconeogenesis would be best performed on the amino acids glycine, alanine and serine which would yield three carbon fragments, but these amino acids are drained from the available pool by the need to dispose of nitrogen in uric acid (McCabe & Bursell 1975). The other amino acids produce two or four carbon fragments which it would be energetically uneconomical to convert to carbohydrates.

Amino acids could also be converted to lipids to provide an excellent energy source which could be readily stored in the body. Mobilization of lipids in the haemolymph is difficult, but it has been overcome by all insects. In some insects mobilization is so efficient that lipids form a fuel for flight, but this is not the case in Diptera.

So the direct use of an amino acid as a fuel source may be an energetically useful alternative in insects with a high proportion of protein in their diets. This would be particularly true if an amino acid is available which is a good energy source, is low in nitrogen atoms to minimize the energy used for its excretion, and which is highly soluble to permit easy translocation to the flight muscles during times of high demand. Figure 6.15 shows that proline is not only highly soluble, but also has a high net energy yield carrying only a single nitrogen atom per molecule. It has been firmly established that tsetse flies do indeed use proline as the prime energy source for flight (see Bursell 1981). The energy is supplied by the partial oxidation of proline to alanine in the flight muscle, with the subsequent reconversion of alanine to proline in

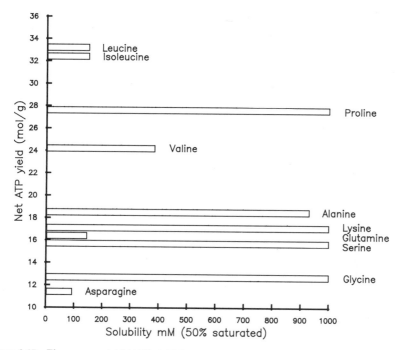

Figure 6.15 The energy yield (ATP yield from total oxidation of the amino acid minus the energy required to detoxify their nitrogen) is expressed against water solubility for a range of amino acids which might potentially be flight energy substrates. (Drawn from data in Bursell 1981.)

the fat body at the expense of stored triglyceride (Fig. 6.16). The synthetic pathway in the fat body cannot keep pace with the rate of proline oxidation occurring during flight and so tsetse flies are restricted to fairly short flight periods. The 'proline battery' is then recharged during the enforced rests between flights.

6.7 AUTOGENY

Autogeny, the capacity to produce at least one egg batch without the need for a blood meal, is found in several blood-feeding insects including some mosquitoes, ceratopogonids, sandflies, blackflies and tabanids. Autogenous insects carry over the nutritional requirements to produce the egg batch either largely or wholly from the immature feeding stages. The adult female may supplement these reserves from several non-blood food sources (see below). Both facultative and obligate autogeny are known. Facultatively autogenous insects have the choice of producing an egg batch autogenously or of taking a blood meal and producing a larger

105

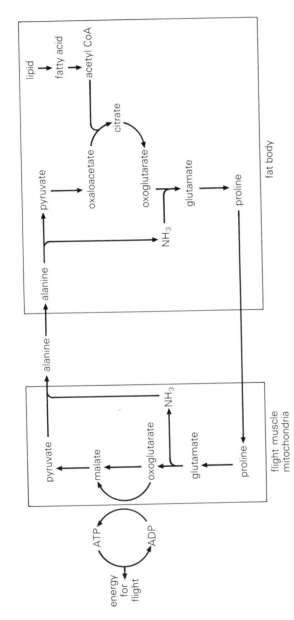

Figure 6.16 Proline is used as the energy source for flight in the tsetse fly. The metabolic pathways involved in both the derivation of energy from the amino acid and the subsequent recharging of the 'proline battery' from triglycerides in the fat body are shown. (Redrawn from Bursell 1977.)

egg batch. This flexibility allows the insect to adjudge whether autogeny is the best strategy considering both the quality of its larval feeding and the availability of hosts for the adult. If no hosts are available to the adult some eggs can still be produced. If hosts are available the adult female can choose to blood-feed and many more eggs can be produced.

Facultative autogeny is clearly a useful strategy; insects in which autogeny is obligatory do not have this choice. Even if feeding was poor for the immature stages of obligate forms they are nevertheless obliged to produce an autogenous egg batch before blood-feeding. This usually means less eggs in the first egg batch. Some insects appear to have gone further still, and are unable to take blood with egg production being entirely reliant on larval feeding. So autogeny, in one of its various forms, may be a useful adaptation for those insects where the quality of larval and/or adult feeding is unpredictable.

The selection pressures which lead to the development of autogeny can operate at the level of either the adult or immature stage of the insect. Let us look at the larval stage first. Density-dependent effects on the larvae of the pitcher plant mosquito, *Wyeomyia smithii*, determine the degree of autogeny displayed by the adult female. In the northern part of its range the larvae of this mosquito are never overcrowded in their very specialized habitat, the leaf pitchers of the purple pitcher plant (Bradshaw 1980, 1983), but in the southern part of their range crowding does occur (Bradshaw & Holzapfel 1983). Because they can acquire sufficient reproductive resources during larval feeding, the northern females produce several egg clutches autogenously (Hudson 1970). In their southern range, where larval feeding is poorer because of crowding, the females require blood to produce the second and subsequent egg batches (Bradshaw 1980).

An example of selective pressures operating on the adult stage is given by two populations of *Aedes taeniorhynchus* studied on the Florida Keys (O'Meara & Edman 1975). One population, on Big Pine Key, had easy access to hosts (deer), while the second population on Flamingo Key had few potential hosts available. Not surprisingly, the engorgement rates of mosquitoes sampled on Big Pine Key (20.9%) were far higher than those on Flamingo Key (8.1%). When the populations were investigated for the presence of autogenous individuals it was found that the Flamingo Key population carried a significantly higher percentage of autogenous forms and that these showed higher fecundity than the autogenous forms from Big Pine Key. From this and other similar studies, it seems that availability of blood sources for the adult female can be a factor influencing the levels of autogeny in a population (O'Meara 1985, 1987). But in many localities closely related autogenous and anautogenous forms are sympatric. So while the availability of nutrients may be a major factor in the expression of autogeny, as shown in the two examples given above, it cannot entirely explain the occurrence of autogeny.

Table 6.4 Three types of female *Ae. taeniorhynchus* have been identified in terms of egg development; autogenous females (1), females which are autogenous if mated (2) and anautogenous forms (3). But this pattern is influenced by the feeding success of the larval stage, as illustrated in this table. (Redrawn from O'Meara 1985.)

Larval diet	Type of female 1	2	3
high	autogenous	autogenous if mated	anautogenous
intermediate	autogenous if mated	autogenous if mated	anautogenous
low	anautogenous	anautogenous	anautogenous

Autogeny in mosquitoes can be triggered by mating; a component of male accessory gland secretion is the triggering substance (O'Meara & Evans 1976). This phenomenon can be seen in *Ae. taeniorhynchus* where the situation is a complex interaction between the genotype of the mosquito, mating and larval diet (Table 6.4) (O'Meara 1979, 1985). Male-induced autogeny usually leads to the production of relatively few eggs (*c*. 25) compared to populations displaying non-male-induced autogeny (*c*. 65) or blood-fed females (*c*. 150) (O'Meara & Evans 1973). Females which can be stimulated into autogeny by mating actively seek a blood meal which substantially increases their reproductive success. Autogeny is a back-up strategy in which a few eggs are still produced even if the search for a host is unsuccessful. In Hemiptera (but not in the mosquitoes described above) the mated male may provide other factors (possibly nutrients) which assist in the development and increase the size of the egg batch. This is shown in both *Cimex* and *Rhodnius* females, which will produce up to 25% less eggs if mated with a previously unfed male compared to a fed one (Buxton 1930, Usinger 1966).

Interestingly, the male insect is rarely considered in discussions of autogeny possibly because much of the work on autogeny has used species where only the female feeds on blood. Like females, the male insect's capacity for successful reproductive activity can depend on their feeding history. Unfed male lice (Gooding 1968) or male *Glossina morsitans*, which will attempt to mate but are incapable of inseminating females (Foster 1976), can be compared with both *Cimex* and *Rhodnius* males which can successfully mate before they are fed (Usinger 1966, Buxton 1930). The factors influencing these different evolutionary choices remain unexplored.

Some autogenous insects can increase the number of eggs they lay by sugar-feeding. Mosquitoes, for example, can efficiently convert sugar to fat, and feeding on sugar meals alone leads to the accumulation of fat in the body (Van Handel 1984), especially in facultatively autogenous mosquitoes (O'Meara 1985) where sugar-feeding may be used to boost the number of autogenously produced eggs (Table 6.5). The sugar is obtained from a variety of sources including flower nectaries, overripe fruits and aphid honeydew. Not all insects are capable of efficient conversion of sugars to fats. *Stomoxys calcitrans* does not show a net increase in the fat reserves of the body after sugar meals (Venkatesh *et al.* 1981, Venkatesh & Morrison 1982) and so sugar meals will make little direct contribution to female fecundity. Sugar may still make an indirect contribution to fecundity by increasing the energy available to the insect; for example, the energy available from the sugar meal increases the insect's flight capacity and its chances of finding a host. It will also lower the demand on the female to use part of the blood meal as an energy source, thereby freeing more resources for egg development.

Feeding on more 'watery' foods (for example nectar) in addition to blood is common in several blood-sucking insect groups, including mosquitoes, blackflies, tabanids and blood-feeding muscoids. However, a blood meal is essential for the successful development of large egg batches in most of these insects. So it is important to these insects that they are not prevented from taking a blood meal because they have just fed on sugar solution. To avoid this, the insects display a dual sense of hunger; in other words despite feeding to repletion on sugar-water these optionally blood-sucking insects will still take a blood meal if it becomes available. The reverse is also true. Recently blood-fed *Aedes aegypti*, which had previously been water-starved, will still probe hot water-soaked pads (Khan & Maibach 1971). This dual sense of hunger is common to those Diptera, blood feeders and non-blood feeders alike, which will feed on sugar solutions but which require a protein meal before they can produce eggs. Regulation of this dual sense of hunger has been studied in the non-blood-feeding blowfly, *Phormia regina*, (Belzer 1978a, b, c, 1979). It has been shown that sugar and protein intake are regulated in two separate ways. Negative feedback from the recurrent

Table 6.5 Some mosquitoes can use sugar meals (10% sucrose in this experiment) to increase the number of autogenously produced eggs. (Adapted from O'Meara 1985.)

Species (strain)	Treatment	Female _n_	% autogenous	Eggs per female (mean ± S.E.)
Aedes taeniorhynchus	unfed	20	90.0	40.8 ± 4.6
(flamingo)	sugar	20	85.0	63.7 ± 7.4
Aedes bahamensis	unfed	20	100.0	53.5 ± 2.1
(Grand Bahama)	sugar	20	100.0	61.0 ± 1.7

nerve on the foregut regulates sugar intake, while feedback from stretch receptors in the abdomen controls the intake of protein. This bicameral control system enables the dual sense of hunger because the insect's hunger for sugar is satiated well before complete distension of the abdomen. This leaves room for a protein meal which is stopped following full stretching of the abdomen.

The efficient use of this dual sense of hunger is also aided by the structure of the gut in Diptera which are optional blood feeders (Fig. 6.1). Under most circumstances sugar meals are sent to the crop and only regurgitated into the midgut as required. This leaves the midgut free to accept an immediate blood meal should one become available. So the insect can bypass the sugar meal and immediately digest the more important protein meal. Diversion of sugar to the crop may possibly have other functions. Nectar contains inhibitors of insect proteinases (Bailey 1952), so the sudden arrival of a full nectar meal in the midgut might seriously impair enzyme activity. Diverting sugar to the crop and gradually regurgitating it in small packets may thus protect the enzymes needed for digestion of the blood meal. Sugar storage in the crop may also be important for water conservation in the insect because the crop is cuticle-lined and no absorption takes place from it.

Using this dual sense of hunger is clearly an advantage to the insect when the availability of blood meals is unpredictable. The sugar meals tide the insect over between blood meals, but do not seriously impair the insect's capacity to take a blood meal should one become available.

CHAPTER SEVEN

Host–insect interactions

Every animal needs to maintain a steady internal environment in order to carry out the various physiological processes which together allow life to continue. The process is known as homeostasis. One of the key organs in homeostasis is the animal's surface covering, be it skin, cuticle or whatever. In mammals and birds the surface covering has become extended to incorporate an outer insulating layer of hair or feathers. These layers have proved to be excellent homes for many different species of permanent and periodic ectoparasitic insects providing them with a relatively constant environment in which to spend their days.

Vertebrate skin is formed of an inner dermis and an outer epidermis (Fig. 7.1). The thickness of these two layers, and the ratio between them, varies considerably on different parts of the body. The dermis is a connective tissue layer carrying the services of the skin, the blood, lymphatic, and nervous systems. Embedded in the dermis are the acini of the gland systems which open onto the skin's surface. There are two basic types of skin-associated gland in mammals. The sebaceous glands produce an oily secretion called sebum which helps prevent the skin and

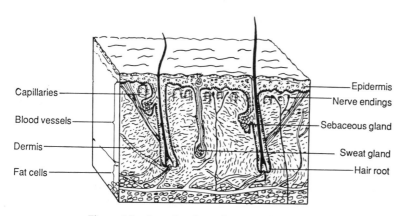

Capillaries

Blood vessels

Dermis

Fat cells

Epidermis

Nerve endings

Sebaceous gland

Sweat gland

Hair root

Figure 7.1   A section through mammalian skin.

111

hair from drying up and which may contain a bactericide. The sweat glands (which are not present in carnivores or rodents) produce a watery secretion used in temperature regulation and in maintaining water and salt balance. Only the uropygial (preen or oil) gland is present in birds. It is located in front of the tail and is used to provide oil for preening. The principal function of preening is to maintain the condition of the feathers and in particular their water-repellent properties.

The epidermis is formed of epithelial cells; this layer does not have nerves, blood or lymph vessels within it. The epidermal cells lying at the dermis–epidermis junction divide and give rise to a stream of new cells which gradually pass through the various levels of the epidermis, eventually arriving at the skin's surface. During this migration they synthesize and store increasing amounts of keratin before eventually dying and forming the outer keratinized skin layer. They are ultimately lost from the surface of the skin as dry, scaly flakes. Keratinized cells are also used to manufacture hairs, feathers and nails. Hair and feathers serve a variety of useful functions such as extra mechanical protection, insulation, coloration (for camouflage or communication purposes), locomotion or as an aid to buoyancy.

7.1 INSECT DISTRIBUTION ON THE SURFACE OF THE HOST

Permanent and periodic ectoparasites are distributed among the hair and feathers in a non-random fashion. One of the major influences on their distribution is the variation in the microclimate in different parts of the body covering. The most important microclimatic factor is the temperature gradient found in the covering; this has both a vertical and lateral component. The degree of shade and the humidity level in different parts of the hair or feather covering also appear to be important in some instances.

The temperature in the hair or feather covering is determined by ambient and core body temperature, the thickness of the covering and the degree of exposure to the sun's rays. Insulation and ambient temperature are not the only variables here; core body temperature also varies with the health of the animal, its activity patterns, age and species (birds commonly have body temperatures over 40°C, while marsupials fall in the range 30–35°C). By altering local blood flow the animal can also alter the temperature of selected parts of its body surface to suit particular circumstances. This means that the temperature of the body's extremities (tail, feet, nose, etc.) are often nearer to ambient than core body temperature. Over the trunk of the body skin temperature is usually maintained nearer core body temperature. Here the density of hairs or

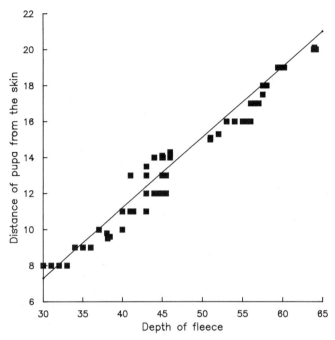

Figure 7.2 Fleece thickness in sheep affects microclimate in the fleece which in turn controls the distance from the skin at which pupae of *Melophagus ovinus* are attached. (Redrawn from Evans 1950.)

feathers, their length and orientation to the skin (which the animal can alter) are more normally the factors determining the temperature rather than alterations in the local blood supply. Let us look at some examples of the effect of temperature gradients on insect distribution in the body covering.

The sheep ked, *Melophagus ovinus*, migrates vertically within the sheep's fleece according to fluctuations in temperature (Evans 1950). This has been shown by looking at the larviposition behaviour of the ked. As the thickness of the fleece varies the fly moves nearer or further from the skin till it finds the right temperature for puparial deposition (Fig. 7.2). The distribution of lice on sheep has also been partly attributed to temperature variation. *Linognathus ovillus* and *L. pedalis* both prefer hairs to fleece and both require similar, comparatively low temperatures for successful reproduction. In consequence *L. ovillus* is most common on the face of sheep, while the foot louse, *L. pedalis*, is restricted mainly to the legs (Murray 1963). An extreme example of temperature-dependent site selection is seen in the anopluran louse, *Lepidophthirus macrorhini*, which is found on the southern elephant seal, *Mirounga leonina*. These lice are inactive below 5°C and are most active at 20–30°C. This essentially

Table 7.1 The choice of feeding site of *Aedes triseriatus* on eastern chipmunks and grey squirrels is influenced by body-hair length and density. The different feeding patterns on the two hosts reflects the differences in hair cover between them. (From Edman *et al.* 1985.)

Host	Back	Ear	Eyelid	Foot	Nose	Chi²
eastern chipmunk	0	27	13	2	8	$p<0.01$
mean hair density per 9 mm²	(1583)	(153)	(254)	(49)	(58)	
mean hair length (mm)	(10.5)	(1.6)	(2.9)	(2.1)	(0.9)	
squirrel	0	19	11	16	4	$p<0.01$
mean hair density per 9 mm²	(454)	(120)	(59)	(81)	(119)	
mean hair length (mm)	(13.6)	(2.2)	(2.0)	(3.8)	(1.4)	
chi² test		N.S.	N.S.	$p<0.01$	N.S.	

limits their distribution to the hind flippers because onshore, with an air temperature of 1.8°C, the temperature of the back of the seal is 6°C and that of the flippers is about 30°C. But the seal only comes ashore for two brief periods (three to five weeks) each year. At sea the flippers are still the chosen site for the lice because they remain, periodically at least, the warmest skin areas as they are used to dissipate heat during bouts of strenuous activity and may be warmed at other times by being held out of the water. It is presumably during these warmer periods that the louse feeds while at sea (Murray & Nicholls 1965).

Anyone who keeps a bird or a dog, or has looked carefully at themselves naked in the mirror, knows that the texture, size, density and colours of feathers or hairs show considerable variation from one body area to another. Such variations in the physical nature of the habitat are a second factor leading to the concentration of permanent and periodic ectoparasites in specific areas. For example, the louse, *Haematopinus asini*, deposits its eggs on the coarse hairs of the tail, legs and mane of the horse but not on the finer body hair. The non-blood feeding louse, *Damalinia equi*, is smaller than *H. asini* so it cannot use these large diameter hairs; therefore it is confined to the horse's finer body hair (Murray 1957). A similar distribution is seen in the cattle lice, *Haematopinus eurysternus* and *Damalinia bovis* (Matthysse 1946).

The feeding sites of temporary ectoparasites are also affected by the varying degree of cover and the different skin thicknesses on different parts of the host's body (Table 7.1) (a quick rule of thumb suggests that the thicker the overlying coat the thinner the underlying skin). Different species of tabanid select markedly different landing sites on cattle (Fig. 7.3) and a positive correlation has been shown between the hair depth at the landing site and the length of the mouthparts of the tabanid choosing to land there (Mullens & Gerhardt 1979). Another example is furnished by the distribution of *Lipoptena cervi* on the host animal. The total span of the tarsi of this hippoboscid is about 0.22 mm, limiting the fly to host regions with hair of this diameter or less, such as the deer's groin, which is one of the fly's favoured sites (Haarlov 1964). A second factor determining site selection by *L.cervi* appears to be skin thickness. The mouthparts are only 0.9 mm long and the fly needs to feed where vascularization of the dermis is nearest to the skin's surface. Clearly the skin areas which have the greatest degree of vascularization near to the surface are probably also the warmest; thus the groin, which is the favoured site for

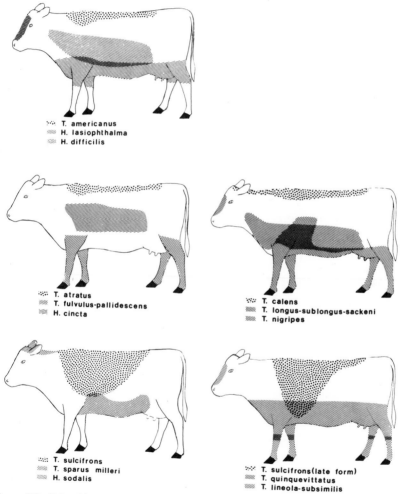

Figure 7.3 Tabanids show a marked preference for different landing sites on the host. Some of these are illustrated here. There is a correlation between the length of the mouthparts of the various species and the thickness of the coat at the various landing sites ($r = 0.5513$). (Redrawn from Mullens & Gerhardt 1979.)

L. cervi, is the warmest skin area on the deer. Heat is likely to be an important factor guiding such insects to their preferred site.

The degree of protection offered by a particular site is another important physical characteristic affecting ectoparasite distribution. Thus, no lice are found on the heads of British grebes or divers where the head feathers offer little or no protection for the insects when the birds are underwater (Rothschild & Clay 1952). Similarly, some aquatic mammals, such as the coypu, trap air in certain parts of the pelage when they are diving. The ectoparasites of such animals are often restricted to these parts of the body which remain dry throughout the dive (Newson & Holmes 1968).

Ectoparasites can adapt to a more or less complete loss of skin covering. For example, elephants (Mukerji & Sen-Sarma 1955) and pigs (Henry & Conley 1970) have only sparse body hair; the lice found on these animals have adapted to living in folds in the skin.

7.2 MORPHOLOGICAL SPECIALIZATIONS FOR LIFE ON THE HOST

As with all living organisms, blood-sucking insects are continuously evolving to better fit their ever-changing niche. Many of the characters adopted are subtle and difficult to ascribe to the blood-sucking way of life while others are more clear-cut, particularly those of the permanent ectoparasites with their highly specialized life-style. The most obvious specializations of blood-sucking insects, their mouthparts, are dealt with separately in Section 5.5.

Blood-sucking insects are far smaller than their hosts. There is a rough correlation between the size of permanent ectoparasites and that of their hosts which suggests there may be an optimum size ratio, but the adaptive significance of the relationship has not been explained (Kim 1985). In the case of the permanent and periodic ectoparasites it is obvious that small size is a great advantage to the insect in helping it escape the host's grooming activities. In the temporary ectoparasites small size helps them to approach and escape with their blood meal, unnoticed by the host. However, not all blood-sucking insects are 'small'. When viewed from a human perspective some of the larger tabanids and triatomine bugs are in anyone's eyes substantial insects. But the potential problems caused by their unusual bulk are acceptable only because they are offset by other factors in their relationship with the host. For example, an adult *Triatoma infestans* is about an inch in length, but because of its exceptionally stealthy approach to the host, commonly a sleeping human, it normally escapes with a full blood meal despite the fact that it commonly feeds around the face. Large tabanids prosper

116

either because the host, commonly a large herbivore, is relatively insensitive to the attack or is unable to defend itself effectively.

Many blood-sucking insects have clearly modified shapes to suit a particular life-style. The lateral flattening of fleas and some streblids and the dorsoventral flattening of lice, polyctenid bugs and hippoboscid flies is a morphological adaptation to life in the covering layer of their hosts. Flattening of these insects allows them considerable freedom of movement among hair or feathers, and also allows adpression against the host or its covering. Both of these adaptations help the insect to evade the host's grooming activities. The flattening may even allow the insect to more regularly avoid the tines of any comb-like device used in grooming. The flattening of the triatomine and cimicid bugs allows them to retreat during the intermeal period, to the safety of cracks and crevices in the home of the host.

Wings may also be a barrier to rapid movement within the covering of the host. Ectoparasites have dealt with this problem in several ways. Some, such as the hippoboscid *Melophagus ovinus*, have lost their wings entirely and rely on other means of transfer between hosts. Other forms lose their wings once the host is found. This may happen by the progressive abrasion of the wings as occurs in some moths, or by the deliberate shedding of part or all of the wing as occurs in females of the streblid *Ascodipteron* and in hippoboscids such as *Lipoptena*. As well as permitting easier passage through the covering of the host, loss of the wings also prevents the insect undertaking flights which may cause it to lose contact with the host. The female streblids of the genus *Ascodipteron* go even further. On contacting a host they shed not only their wings but also their legs. They then burrow deeply into the host's skin (during which they may lose the halteres as well) so that only the rearmost abdominal segments are exposed (Marshall 1981).

Flattening is a general adaptation to life in the host's covering, but more subtle adaptations are often seen suiting the permanent ectoparasite to particular parts of its host's body. Consequently, different species inhabiting similar sites on an animal often adopt a similar shape and size; conversely large morphological differences may be seen between related species living on different parts of the same animal. Such subtle adaptations can often be seen in the surface covering of ectoparasites. Commonly the cuticle of permanent ectoparasites is covered with spines and bristles which may be aggregated together in the form of combs. These cuticular extensions are seen in the streblids, but are particularly well developed in the polyctenids, nycteribiids and fleas; other ectoparasites from the Anoplura, Staphylinidae, Pyralidae and Hippoboscidae possess analogous cuticular structures (Marshall 1981). These cuticular extensions are usually associated with delicate appendages such as antennae, or weaker areas in the insect's cuticle such as the intersegmental membranes. Clearly the type and degree of development of the cuticular extension adopted by the ectoparasite are partly dependent on the type of host it is associated with. This can be seen in the remarkable degree of convergent evolution which has occurred among distantly related ecto-parasites on different host types. At a macro-level this is shown in bird-infesting forms which tend to have much longer, more slender bristles than mammal-infesting close relatives. The nature of the association is also important. Nest-dwelling fleas which visit the host only briefly to feed tend to have a reduced cuticular embellishment compared to fleas which live for long periods in the host's fur. But the association between host type and the form of the cuticular covering of the insects ectoparasitic upon it can be traced down to a much more detailed level. Traub (1985) states:

The chaetotactic modifications may be so diagnostic that infestation of a shrew can be recognised merely by examining the spines of the flea, irrespective of the taxonomic standing or geographic location of the flea.

The function of the tremendous array of combs, bristles and spines present on ectoparasites is the subject of some debate. It has been proposed that the combs are used for attachment or to help prevent dislodgement (Humphries 1967, Amin & Wagner 1983, Traub 1985) which could be achieved by interlocking with host hair, particularly when there is backward movement of the insect. It has also been argued that the combs protect the delicate body regions (Smit 1972, Marshall 1980). It seems improbable that the remarkable degree of convergent evolution in the shape, size and spacings of combs which is shown by poorly related fleas on the same host, or the correlation of the spacing of the cuticular extensions with the size of the host hair (Fig. 7.4) (Humphries 1967, Lehane & Lenka, in prep.) would be seen if the cuticular embellish-ments had only a protective function. In addition there is a good correlation between the degree of development of these cuticular embel-lishments and the comparative risk to the flea if it lost contact with the host. So, the bristles, spines and combs are developed to the highest degree in ectoparasites of nocturnally active, flying or tree-dwelling hosts while they are least developed in ectoparasites of diurnally active, surface-dwelling communal forms (Traub 1985). Clearly the danger to the flea in losing host contact in the former case is far greater than in the latter, again indicating an anchorage function for the combs. On the other hand these embellishments are commonly associated with delicate parts of the insect's body, such as antennae, mouthparts or intersegmen tal membranes, pointing to a protective function (Marshall 1981) however, there is no biological reason why the embellishments could no

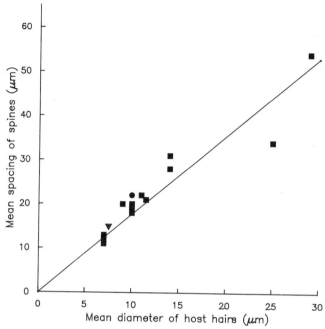

Figure 7.4 There is a significant correlation between the spacings of the spines on ectoparasites and the width of the host's hair. This suggests that the spines can interlock with the hair helping prevent the dislodgement of the parasite. ■, 15 spp. of flea; ●, *Platypsyllus castoris*; ▼, *Nycteribia biarticulata*. (Redrawn from Humphries 1967.)

perform both functions. The association with the weaker points of the cuticle suggests a possible protective origin with a later adaptive evolution of the size, shape and spacing of the cuticular extensions to meet the anchorage requirements demanded by the host's covering and life-style. The evidence suggests that the combs, bristles and spines seen today serve the dual purpose of attachment and protection.

An urgent problem which must be faced by all insects is water loss. The lipids of the cuticle play a vital role as a waterproof covering. Abrasion of the surface layers of the cuticle may cause a dramatic and fatal increase in water losses. Ectoparasites, particularly those which move rapidly through closely packed hair, are at risk because of the abrasive character of their surroundings. Waterproofing of the cuticle depends not only on the outer wax layer but also on lipid in the other constituent layers of the cuticle. For this reason, and because of the thinner separating layer between the haemolymph and the air, water loss from an abraded thin area of cuticle like an intersegmental membrane is probably considerably greater than losses from abraded areas of thicker cuticle. Combs, spines and bristles which guide hair and feathers away from these membranes may well have an important role in protecting the insect from desiccation.

119

Blood-sucking insects, particularly the permanent and periodic ectopa-rasites, have developed specific means of attachment to the host. Tarsal claws are almost universal and are used for gripping the hair or feathers of the host animal. Hippoboscids possess paired tarsal claws each of which operates against a basal thumb. Ornithophilic species tend to have lighter, more deeply cleft claws than mammophilic forms (Kim & Adler 1985). There are similar arrangements in nycteribiids (Theodor 1967) and the legs of anopluran lice also possess very well-developed claws. Some permanent ectoparasites effect an even stronger attachment to the host. The sticktight fleas, such as *Echidnophaga* spp., have anchoring mouth-parts which allow them to attach firmly to the host for long periods. Burrowing fleas such as the chigoe flea of man, *Tunga penetrans*, actually tunnel into the skin (it is possible that the host immune response to the attached flea is of more importance in the embedding of the flea than the burrowing activities of the flea itself; Traub 1985). These burrowing forms are commonly neosomic (Audy *et al.* 1972), showing a tremendous enlargement of the abdomen compared to the rest of the body after attachment of the adult to the host. Streblids of the genus *Ascodipteron* also attach firmly to the host and are neosomic; the adult females shed their wings and their legs once the bat host has been contacted, then burrow deeply into its skin.

To ensure that their offspring remain in contact with the host, lice possess special cement-producing glands. These permit the female to glue her eggs to the host's hairs or feathers or, in the case of the human body louse, *Pediculus humanus*, to the host's clothes. One flea, *Uropsylla tasmanica*, is also known to glue its eggs to the fur of the host (Dunnet 1970), and females of the wingless hippoboscid, *Melophagus ovinus*, glue their viviparously produced offspring to the fleece of the sheep.

As mentioned above, hair and feathers form an abrasive and intrusive environment. Ectoparasites, particularly the permanent ectoparasites, have adapted their body shape to minimize the damage that the covering causes. For this reason a reduction in the size of antennae and/or their protection in refuges is common in permanent ectoparasites. It is seen in the anopluran lice of mammals where the more highly specialized forms, such as the Linognathidae, have reduced the number of antennal segments even below the five seen in other Anoplura. Robert Hooke (1664) first described the antennal groove into which the flea's antennae can be lowered and raised again as required. Nycteribiids and streblids also hide their antenna in grooves; in hippoboscids the delicate arista of the flattened antennae can be hidden in a groove of the larger second antennal segment. Like the antennae, the potentially vulnerable mouth-parts are also usually protected in some fashion. This may be achieved by folding them back under the head as occurs in Polyctenidae, by folding them under and partly retracting them into the head as occurs in

Siphonaptera, by retracting the delicate components fully into the protection of the head and its snout-like extension as occurs in Anoplura, or by heavy cuticularization of the complete head and thorax as occurs in hippoboscids. In nycteribiids the head is bent back and lies protected in a groove on the dorsal surface of the thorax.

Adaptation of the sensory organs occurs in response to the blood-sucking habit. Reduction in the size and protection of the antennae has been discussed above. Loss of eyes is common in permanent ectoparasites, particularly the smaller forms. Eyes are greatly reduced or absent in polyctenids, apterous hippoboscids, nycteribiids, streblids, and anopluran lice. This may be another adaptive response to the abrasive surroundings of these insects, permitting the thickening of the cuticle of the head capsule and so minimizing abrasion damage. The loss of receptors is permissible because of the continued close association of the insect with the host. If we look at chemoreceptors we can see another pattern emerging as a response to the blood-sucking habit. This is illustrated in the mosquitoes where the female bears between two and four times the number of chemoreceptors as the non-blood-feeding male (Chapman 1982). This sexual dimorphism may be a direct result of their different feeding habits.

7.3 HOST IMMUNE RESPONSES TO INSECT SALIVARY SECRETIONS

After a few feedings by a particular blood-sucking insect species on a host animal, a pruritic, red weal will start to appear at biting sites. This is the basis of most peoples earliest awareness of blood-sucking insects – their bites itch! . . . but what causes the itch?

Three possibilities immediately present themselves: (a) the response is a localized traumatic reaction to the injury caused by the insect's mouthparts; (b) it is a response to a toxin introduced into the wound in the insect's saliva; or, (c) it is an immune response to an antigen in the saliva. It was shown not to be a response to mechanical injury in a series of experiments in which the salivary glands of mosquitoes were surgically severed. These mosquitoes were then unable to introduce saliva into the wound during feeding and no host reaction occurred when they subsequently fed on a sensitized host (Hudson et al. 1960). So the saliva causes the problem. Accumulated weight of evidence, largely based around the passive transfer of reactivity from sensitized to naive hosts, subsequently showed that the response was immunologically based and was due to an antigen rather than a toxin in the saliva.

There has been a tremendous development in the understanding of the vertebrate immune response in the 25 years since we began to

121

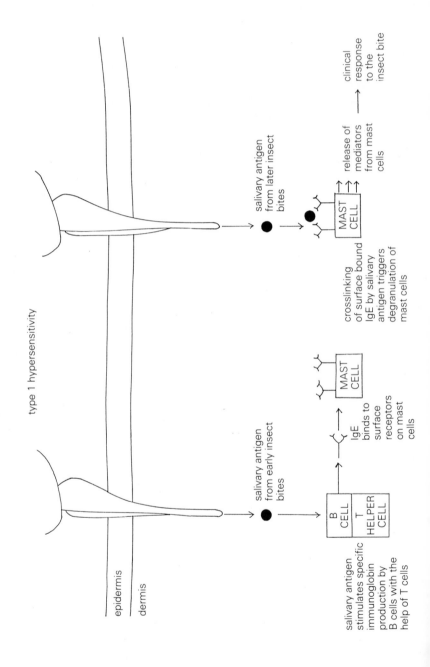

type 1 hypersensitivity

epidermis

dermis

salivary antigen from early insect bites

salivary antigen stimulates specific immunoglobin production by B cells with the help of T cells

B CELL

T HELPER CELL

IgE binds to surface receptors on mast cells

MAST CELL

salivary antigen from later insect bites

crosslinking of surface bound IgE by salivary antigen triggers degranulation of mast cells

MAST CELL

release of mediators from mast cells

clinical response to the insect bite

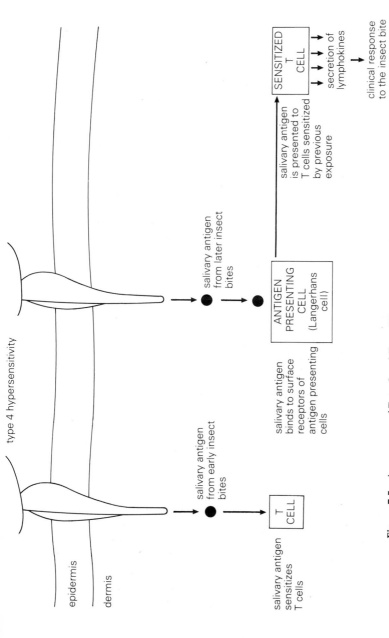

Figure 7.5 A summary of Type I and Type IV immune responses to insect bites.

123

appreciate the immunological basis of the host response to insect bites; however, very little of this work has been carried out on vertebrate reactions to insect bites. In consequence, while some of the details of the response are becoming clear, much remains to be discovered. The picture that will emerge after further research will probably be far more complex than the one currently available.

Some of the details of the vertebrate response to insect saliva are known and it seems that the vertebrate reaction often follows a five-staged career (Nelson 1987):

(1) No response.
(2) A delayed (Type IV) immune response with associated pruritis (itchiness).
(3) An immediate (Type I) immune response followed, 24 – 48 h later, by a delayed (Type IV) immune response with both stages having an associated pruritis.
(4) Only the immediate (Type I) immune response with an associated pruritis.
(5) No response.

(A detailed discussion of immunological mechanisms is outside the scope of this book but the basic details of Type I and Type IV reactions are given in Figure 7.5, and the main factors leading to pruritis are outlined in Figure 7.6.)

The various stages in the sequence of response to insect bites given above were first described in some depth for the interaction of guinea-pigs and fleas (Larrivee et al. 1964). The details of the histopathology seen in this interaction, which are outlined in Box 7.1, are not standard for every insect bite, but are a good general guide to the type of response which has been reported in the few thorough investigations carried out to date. Much more work in this area is required before a clear picture will emerge; thus although it is known that basophils play a major part in the vertebrate reaction to tick bites (Wakelin 1984), their role in responses to insect bites is still largely unexplored.

The number of bites required before the onset of each of the five stages of the response are known to vary considerably among different individuals, with different insect–vertebrate combinations and with the conditions under which the study is made (Nelson 1987). The timing seen in the highly controlled guinea-pig and flea study cannot be used as a definitive guide to reaction times in the real world where widely varying responses occur. As a rough guide, regular exposure to bites usually ensures stage 5 is reached within two years. With irregular exposure each stage can be very protracted and stage 5 may never be reached. Even when it does occur the lack of response seen in stage 5 may

124

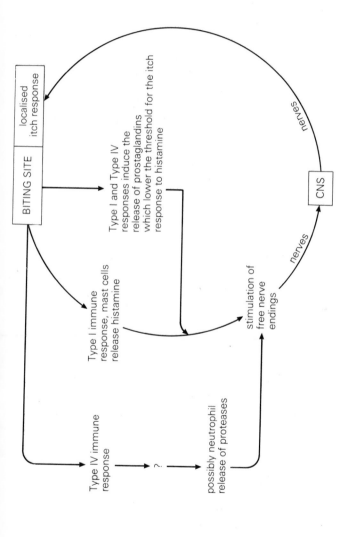

Figure 7.6 Type I and IV immune responses may follow an insect bite and both are associated with an intense itchiness (pruritis). The mechanisms leading to the itch are summarized.

BITING SITE

localised itch response

Type I and Type IV responses induce the release of prostaglandins which lower the threshold for the itch response to histamine

Type I immune response, mast cells release histamine

Type IV immune response

?

possibly neutrophil release of proteases

stimulation of free nerve endings

nerves

nerves

CNS

125

Box 7.1 Histopathology of the various stages in the sequence of host response to insect bites

---

The various stages in the sequence of response to insect bites were first described in some depth for the interaction between guinea-pigs and fleas (Larrivee *et al.* 1964); the details of the histopathology are given here.

Stage 1 lasts about five days and the animal shows no response to the flea bites during this period.

Stage 2 begins about six days after the fleas' first feed on the guinea-pig. Following feeding there is no immediate immune response; but 24 h after the blood meal a clear response is visible; the bite site has been infiltrated with monocytes and lymphocytes and, to a lesser extent, with eosinophils.

Stage 3 begins about nine days after the fleas' first feed on the guinea-pig. The immediate immune response to insect-feeding now becomes apparent and 20 min after the blood meal has been taken the bite site is invaded by large numbers of eosinophils. The delayed response is still occurring and by 24 h after the bite the eosinophils have been largely replaced by monocytes and lymphocytes.

Stage 4 begins about 50 days after the fleas' first feed on the guinea-pig. The immediate response is in operation and the bite site is rapidly infiltrated by eosinophils. The delayed response has now largely disappeared and very few monocytes or lymphocytes appear in the wound.

Stage 5 begins about 80 days after the fleas' first feed on the guinea-pig. Desensitization to the bites occurs and almost no cellular response to the bite can be seen.

---

be restricted to particular zones of the body which are receiving regular bites. If the biting insect moves just a few inches from this zone it may still induce a full reaction at the new biting site.

There are other factors defining the nature of the immune response and its timing in addition to the host's degree of prior exposure to antigen and the regularity of that exposure. These factors include the age and nutritional status of the host, the nature of the antigens introduced by the insect, and the route by which they are introduced (i.e. pool versus vessel feeders). There is also a genetic component in host responsiveness. In any given host population there is usually a spectrum of reactivity to bites. At one extreme, a few of the individuals being bitten may die from anaphylactic shock, others may develop massive oedema or an intense pruritis, while at the other extreme some individuals may fail to respond to bites at all; most hosts will fall somewhere between. Work on mice infested with the louse, *Polyplax serrata*, has illustrated the variation in resistance which can occur (Clifford *et al.* 1967); this work is outlined in Figure 7.7.

In some brave research, in which mosquitoes were allowed to feed on

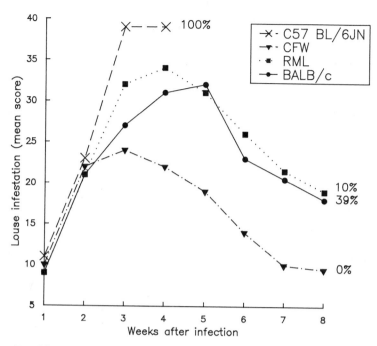

Figure 7.7 Mice prevented from grooming can only control infestations of lice (*Polyplax serrata*) by immunological means. By infesting different strains of mice (see key) the genetically determined variability in the immunological responses of the host becomes clear as some strains develop greater infestations than others. Mortality (%) is greatest in those strains developing the highest infestations. Louse infestations are quantified as an average score. This is an arbitrary scale from 0 to 40 where 0 = no lice; 10 = rare; 20 = few to moderate; 30 = many; 40 = very many. (Redrawn from Clifford *et al.* 1967.)

the investigator, the time of onset of the pruritis associated with the immediate response (stages 3 and 4) was established to be about three minutes (Gillett 1967). Because of the alerting effect of the pruritis on the host, the mosquito is well advised to have completed the meal inside these three minutes and to have left the scene of the crime. Those which are still feeding at that time have an increased chance of being swatted (a most efficient agent of natural selection) or of being disturbed before a full meal has been taken (which may affect reproductive success). In this way the immediate response, and the irritability resulting from it, are likely to be strong factors selecting haematophagous insects for rapid feeding, especially if the host animal is an efficient groomer. This view is supported by comparative work on the rapidity of feeding of wild and colonized mosquitoes whose regular hosts include primates. The colonized mosquitoes had been fed on restrained hosts over a period of three years prior to the experiments which had effectively removed the selective pressure for rapid feeding. It was found during the experiments that these mosquitoes would often fail to feed within the 'safe period' (i.e. in the three minutes before the onset of the immediate response). In contrast wild mosquitoes rarely failed to complete feeding within the 'safe period' (Gillett 1967). Although these results are intellectually appealing the situation is not clear-cut because in more recent laboratory work it proved impossible to select *Aedes aegypti* for feeding rate (Mellink 1981).

It is interesting to consider the concept of a safe period when, as happens under field conditions, a series of temporary ectoparasites attack a host over a period of time. In these circumstances only the first insect to feed has a safe period; subsequent insect attacks are likely to be met by an irritable host. As argued above, an irritable host is likely to be dangerous, possibly killing the insect or impairing its reproductive success by interrupting its feeding. This series of events are to the reproductive advantage of the first insect that managed to feed within the safe period, because its offspring will have fewer competitors. In this sense stimulating an immune response in the host could be seen, not as an unfortunate consequence of the injection of anti-haemostatic factors in saliva, but as a positive advantage to the insect which can arrive early enough and which can feed quickly enough.

Not all blood-sucking insects complete a blood meal in under three minutes. For example, adult triatomine bugs may take 20 min to complete a meal. Whether they cause an immediate hypersensitivity response and how they cope with any host irritability caused by it is not known. The consequences of host irritability are considered further below

7.4 BEHAVIOURAL DEFENCES OF THE HOST

Given the correct microclimatic and physical conditions, insects usually choose to feed (and reside in the case of permanent and periodic ectoparasites) at host body sites where they will be least disturbed. Because the host grooms, preens or shows a variety of other defensive behaviours towards ectoparasites this means, in practice, that ectoparasites are usually found in only a restricted part of their potential range. This is most obviously seen in permanent and periodic ectoparasites: the distribution of the louse, *Polyplax serrata*, on the mouse is a well-described example. Normally, an infested mouse bears about 100 lice which are congregated on its head and neck. The grooming of the head and neck is performed by the toes whereas the rest of the body is groomed by using its two lower incisors in a comb-like manner. If the mouse is prevented from grooming with its teeth, lice move onto the body and increase in numbers to over 2000 in three weeks. Clearly grooming with the teeth is a more efficient method of removing lice than grooming with the toes. Grooming with the toes does nevertheless kill lice. This is shown by the increased louse populations on the head and neck when the mouse is prevented from using its feet for grooming (Murray 1987).

Mutual grooming is another important factor limiting louse numbers on mice. In stable hierarchies of mice, prevented from grooming with their feet, louse burdens on the head and neck areas were still kept low because mice groomed each other using their teeth (Lodmell *et al.* 1970). Interestingly, if the hierarchy of the group was not stable, grooming did not control louse numbers and mice died from the excessive burdens of lice which developed. Mutual grooming is also seen in social birds such as penguins (Brooke 1985) and, of course, in the primates including man.

Grooming, as a means of permanent ectoparasite control, decreases in efficiency with increasing host size. A healthy mouse can restrict lice to the head and neck. On the rat, *Rattus norveigicus*, lice are to be found widely distributed on the trunk. On the vole, *Microtus arvalis*, which is intermediate in size between the mouse and the rat, lice are again restricted to the neck and head with another colony at the base of the tail (Murray 1987). But, although larger animals may be unsuccessful in restricting parasites to particular parts of the body, grooming by the host is still important in restricting the total number of ectoparasites on the body as a whole. For example, the ox can use the coarse surface of its tongue as a comb in grooming and can significantly reduce the numbers of lice on its body as a consequence (Lewis *et al.* 1967).

When the host moults the number of ectoparasites falls significantly. There is some direct parasite loss with the covering itself while changes in the microclimatic and physical characteristics of the insect's environment also reduces parasite numbers. Other insects die due to the increased efficiency of grooming in the thinner outer covering of hair or feathers. Moulting may also occur artificially, as it does each season in the shearing of sheep, where it causes a dramatic fall in the population of the sheep ked, *Melophagus ovinus* (Evans 1950). Moulting takes a variety of forms in different animals. It causes most damage to ectoparasite populations when it is sudden, is widespread on the body surface and involves the loss of a considerable proportion of the body's protective covering. But moulting is not always disastrous for ectoparasites. The lice *Haematopinus asini* and *H. eurysternus* both deposit their eggs on the coarse hairs of the tail, legs and mane but not the finer body hair of their respective hosts, the horse and cow. The non-blood-feeding lice *Damalinia equi* and *D. bovis* are smaller and egg attachment is confined to the finer hair covering of the horse and cow, respectively (Matthysse 1946, Murray 1957). Both *Damalinia* spp. are seriously affected by moulting, as large numbers of eggs will be shed along with the body hair of the animals, but the two *Haematopinus* spp. are hardly affected because, unlike the body hairs, the long hairs of the mane and tail are not shed wholesale in the moult.

Even though permanent ectoparasites are normally best adapted for life on one particular part of a host animal, and are restricted further by the grooming activities of that animal, given the right circumstances they can often spread widely over the animal's surface; this is often seen in sick or injured hosts. In such animals ectoparasites are dispersed over an unusually large host body area, often in abnormally high numbers. The principal reason for this is probably the inability of the host to groom itself efficiently, but delay in moulting and the possibility that an unhealthy host is more attractive or available to the insect (see Section 8.4) may also be contributory factors.

The number of temporary ectoparasites which successfully feed on a host is also affected by host defensive activity. One detailed study has been conducted on the feeding success of mosquitoes on a selection of ciconiiform birds (herons and egrets). The birds were held overnight in test cages containing mosquitoes and the feeding success of the insects was determined the following morning. It was clear from the results that the black crowned night heron and green heron were bitten far more frequently (three to eight times more) than the other five species used. This variability of biting frequency was not related to size, colour, weight or smell of the birds. Experiments in which restrained birds were exposed to mosquitoes made it clear that different levels of anti-mosquito behaviour among the different birds determined which species were bitten most often. The types of anti-mosquito behaviour seen in these birds is summarized in Table 7.2. Foot-pecking and foot-slapping appeared to be most effective in species where mosquitoes attack the exposed leg (Weber & Edman 1972). The five species which achieved a degree of success in protecting themselves from mosquito attack showed an average of about 3000 movements per hour, so they were virtually in perpetual motion! The most frequently bitten species, the green heron and black crowned night heron, also displayed anti-mosquito activity but at the lower level of about 650 movements per hour (Weber & Edman 1972).

These studies also showed that host anti-mosquito behaviour could have a significant impact on the amount of blood which a feeding mosquito obtained. Less than 2% of the mosquitoes feeding on the black crowned night heron or the green heron failed to get a full blood meal but between 15 and 31% of those feeding on bird species that showed more efficient anti-mosquito behaviour obtained less than half of a complete blood meal.

The amount of host defensive behaviour seen is directly related to the number of mosquitoes attacking the host (Edman *et al.* 1972, Waage &

Table 7.2 The anti-mosquito behaviour of a range of ciconiiform birds, showing that different host species display various types and degrees of defensive behaviour against blood-sucking insects. (From Edman & Kale 1971.)

Anti-mosquito behaviour	Night heron	Green heron	Little blue heron	White ibis	Louisiana heron	Cattle egret	Snowy egret
using head and bill							
head shake	+	+	+	+	+	+	+
head rub (body)			+	+	+	+	+
bill snap or jab	+	+	+	+	+	+	+
bill rub (body)			+	+	+	+	+
bill rub (legs, feet)			+	+	+	+	+
bill rub (perch)			+				
bill peck (body)			+	+	+	+	+
bill peck (legs, feet)			+	+	+	+	+
bill peck (perch)			+		+		
using legs and feet							
foot shake				+			
foot stamp (perch)			+		+	+	+
foot slap (other foot)			+		+	+	+
head scratch				+	+	+	
using body							
wing flip or flap			+		+	+	+
body fluff				+		+	+

Nondo 1982). In consequence, the number of mosquitoes successfully obtaining a blood meal and the amount of blood obtained by a population of mosquitoes is regulated by the number of mosquitoes attacking the host and the efficiency of the host's anti-mosquito mechanisms. Interactions such as this may constitute an efficient density-dependent means of limiting the size of blood-sucking insect populations, particularly those closely associated with a single host species (Schofield 1982) (see Section 7.5).

Mammals also engage in defensive behaviour against temporary ectoparasites. Most of us have seen cows flicking their ears or swishing their tails in response to the attentions of the large numbers of insects which are attracted to them. The number of ear flicks and tail swishes are directly related to the number of flies which are present on the cattle (Harris et al. 1987) and these mechanisms, together with head swings, leg kicks and shuddering of the skin, are an effective means of reducing annoyance from these insects (Fig. 7.8) (Warnes & Finlayson 1987). The efficiency of these mechanisms in reducing fly attack can be considerable as can be seen from field work in Zimbabwe, where 15 times more tsetse flies fed on sedated goats unable to show defensive behaviours than from normal animals. Responses of game animals to tsetse and other

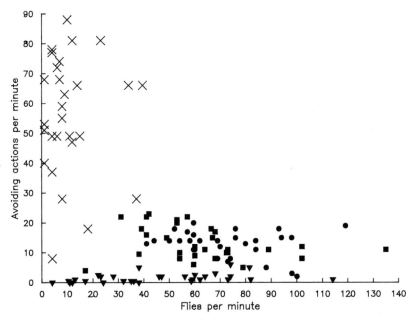

Figure 7.8 Calves displaying the highest levels of behavioural defences have the fewest number of stableflies settling on them. Different calves are represented by different symbols. (Redrawn from Warnes & Finlayson 1987.)

biting flies are also directly correlated with the number of flies attacking the animal. The responses seen include neck shuddering, tail lashing and scraping of the body with the hooves (in giraffe) and distressed lions rolling on their backs, hiding in hyaena holes or climbing trees to avoid attacks from *Stomoxys* spp. (Kangwangye 1977).

As well as these physical attempts to ward off the insect, large mammals show aggregation behaviour as a defence against temporary ectoparasites. Caribou and reindeer populations are gregarious throughout the year, but in the post-calving season the formation of particularly large herds occurs. These large aggregations of animals coincide with the seasonal peaks of blood-sucking insects on these northern ranges. The attack rate from blood-sucking insects on these herded rangifers is about ten times higher for animals on the periphery of the herd compared to those at the centre (Breev 1950). When carbon dioxide baited silhouette traps were substituted for real animals, and placed in either a herded pattern or to represent individual animals, insects were found to be more attracted to individual traps. Also, even though the 'herd-size' was only 24, the traps on the periphery received more attention from blood-sucking insects than did traps at the centre of the 'herd' (Helle & Aspi 1983). Temporary ectoparasites appear to be attracted to single animals with a good area of clear space around them, a situation where they are less likely to be crushed or swatted. It also provides easier access to the lower halves of the animals which are the preferred feeding sites of many temporary ectoparasites. Indeed, if insect attack is particularly high, cattle crowd closely together and eventually lie down exposing only their backs to the attackers. As a final response to intense insect activity cattle will stampede. The efficiency of herding as a mechanism for reducing the attentions received from blood-sucking insects may be increased by the movement of the herd onto selected sites. At times of peak fly activity Camargue horses aggregate at sites known locally as chomadous. These are sites exposed to maximum wind velocity, which acts to minimize insect activity. Similar sites are chosen by cattle subjected to severe attack by blackflies. Reindeer also congregate at specific areas known as tanders and it is suspected that local climatic or biotic factors at these sites minimize insect activity.

Laboratory studies have also shown that infection of the host can have an impact on the number of successful attacks by temporary ectoparasites. Many rodent species show highly efficient anti-mosquito behaviour and for this reason are only rarely fed on by mosquitoes. It is also known that rodent populations may be enzootic for various mosquito-borne diseases such as the arboviral disease, Venezuelan equine encephalitis. How can these seemingly conflicting pieces of information be reconciled? Under experimental conditions the mouse, *Mus musculus*, displays a series of very efficient behavioural mechanisms that prevent feeding by mosquitoes. Malaria-infected mice show the periodic peaks of parasitaemia which are typical of this disease. During, and particularly just after, these peaks the mouse is less able to defend itself against mosquitoes which feed readily from it (Fig. 7.9). These feeding opportunities for the mosquito, which occur a day or two after peak parasitaemia in the mouse, coincide with a peak in the number of gametocytes (the stage of the malaria parasite which is infectious for the mosquito) in the mouse's blood (Day & Edman 1983). In other words, the malaria parasite appears to have modulated the behavioural activity of the host in such a way as to maximize its chances of transmission by the vector. But how does the uninfected mouse become infected in the first place if it is only rarely fed upon by mosquitoes? Mice often eat attacking mosquitoes and it has been shown that they can become infected with both malaria (Edman et al. 1985) and La Crosse virus (Yuill 1983) in this way. The chances are that this is the normal means of transmission and that mice are only rarely infected by receiving an infective bite. Clearly if these sorts of relationship occur in the field they will be important not only in the epidemiology of the particular diseases concerned, but also in making us think more deeply about the ways in which disease transmission occurs.

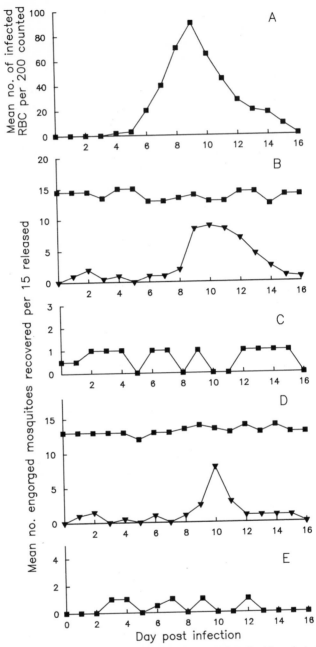

Figure 7.9 Feeding of mosquitoes on restrained mice infected with malaria is continuous and shows no peaks (B, upper plot, D, upper plot). Mosquito feeding on unrestrained mice infected with malaria shows clear peaks (B, lower plot, *Aedes aegypti*; D, lower plot, *Culex quinquefasciatus*) which occur just after the occurrence of maximum parasitaemia (A). Virtually no feeding occurs from uninfected, unrestrained mice (C, *Aedes aegypti*; E, *Culex quinquefasciatus*) which show efficient anti-mosquito behaviour. (Redrawn from Day & Edman 1983.)

Other behavioural activities of the host may also affect ectoparasitic insects. Many birds take dust baths and some anoint themselves with the defensive secretions of certain ants (Formicidae), a practice known as anting. Both of these behaviours have been reported to kill ectoparasites, but whether this is their prime purpose is unknown. Mud-bathing, which many large mammals enjoy, partially protects them from temporary ectoparasites as the insects are probably unable to bite through the dried crusts of mud seen on such animals. Immersion in water, as seen by hippopotami, is also a very efficient means of avoiding blood-sucking insects.

7.5 DENSITY-DEPENDENT EFFECTS ON FEEDING SUCCESS

Population limitation may be achieved by single, but more normally by multiple, factors and the limiting mechanisms that operate may change in time and space. These factors are commonly subdivided into density-dependent and density-independent categories. Separating the effects of one from the other, under natural conditions, is often a complicated process, not least because density-dependent factors will always influence the fitness of individuals to withstand density-independent pressures. But to generalize, in rapidly fluctuating environments or environments which the insect finds harsh, density-independent factors such as temperature and humidity are often of most importance in limiting population numbers. In more congenial circumstances, where populations are likely to expand more smoothly and continuously, insect populations are more likely to be limited by density-dependent factors such as competition for food or space, or by increased exploitation of the population by parasites or predators. Blood-sucking insects show a range of life-style strategies. At one end of this spectrum are insects such as the mosquitoes which are r-selected, having high reproductive rates adapted to maximize the instantaneous rate of population increase in unstable habitats and showing strong dispersal capacity. At the other end of the spectrum are those insects, like reduviid bugs and tsetse flies, which are K-selected, have low reproductive rates, and which are adapted to succeed under highly competitive conditions in stable habitats. To generalize once again, density-dependent effects are likely to be more important in the limitation of population size in K- rather than r-strategists. Because of the approach taken in this book, I intend to discuss only density-dependent effects on feeding success as these present some circumstances peculiar to the blood-sucking habit.

Food availability can be a density-dependent factor limiting population size in blood-sucking insects as it can in all other animals. In many, if not most, circumstances blood-sucking insects will differ from most other animals in the nature of their density-dependent control by food availability. This is because it is not normally a shortfall in the sheer physical quantity of food (i.e. blood) which limits their population growth, but the increasing difficulty of obtaining this food as the density of the insect population increases. There are two clear ways in which this may happen, which both depend on the increased stimulation of the host's immune system as the number of attacking insects increases.

The first way the host's immune system can influence the feeding success of insects is through acquired resistance. The subject of acquired resistance is at the early stages of investigation and has only been clearly demonstrated for a small number of insect–host combinations (Nelson 1987). All the convincing examples are permanent ectoparasites. When acquired resistance is seen in a host the immune response appears to bring about a change in the physical nature of the feeding site such that the insect finds it more difficult to feed. The circulating antibodies, produced against the salivary antigens of the insect, appear to have little impact (Fig. 7.10). Acquired resistance rarely, if ever, reaches the status

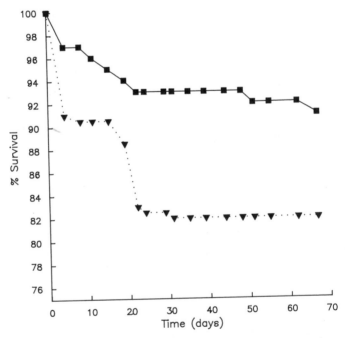

Figure 7.10 When tsetse flies are continually fed at high densities on the previously exposed ear of a rabbit (▼), their percentage survival decreases compared to flies fed on the naïve ear of the same rabbit (■). This suggests that localized immune responses and not circulating antibodies are responsible. (Redrawn from Parker & Gooding 1979.)

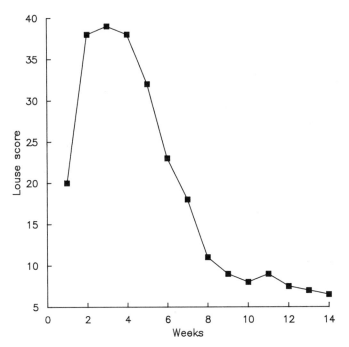

Figure 7.11 A hindfoot-amputated mouse infected with the louse, *Polyplax serrata*, will show localized acquired resistance and the numbers of lice will fall with time. This type of cycle can only be produced on an infested area of a mouse once in its lifetime. The decrease in the number of lice is caused by starvation. (Redrawn from Nelson *et al.* 1977.)

of full immunity. It normally results in just a lowering of the numbers of insects which can maintain themselves on the host rather than their complete elimination.

A characteristic pattern of population growth and decline is seen when permanent ectoparasites move onto a naive host which then develops an acquired resistance (Fig. 7.11). For the first four weeks following infestation of a hindfoot-amputated mouse with *Polyplax serrata*, the numbers of lice on the mouse increase. During this time the skin becomes infiltrated with large numbers of lymphocytes, eosinophils and neutrophils which peak in the second week and decline thereafter. These changes are accompanied by hyperplasia of the skin and vasoconstriction. This immune response to the presence of the lice then leads to the development of acquired resistance which appears over the next eight weeks. Acquired resistance is characterized by an increase in the numbers of lymphocytes and monocytes in the skin as well as increasing numbers of fibroblasts and mast cells. As these changes occur so the numbers of lice on the hindfoot-amputated mouse decrease until eventually very few, or even none, remain. This second phase of response, which coincides with the decline in the numbers of lice on the host, is not considered part of the immune response because polymorphonuclear granulocytes are low throughout. It is a chronic response resembling the response to topically applied chemical irritants. Further evidence that this is an acquired host response, induced by the feeding activity of the lice, is shown by the fact that the degree of resistance expressed is directly correlated to the intensity and duration of the infestation on the hindfoot-amputated mouse. The response of the mouse to the lice also appears to include some progressive, physiological adaptation to the increasing burden, because the sudden transfer of large numbers of lice from an infested mouse to a healthy, naive, hindfoot-amputated mouse can lead to the death of the new host within 24 h, possibly due to toxic or anaphylactic shock.

Further work on the louse–hindfoot-amputated mouse combination showed that when resistant skin from sensitized mice was grafted onto naive, athymic mice it continued to show resistance to louse feeding. Lice fed happily on adjacent skin, or skin grafted from non-resistant mice, but not on the grafted resistant skin (Bell *et al*. 1982, Nelson & Kozub 1980). So, acquired resistance is a local phenomenon restricted to the area of skin exposed to feeding lice. The operating factor seems to be the impairment of louse feeding caused by the restriction in blood flow to that skin region.

What is the importance of this localized response to the host? Permanent ectoparasites tend to accumulate in the places which the host finds the most difficult to groom. The increasing development of acquired resistance in that area gradually reduces louse numbers because the least fit lice show reduced fecundity and increased generation time or die of starvation or are driven out onto groomable areas of the body. In this way numbers in the ungroomable, favoured sites oscillate around a critical density at which the localized immune response begins to appear. So, acquired resistance is a means of limiting ectoparasite numbers in ungroomable regions of the body.

The effect of acquired resistance on temporary ectoparasites is less clear. There certainly can be effects on the feeding insect but these have only been demonstrated in the laboratory under conditions of intense, continuous challenge. When colonized tsetse flies are routinely fed on rabbit's ears, a stage is commonly reached where the blood supply to the ear is significantly reduced (a similar effect has been seen with mosquitoes feeding on the ears of mice; Mellink 1981). It is known that the response is localized in the exposed ear (Fig. 7.10) and it is not correlated with the levels of circulating antibody. The response in the ear is associated with increased mortality of feeding tsetse flies (Fig. 7.10) and with lowered pupal weights (Parker & Gooding 1979). The obvious explanation for the observed effects on these insects is that they are taking smaller blood meals, but this is not the case and the underlying causes are not clear. It is also not clear if this evidence can be used to argue for naturally occurring acquired resistance to temporary ectoparasites, because the artificially high feeding levels seen in these experiments are unlikely to occur in field situations. No record of a naturally occurring acquired resistance to temporary ectoparasites has yet been recorded.

The second way that the immune response can diminish the feeding success of blood-sucking insects is through its stimulation of increased defence behaviour by the host (see above). The level of defensive behaviour in the irritable host is dependent on the density of attacking insects and is provoked by the pruritis induced by insect bites. In this way the immune response limits the feeding success of the insect population as a whole, even though early assailants may obtain a full blood meal before pruritis and defensive behaviour are stimulated. Reduced feeding success in the insect population probably leads to increased generation times, reduced fecundity and increased mortality. An effect of this kind has been recorded under experimental conditions

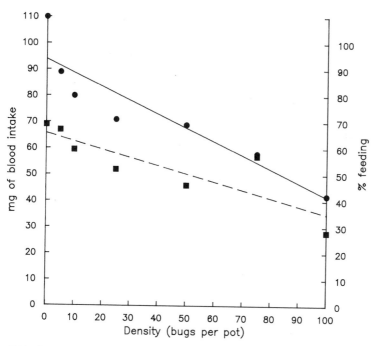

Figure 7.12 Increasing the density of fifth-instar nymphs of *Triatoma infestans* leads to a decrease in both the intake of blood (●) and the number of bugs successfully feeding (■). This is a factor in the density-dependent limitation of bug populations. (Redrawn from Schofield 1982.)

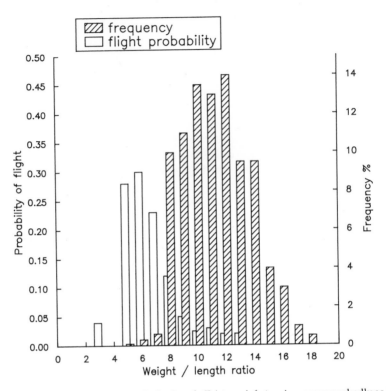

Figure 7.13 The probability of flight for female *Triatoma infestans* increases markedly as the bug gets hungrier and its weight–length ratio decreases. The frequency of weight–length ratios of a population of bugs taken from rural houses in Brazil shows that flights would be expected under natural conditions. Such flights may serve to reduce population density in the house, increasing the feeding success of the remaining individuals and regulating population density. (Redrawn from Lehane & Schofield 1982.)

in the reduviid bug, *Triatoma infestans*, (Schofield 1982). As the number of bugs in a population increases 'scramble competition' comes into operation, the number of bugs successfully feeding and the amount of blood they are ingesting declines (Fig. 7.12). This leads to an increase in the development times of the immature stages, a decrease in fecundity and an increase in the likelihood of adult flight (Fig. 7.13) (Lehane & Schofield 1981, 1982). Each of these effects serves to adjust the population to a lower stable density which occurs, in this instance, without an increase in mortality. There is evidence from South America that the size of reduviid bug populations in dwellings shows a strong positive correlation with the number of occupants (see Schofield 1985), so these limiting mechanisms may be operating in the field. There is also field evidence that high densities of tsetse flies may lead to decreased feeding efficiency in these insects (Vale 1977).

Under field conditions the relationship between feeding success and the number of attacking insects could be very complex. We know there are interspecific differences in host tolerance to insect biting. There are also intraspecific differences which can depend on host health, age and other factors. Tolerances also probably change in a complex manner with changing insect density. There are likely to be interspecific effects on feeding insects when mixed insect species are feeding on the same animal. For example, large numbers of insects present on a host may possibly disturb each other when feeding. Extensive field work is required before we will begin to clarify how important density-dependent effects are on feeding success as a means of limiting blood-sucking insect populations. The most likely occasion where density-dependent feeding success may act as a factor limiting population size is when the insects have access to only a limited number of host animals and/or a limited number of host species. Such circumstances regularly arise for arctic insects, and for nest-dwelling forms (including the domestic bug populations discussed above). Studies on these forms are most likely to improve our knowledge in this area.

The life histories of many blood-sucking insects (for example mosquitoes and blackflies) have the effect of minimizing the intraspecific competition for blood and as a consequence maximizing the reproductive potential of the available females. This is achieved by having larval forms which feed on non-blood food sources, which is the case for most holometabolous blood-sucking insects. Many blood-sucking forms take this further and reserve blood feeding for the adult females only, males normally feeding from a variety of sugar sources. Both strategies limit the attention paid to the host, the dividend for the insect being the reduced response it promotes from the host and the increased chance of successful feeding by the mature females. But not all blood-sucking insects adopt this dual feeding strategy. In hemimetabolous insects, such as lice and the reduviid bugs, both adult and juvenile stages compete for the same blood food sources. Although this is clearly a successful life history strategy for these species, it must limit the theoretical numbers of successful adult females compared to a situation in which juvenile stages and males are nourished by an alternative food.

As mentioned above, in most circumstances the total volume of available blood is not a limiting factor on the population size of blood-sucking insects, but blood is not a resource which is evenly spread throughout an ecosystem; it is collected together in 'packets' (hosts). In some situations, where hosts are scarce, the difficulty of finding such a 'packet' may be a factor limiting population size. In circumstances where the availability of a blood meal is unpredictable, many blood-sucking insects cease to rely on the adult obtaining a blood meal in order to lay eggs. Instead they concentrate on maximizing larval nutrition and carry over reserves to the adult stage permitting them to produce a batch of eggs autogenously (see Section 6.7).

Man's manipulation of the planet has had a profound impact on the availability of food for blood-sucking insects. This in turn has had a tremendous influence on the distribution and abundance of blood sucking insects and on the epidemiology of the diseases they transmit. A good example is furnished by the changing feeding patterns of the anopheline mosquitoes of Europe discussed in Section 3.1. Another example is the growth of unplanned tropical urbanization which has greatly increased the available food resource for some blood-sucking insects. Partly as a result of this, populations of insects such as the yellow fever mosquito, *Aedes aegypti*, have boomed. Originally thought to be a rot hole-breeding species limited to the forests of East Africa, this mosquito has followed man into unplanned urban developments throughout the tropics where, along with other exploiters of these conditions such as *Culex quinquefasciatus*, it is responsible for a consider able degree of disease transmission.

CHAPTER EIGHT

Transmission of parasites by blood-sucking insects

Like all other organisms, blood-sucking insects have their own array of parasites. Many of these parasites are common to a range of different insects, blood-sucking and non-blood-sucking alike. Others, which are also parasites of the vertebrate host, are peculiar to blood-sucking insects and the relationship depends on the blood-sucking habit for its existence. Because of the approach taken in this book, it is the relationships between blood-sucking insects and this latter group of parasites on which I will concentrate in this chapter.

8.1 TRANSMISSION ROUTES

Table 8.1 shows that blood-sucking insects are responsible for the transmission of many important disease-causing organisms. At its simplest, transmission may simply involve the insect as a mechanical bridge between two vertebrate host species. At its most complex, transmission involves an obligatory period of replication and/or development by the parasite in the vector insect. A division is often drawn between 'mechanical' and 'biological' transmission.

Mechanical transmission is said to occur when the blood-sucking insect is no more than a flying pin, transferring pathogens from one vertebrate host to another on contaminated mouthparts. More rarely other parts of the insect's body are involved as happens in the deposition of *Francisella tularensis* in the faeces of fleas. Relatively little work has been carried out on the relationships grouped together as mechanical transmission and the possibility of more complex interactions should not be ignored. For example, it has been suggested, on the basis of epidemiological evidence, that some mechanism for the concentration of parasites in the vector's mouthparts may occur in the mechanical transmission of trypanosomes such as *Trypanosoma vivax viennei* and *T. brucei evansi*

Table 8.1 Some of the most important associations of disease-causing organisms carried to man and other animals by blood-sucking insects: (a) viruses, (b) rickettsia and bacteria, (c) protozoa, and (d) nematodes.

	Major vectors	Major hosts	Geographical distribution
(a) VIRUSES			
Semliki Forest Virus	mosquitoes	man	N. Africa, S. Africa, Amazon, Phillipines, Malagasy
Chikungunya	mosquitoes	man, monkeys	India, Africa, E. Asia
Venezuelan equine encephalitis	mosquitoes	man, equines, rodents, birds	S. America, Southern USA, Europe
Western equine encephalitis	*Culex tarsalis*	man, equines, birds, reptiles, amphibians	America, Czechoslovakia, Italy
yellow fever	*Aedes aegypti*, several other mosquito species	man, monkeys	Africa, tropical America
dengue	*Aedes aegypti* *Aedes* spp.	man	S.E. Asia, Caribbean
St. Louis encephalitis	*Culex tarsalis* *Culex* spp.	man, birds, rodents, bats	America
Japanese encephalitis	*Culex tritaeniorrhynchus* *Culex* spp.	man, equines, pigs, birds	S.E. Asia
Murray Valley encephalitis	*Culex annulirostris* mosquitoes	man, birds	E. Australia, Phillipines, New Guinea
Wesselsbron	mosquitoes	man, sheep	Central and S. Africa, Thailand
Ilheus	forest mosquitoes	man, monkeys, birds	Central and S. America
West Nile Virus	*Culex* spp.	man, birds	Central and N. Africa, India, Mediterranean, USSR
sandfly fever	*Phlebotomus papatasii*	man	Mediterranean, Near East, India, Sri Lanka, Southern China, Central Asia
African Horse sickness	*Culicoides* spp.	equines	South, Central and East Africa, India

144

bluetongue	*Culicoides* spp.	sheep, rodents	world-wide
Rift Valley Fever	*Aedes* spp. / *Eretmopodites* spp. / *Ochlerotatus* spp.	man, domestic and feral animals	Central, E. Africa and Southern Africa
California encephalitis	*Aedes* spp. / *Culex* spp.	man	N. America
myxomatosis	mosquitoes / *Spilopsyllus cuniculi* / *Culicoides* spp.	rabbits	America, Europe, Australia

(b) RICKETTSIA AND BACTERIA

Rickettsia prowazekii	*Pediculus humanus humanus* / *Pediculus humanus capitis*	man, possibly cycles in domesticated animals	world-wide
Rickettsia typhi	*Leptopsylla segnis* / *Xenopsylla cheopis* / *Nosopsyllus fasciatus*	man, rats, mice	world-wide
Rochalimaea quintana	*Pediculus humanus humanus*	man	Europe, Mexico, China, Ethiopia, Algeria
Bartonella bacilliformis	*Phlebotomus verrucarum*	man	S. America
Anaplasma marginale	biting flies	cattle, zebra, water buffalo, bison, antelopes, deer, elk and camel	tropics and subtropics
Eperythrozoon suis	*Haematopinus suis*	pigs	world-wide
Borrelia recurrentis	*Pediculus humanus humanus*	man	Ethiopia, Eritrea
Francisella (Pasteurella) tularensis	*Chrysops* spp. / *Tabanus* spp.	man, rodents, rabbits, birds	N. Hemisphere
Yersinia (Pasteurella) pestis	*Xenopsylla cheopis* / *Synosternus pallidus* / *Nosopsyllus fasciatus*	man, rodents	Manchuria, S.E. China, Thailand, Java, Burma, E. Africa, India, Iran, Malagasy
Treponema pertenue	*Aedes aegypti*	man	northern South America

Table 8.1 (Continued)

	Major vectors	Major hosts	Geographical distribution
(c) PROTOZOA			
Plasmodiidae			
Plasmodium falciparum	Anopheles spp.	man	tropics, subtropics and some temperate countries
Plasmodium ovale	Anopheles spp.	man	tropical Africa
Plasmodium vivax	Anopheles spp.	man	Some temperate and tropical areas worldwide
Plasmodium malariae	Anopheles spp.	man, chimpanzees	tropical Africa, India and S.E. Asia
Plasmodium gallinaceum	Aedes spp. Armigeres spp.	domestic fowl	India
Plasmodium relictum	Culex spp. Anopheles spp. Aedes spp.	pigeons, anatidae, passerines	world-wide
Plasmodium cynomolgi	Anopheles spp.	monkeys, macaques, occasionally man	Asia
Plasmodium knowlesi	Anopheles hackeri	monkeys, macaques, occasionally man	S.E. Asia
Plasmodium berghei	Anopheles spp.	tree rat	Central Africa
Haemoproteus columbae	hippoboscids	various birds particularly pigeons and doves	world-wide
Leucocytozoon simondi	Simulium spp.	ducks and geese	N. America, Europe, Asia
Leucocytozoon smithi	Simulium spp.	turkeys	N. America, Europe
Trypanosomatidae			
Leishmania tropica	Phlebotomus spp.	man, dogs, rodents	Asia, Middle East, Mediterranean
Leishmania donovani	Phlebotomus spp.	man, rodents, serval, genet cat, dogs, foxes, jackals	S. America, tropical & N. Africa, Asia, Mediterranean

Species	Vector	Hosts	Distribution
Leishmania brasiliensis	*Lutzomyia* spp.		S. America, Iran
Trypanosoma brucei	*Glossina* spp.	man, rodents, monkeys equines, cattle, sheep, dogs, cats, many wild game animals are reservoirs	Africa from latitudes of 15° N to 25° S
Trypanosoma gambiense	*Glossina* spp.	man, cattle, sheep, goats, horses, dogs, cats, pigs	W. and C. Africa from latitudes of 15° N to 18° S
Trypanosoma rhodesiense	*Glossina* spp.	man, wild game animals are reservoirs	tropical E. Africa
Trypanosoma congolense	*Glossina* spp.	a very wide range of domestic and wild game animals	tropical Africa
Trypanosoma vivax	*Glossina* spp. biting flies	cattle, water buffalo, sheep, camels, goats, horses, antelope, deer	tropical Africa, Caribbean, Central and S. America
Trypanosoma uniforme	*Glossina* spp.	sheep, goats, cattle, antelope	tropical Central Africa
Trypanosoma simiae	*Glossina* spp. biting flies	warthog, pigs, camels	tropical E. and Central Africa
Trypanosoma suis	*Glossina* spp.	pigs	Zaire
Trypanosoma evansi	biting flies	camels, horses and dogs, a wide range of domestic and feral animals are reservoirs	India, Far East, N. Africa, Near East, Central and S. America
Trypanosoma equinum	biting flies	equines, dogs, cattle, sheep, goats	Central and S. America
Trypanosoma theileri	*Tabanus* spp. *Haematopota* spp.	cattle	world-wide
Trypanosoma melophagium	*Melophagus ovinus*	sheep	world-wide
Trypanosoma lewisi	*Ceratophyllus fasciatus*	rats	world-wide

Table 8.1 (*Continued*)

	Major vectors	Major hosts	Geographical distribution
Trypanosoma rangeli	reduviid bugs	man, dog, cat, opossum, monkey	S. America
Trypanosoma cruzi	reduviid bugs	man, opossum, armadillo, wide range of feral and domestic reservoir hosts	S. America
(d) NEMATODES			
Onchocercidae			
Onchocerca volvulus	*Simulium* spp.	man	tropical Africa, Central America
Onchocerca gutturosa	*Simulium* spp. *Odagmia* spp. *Friesia* spp.	cattle, buffalo	world-wide
Onchocerca gibsoni	*Culicoides pungens*	cattle, zebu	Asia, Australasia, Southern Africa
Onchocerca lienalis	*Simulium* spp.	cattle	Australia, N. America
Onchocerca cervicalis	*Anopheles* spp. *Culicoides* spp.	equines	world-wide
Filariidae			
Wuchereria bancrofti	*Culex* spp. *Aedes* spp. *Mansonia* spp. *Anopheles* spp.	man	tropics, subtropics and some temperate countries
Mansonella ozzardi	*Culicoides furens* *Simulium amazonicum*	man	S. America, Caribbean
Brugia malayi	*Aedes* spp. *Mansoni* spp. *Anopheles* spp.	man, leaf monkeys, cats, dogs, civet cats, pangolins	India, S.E. Asia

148

Species	Vector	Host	Distribution
Brugia pahangi	*Mansonia* spp. *Armigera* spp.	cat, dog, civet cat, leaf monkeys, tigers, slow loris	Malaysia
Brugia patei	*Mansonia* spp. *Aedes* spp.	cats, dogs, genet cat, bush baby	Africa
Brugia timori	*Anopheles* spp.	man	Indonesia
Loa loa	*Chrysops silacea* *Chrysops dimidiata*	man, baboons, monkeys	W. and C. Africa
Dirofilaria immitis	mosquito spp.	dog, fox, wolf, cat and occasionally man	tropics, subtropics, some temperate countries
Dirofilaria repens	mosquito spp.	dogs and occasionally man	Europe, Asia, S. America
Parafilaria multipapillosa	*Haematobia atripalpis*	equines	E. Europe
Ornithofilaria fallisensis	*Simulium* spp.	anatid birds	N. America
Elaeophora schneideri	*Hybomitra* spp. *Tabanus* spp.	sheep, deer, elk	N. America
Setariidae *Setaria equina*	*Aedes* spp. *Culex* spp.	equines	world-wide
Setaria labiatopapillosa	*Anopheles* spp.	cattle, deer, giraffe, antelope	world-wide
Dipetalonema reconditum	*Ctenocephalides* spp. *Pulex* spp.	dogs	N. America, Africa, S. Europe
Dipetalonema evansi	*Aedes detritus* *Culicoides* spp.	camels	Egypt, Far East, E. USSR
Dipetalonema streptocerca	*Culicoides* spp.	man	C. and W. Africa
Dipetalonema perstans	*Culicoides* spp.	man, anthropoid apes	Africa, S. America
Stephanofilaria stilesi	*Stomoxys calcitrans*	cattle	USSR, N. America
Spiruridae *Habronema majus*	*Stomoxys calcitrans*	equines	world-wide

(Wells 1982). As our understanding of the subtleties and complexities of the interactions between parasite and vector increases, the collection of parasites placed in the 'mechanical' group progressively decreases.

Clearly any blood-sucking insect is a potential mechanical transmitter, but, because the pathogens cannot survive for long outside the host's body, insects which habitually take a succession of partial meals from several vertebrate hosts are probably the most efficient mechanical transmitters. Larger insects with painful bites, such as biting muscids and tabanids, are often disturbed by the vertebrate host before completing their meal. As they commonly feed on large herbivores found in groups and are highly mobile insects, the unfinished meal is often quickly completed on a second vertebrate host. Another important factor in mechanical transmission is the amount of blood which can be transferred between animals by the insect; because, for any given level of parasitaemia, the more blood transmitted the greater the chance of disease transmission. Large tabanids, such as *Tabanus fuscicostatus*, may transmit as much as a nanolitre of blood on their sponging mouthparts. Probably for these reasons, tabanids are the mechanical transmitters of trypanosomes such as *Trypanosoma evansi* (= *T. equinum*, *T. hippicum* and *T. venezuelense*) and on occasion *T. equiperdum* (Soulsby 1982) which cause serious and often fatal diseases in equines in many parts of the world. There is also evidence for the mechanical transmission of *Trypanosoma vivax* by the tabanid, *Cryptotylus unicolor*. Smaller insects which have less painful bites and which are capable of moving less blood between vertebrates are also important mechanical vectors. Thus myxoma virus, which causes myxomatosis in rabbits, is mechanically transmitted by the rabbit flea, *Spilopsyllus cuniculi*, in Britain and by mosquitoes in Australia.

Biological transmission is said to occur when there is a biological dependency between the vector and parasite, with the pathogen undergoing a period of propagation and/or growth and development in the insect. Biological transmission is the most common and important means of pathogen transmission by insects.

Each parasite has its own peculiarities of life-style in the insect. Some of these characteristics can be used to subdivide biological transmission into the following categories (Huff 1931):

(a) *Propagative transmission*
Propagative transmission occurs when the pathogen multiplies in the insect host but undergoes no development. This occurs with bacterial pathogens such as *Yersinia (Pasteurella) pestis* where the large numbers of pathogens leaving the flea are essentially the same as those ingested. Arboviruses are usually included in this category although this is not strictly correct as these pathogens acquire a portion of the plasma membrane from the cell in which they were

formed. This coat is species-specific and in this respect the virus which leaves the insect is different from that entering. In essential details, however, such as the nature of the envelope and coat proteins and the genetic information contained in each virus particle, they are members of this category.

(b) *Cyclopropagative transmission*
Cyclopropagative transmission occurs when the pathogen not only multiplies in the insect but also changes its form in some manner. A large number of parasites are included in this group, such as many of the trypanosomes, leishmaniases and the malaria parasites.

(c) *Cyclodevelopmental transmission*
Cyclodevelopmental transmission occurs when the pathogen undergoes a developmental transformation in the vector but does not multiply, which is true of helminths such as the filarial worms.

The details of the sojourn in the insect tend to vary not only from one group of pathogens to another but also from species to species. As space does not allow a detailed discussion of the passage of each important parasite, some generalizations based on major parasite groups will be given. In addition an outline of the main routes taken by parasites transmitted by insect vectors is given in Figure 8.1.

Leishmania have been subdivided into two groups according to their distribution in their sandfly hosts. The subgenus *Leishmania* (Suprapylaria) are limited to the foregut and midgut of their sandfly vectors. The subgenus *Viannia* are found in the vector foregut, midgut and hindgut (Lainson & Shaw 1979, 1987). In the subgenus *Leishmania*, amastigotes are released in the midgut of the sandfly from the vertebrate macrophages ingested in the blood meal. Division of these amastigotes occurs in the blood meal in the first two to three days following feeding. The free-swimming promastigotes produced remain constrained within the peritrophic membrane. As the peritrophic membrane breaks down at the completion of blood-meal digestion, parasites escape and shortened haptomonad-like promastigotes attach to the stomadaeal valve. Colonization of the oesophagus and pharynx by small, rounded, sessile, flagellated paramastigotes follows. There is now increasing evidence that leishmanial promastigotes then undergo metacyclogenesis to produce a final infective metacyclic stage (Sacks 1989). A similar picture is seen in the subgenus *Viannia*, except that the ileum and pylorus regions of the hindgut become colonized with small, rounded, haptomonad-like promastigotes and paramastigotes (Molyneux & Killick-Kendrick 1987). Intracellular forms have been reported in insect guts but their importance is unknown (Molyneux *et al.* 1975).

In all these stages there is a greater or lesser degree of attachment of the various forms to the intestinal wall, some of which is probably mediated

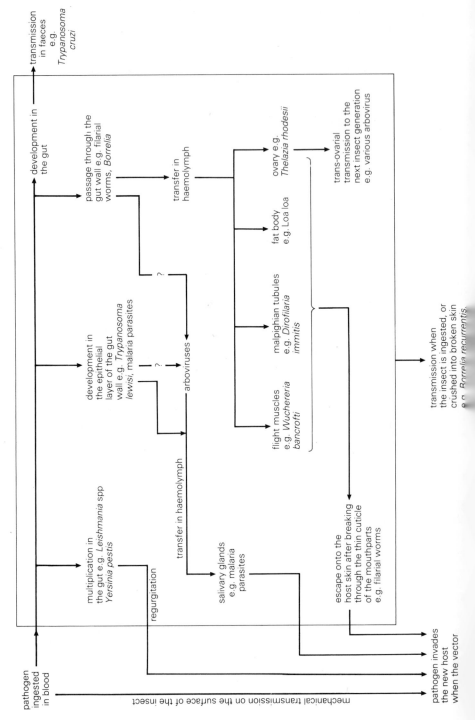

by insect derived lectins. (Lectins, which cross-link molecules possessing suitable carbohydrate residues, may also have a defensive function in insects; see Section 8.7.) Some carbohydrate meals increase the probability that parasites will be transmitted at a subsequent blood meal. This probably arises because of competitive inhibition of the lectins by the carbohydrates in the meal, which means more free-swimming infective forms are available in the proboscis (Dwyer 1974, Molyneux & Killick-Kendrick 1987). It has also been suggested that attached parasites interfere with the functioning of cibarial receptors, decreasing sandfly-feeding efficiency and increasing the frequency and duration of probing. This increases the chances of parasite transmission which may occur even though the insect fails to feed successfully (Killick-Kendrick et al. 1977, Beach et al. 1985).

Trypanosomes transmitted by insects can be divided into two groups based on the route taken out of the vector. Those transmitted via the proboscis are known as the salivaria and those transmitted in the faeces, the stercoraria. The stercoraria consist of parasites such as *Trypanosoma cruzi*, transmitted by reduviid bugs and causing Chagas disease or South American sleeping sickness in humans, *T. lewisi* of rats, transmitted by rat fleas, and *T. theileri*, transmitted to cattle by tabanids. In *T. lewisi* circulating trypomastigotes are ingested in the blood meal of the flea. Rarely, trypomastigotes invade midgut epithelial cells and undergo reproduction (it is uncertain whether this is sexual or asexual) producing further trypomastigotes. These emerge and invade new cells to repeat the cycle. Eventually free trypomastigotes transform into epimastigotes. These pass down the gut, divide, and transform into the infective trypomastigote-stage. These become established in the hindgut and are passed in the faeces. The situation in *T. cruzi* is different with development occurring entirely in the lumen of the gut. Within hours of ingestion of the infective meal stumpy trypomastigotes are formed; 14–20 h after the meal amastigotes appear, which are joined by sphaeromastigotes which divide asexually. These then form promastigotes and later small epimastigotes, which divide to give rise to long epimastigotes which attach to the cuticle of the bug's rectum. These progressively transform into the infective-stage metatrypanosomes about 8 (nymph) to 15 (adult bug) days after the infective blood meal.

The salivaria can be subdivided on the basis of their infectivity to the vector and their site of development. The *Trypanosoma vivax* (subgenus *Duttonella*) group are highly infective to the insect and development is entirely limited to the insect's proboscis and cibarium. Those trypanosomes which are able to anchor themselves to the wall of the proboscis before being swept into the midgut with the blood meal are the originators of the infection. Epimastigotes are formed which divide rapidly to form colonies attached to the mouthpart walls. Some of these parasites

153

detach and pass forward to enter the hypopharynx where they transform to trypomastigote forms. These in turn develop into the infective meta-trypanosomes.

The *Trypanosoma brucei* (subgenus *Trypanozoon*) group have the lowest insect infectivity of the four salivarian subgenera. The final site of development is the salivary glands. The trypanosomes first establish themselves in the gut. They remain in the endoperitrophic space for about four days after the blood meal, then they begin to appear in the ectoperitrophic space. From the early part of the century until the 1970s it was believed that the trypanosomes achieved this by passing down the gut to the rectum, where the peritrophic membrane was torn up by the rectal spines. The trypanosomes were then thought to pass around the open ends of the membrane, pass back up the gut in the ectoperitrophic space, penetrating the peritrophic membrane where it emerges from the proventriculus, an area of the membrane thought to be 'soft'. This theory states that, once back in the endoperitrophic space, the trypanosomes travel up the foregut to the end of the food canal. They then emerge from the food canal and enter the hypopharynx and travel to the salivary glands via the salivary ducts. In the 1970s evidence was produced to suggest an alternative route in which the trypanosomes penetrate the peritrophic membrane to escape to the ectoperitrophic space. They then penetrate and enter the midgut cells (with possibly an intracellular development stage) and finally move, probably through the haemolymph, to the salivary glands, which they enter by penetration. The evidence for both routes is reviewed by Evans & Ellis (1983).

The *Trypanosoma congolense* (subgenus *Nannomonas*) group occupies an intermediate position between the *vivax* and *brucei* groups. It displays an intermediate degree of infectivity for *Glossina*. The final site of development is the hypopharynx, but this follows a developmental period in the midgut. The basic route followed by the trypanosomes in the vector has classically been thought to be similar to that of the *brucei* group. Thus parasites initially develop in the endoperitrophic space, escape from it in the hindgut and pass back up the midgut in the ectoperitrophic space. After burrowing through the 'soft' peritrophic membrane (where it emerges from the proventriculus) they pass forwards to become established in the foregut. Here they attach to the walls of the labrum and some parasites finally enter the hypopharynx which is where the infective metatrypanosomes are found; invasion of the salivary glands does not occur. But this route may have to be re-evaluated in the light of the new route suggested for *brucei* group trypanosomes described above.

The fourth group (subgenus *Pycnomonas*) has only a single representative, *Trypanosoma suis*, which has a very limited distribution.

In the Haemosporidia male (micro-) and female (macro-) gametocytes are taken up in the blood meal by the vector which is commonly a mosquito. Almost immediately the male gametocyte produces numbers of flagellate microgametes by a process termed exflagellation. These fertilize the female gametocyte and the resulting ookinete penetrates the midgut wall. Most Haemosporidia, including the human malaria parasites, then produce a thin-walled oocyst which becomes established on the outer wall of the gut epithelium beneath the basement membrane and muscle layers. Exceptionally, oocysts of *Hepatocystis* break free of the gut, and travel to the head capsule of their *Culicoides* hosts before encysting. Rapid division occurs in the oocyst with the production of fusiform sporozoites sometime between the 4th and 15th day following the infected blood meal. The number of sporozoites produced varies with species; *Plasmodium falciparum* produces about 10 000 (Pringle 1965) and some haemoproteids as few as 30 or 40. The oocyst ruptures releasing the sporozoites, which become distributed throughout the insect's body. Some penetrate the salivary glands and at the time of the next blood meal gain access to the lumen and are passed to the next vertebrate host in the saliva. *Hepatocystis* are believed to take a slightly different route with the sporozoites directly entering the mouthparts.

Adult filarial worms produce live young called microfilariae (L_1). These first stage larvae are ingested along with the blood meal of the vector insect which may be a mosquito, muscid fly, flea, ceratopogonid midge, tabanid or blackfly depending on the species of worm. Microfilariae of some filarial worm species possess a sheath which is the elongated remnant of the eggshell. This sheath is normally shed once the worm is in the vector's intestine. Successful microfilariae usually pass through the intestinal wall within an hour of blood meal ingestion. Once in the haemolymph they pass to one of the insect's internal organs in which they develop through the second larval stage to the infective third stage (L_3). The choice of internal organ for development varies among species of filarial worm. The filarial worms of humans, *Onchocerca volvulus*, *Brugia malayi* and *Wuchereria bancrofti*, all take about 8–12 days to develop in the thoracic flight muscles of the vector. Another filarial worm of man, *Loa loa*, develops in the fat body of *Chrysops* spp. while the filarial worms of dogs, *Dirofilaria immitis* and *D. repens*, both develop in the Malpighian tubules of their mosquito vectors. The infective, third stage larva moves in the haemolymph to the insect's mouthparts. It breaks through the cuticle at the time of the next blood meal to become deposited on the skin of the vertebrate host which the worm enters through the puncture hole left by the insect's mouthparts.

Arboviruses are ingested by the vector in the blood meal. Infection of the insect only occurs if the virus is ingested in sufficient concentration to overcome the insect's gut barrier (Storey 1933, Chamberlain & Sudia 1961). The infection threshold level varies among arboviruses and among insects. The nature of the gut barrier is not understood but may centre around the number of available attachment sites for the virus on the surface of the midgut cells. The gut barrier may be intimately associated with virus–vector specificity as it is often possible to infect otherwise refractory insect species by the direct injection of the virus into the insect's haemocoel. In relation to the gut barrier, it is notable that mosquitoes are more easily infected with arboviruses when they are concurrently infected with microfilariae, which probably assist the passage of the virus through the intestinal barriers (Mellor & Boorman 1980, Turell et al. 1984). Even after successful infection of the gut epithelium the virus may be prevented from infecting the haemocoel by the 'mesenteronal escape barrier' and the salivary glands by the 'salivary gland infection barrier' (Kramer et al. 1981). These barriers are also dose-dependent but the mechanisms involved are again not understood. Even if the arbovirus achieves the infection of the salivary glands, transmission will only occur if the virus overcomes the 'salivary gland escape barrier' (Grimstad et al. 1985).

Once in the haemolymph arboviruses infect other tissues in addition to the salivary glands. Infection of the ovaries is particularly important in some virus–vector associations because transovarial transmission can occur to the next generation of vectors (Watts et al. 1973, DeFoliart et al. 1987). Not only can the female progeny infect new vertebrate hosts but the infected male progeny (non-blood-feeding) may increase the population of infected vectors by venereally transmitting the arbovirus to previously uninfected females (Thompson & Beatty 1977).

8.2 SPECIFICITY IN VECTOR–PARASITE RELATIONSHIPS

Each parasite transmitted by blood-sucking insects is normally associated with a restricted number of vector species. Even mechanically transmitted organisms such as the myxoma virus show a degree of specificity in their vector associations. When myxomatosis escaped from an experimental area of the Murray Valley in Australia in 1950, it first moved along major watercourses and was transmitted by the locally abundant, river-haunting mosquito, Culex annulirostris. In subsequent seasons it managed to move out across vast areas of semi-arid country by using another mosquito, Anopheles annulipes, which survived the harsh local conditions by using rabbit burrows for shelter. In Europe the myxoma virus uses yet another carrier, the European rabbit flea, Spilopsyllus cuniculi. Specificity in this instance is largely determined by details of local ecology, the myxoma virus making use of any appropriate 'flying syringe' for its mechanical transmission from one vertebrate host to another. Relationships where the parasite has a biological dependence on the host insect, are more specific. Thus, the four malaria parasites of man are only transmitted by anopheline mosquitoes, *Culicoides* spp. are the vectors of bluetongue virus and Chagas disease is transmitted by reduviid bugs.

There are several ways in which a specific relationship between a pathogen and its vector insect can be mediated. Of prime importance is vector–pathogen coincidence; except for transovarial transmission, if the insect does not feed on a host containing the parasite it cannot become a vector. So one of the most important factors determining vector–parasite relationships is the host choice made by the feeding insect. Host choice is influenced by geographical, ecological, morphological, behavioural and other considerations and this complex issue has been discussed in detail in Chapter 3.

The importance of physiological factors can be shown under experimental conditions when abnormal combinations of pathogen and vector are contrived. Under these circumstances we find that, like the human malaria parasites which die if ingested by any insect other than an

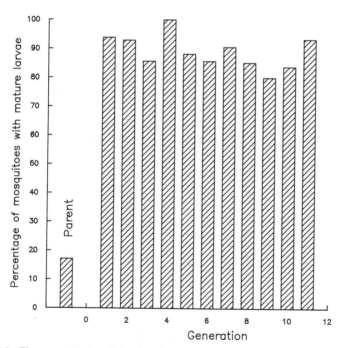

Figure 8.2 The susceptibility of blood-sucking insects for the organisms they transmit can be genetically determined. Here a West African strain of *Aedes aegypti* was selected for susceptibility to the filarial worm *Brugia malayi* over five generations. The selected mosquitoes were then maintained as a colony for several generations without an appreciable loss in susceptibility. (Drawn from data in Macdonald 1962b.)

anopheline mosquito host, most parasites will only develop successfully in a very narrow range of insect species. Physiologically based susceptibility to particular parasites, as well as varying among insect species, also differs intraspecifically. This has been clearly shown in the relationship of mosquitoes with both malaria parasites (Huff 1929, 1931, Ward 1963, Kilama & Craig 1969, Collins *et al.* 1986) and filarial worms (Kartman 1953, Macdonald 1962a, b, 1963). In both associations there is considerable variation in the ability of individual mosquitoes of a given species to permit the development of the parasites; this can range from complete refractoriness to complete susceptibility. Selective breeding experiments have shown that vector susceptibility is usually genetically based (Fig. 8.2), but in two instances extrachromosomal inheritance has been reported. Examples of the mechanisms involved are given below, but the field is still a controversial one and our understanding of it is still far from complete. It should also be borne in mind that parasite populations also display a spectrum of infectivities for available vectors which further complicates the situation found in the field.

Anopheles gambiae is the major vector of human malaria in Africa. Experimentally it can also transmit a range of other malaria parasites of simians, rodents and birds. A strain of the mosquito selected for refractoriness to the simian parasite *Plasmodium cynomolgi* is also refractory to other malaria species (Collins *et al.* 1986) suggesting a common refractory mechanism. Refractory mosquitoes defend themselves against the parasites by encapsulating healthy ookinetes (see Section 8.7) suggesting that the mechanism may be the ability of refractory mosquitoes to recognize the invaders.

In contrast to this broad spectrum mechanism against malaria parasites, work on the refractoriness of mosquitoes to filarial worms shows there are separate genes controlling susceptibility to various parasite species. So while susceptibility of *Aedes aegypti* to *Brugia malayi* (developing in thoracic musculature) is controlled by a single sex-linked, recessive gene designated f^m (Macdonald 1962b, 1963), susceptibility of the same mosquito to another filarial worm, *Dirofilaria immitis* (developing in Malpighian tubules) is controlled by a different sex-linked recessive gene designated f^t (McGreevy *et al.* 1974). Susceptibility of *Aedes aegypti* to *Plasmodium gallinaceum* is controlled by yet another gene, a dominant autosomal allele designated *pls*. From these data it might be thought that each gene is acting on a particular organ system and conferring refractoriness to anything attempting to develop there, but this is not so as can be seen in *Culex pipiens*. A sex-linked recessive gene *sb* controls susceptibility of this mosquito to *Brugia pahangi*, but does not control susceptibility for another filarial worm, *Wuchereria bancrofti*, even though both follow the same route into the mosquito and both develop in the insect's flight musculature. A complete explanation of the mechanisms defining these host specificities awaits our understanding of the nature of the phenotypic expression of these genes. However, from these results, and in contrast to the anti-malaria genes discussed above, it seems clear that these genes are not operating on some generalized defensive mechanism in the insect.

Non-Mendelian inheritance of susceptibility to *Brugia malayi* and *B. pahangi* has been reported in mosquitoes of the *Ae. scutellaris* complex (Trpis *et al.* 1981). It has also been reported to govern the susceptibility of tsetse flies for trypanosomes (Maudlin & Dukes 1985). Susceptibility appears to depend upon the numbers of rickettsia-like organisms (RLOs) which are present in the female parental fly (Maudlin & Ellis 1985). Under natural conditions only about 10% or less of tsetse fly populations are susceptible to trypanosome infection (Jordan 1974). It has been suggested that refractory tsetse defend themselves against invading trypanosomes by using gut-based lectins (see Section 8.7). Lectins bind to specific sugar residues on the surface of more than one trypanosome, thereby linking them together (agglutination). It is possible that in susceptible flies RLOs interfere with these lectins by producing an enzyme called chitinase. This enzyme attacks chitin, releasing glucosamine which binds to specific lectins inhibiting their action on trypanosomes (Welburn *et al.* 1987).

We have discussed susceptibility–refractoriness as if it were an all-or-nothing phenomenon, whereas in fact it is commonly a subtler issue with varying degrees of refractoriness and susceptibility shown by individual insects. In other words, the major genes pinpointed above are likely to have their impact modulated by other factors in the insect. For example, looking again at *Aedes aegypti* which are fully susceptible to *Brugia pahangi* (i.e. f^m/f^m), we find that about a quarter of the larva which enter the thoracic flight muscles die in the first few days (Beckett & Macdonald 1971). There are probably other subtle ways in which vectors respond to infection (Minchella 1985). The fact that absolute refractoriness does not always develop in a vector population exposed to high parasite challenge suggests that adoption of refractoriness by the vector has its costs as well as its benefits. For example, any physiologically based defence mechanism entails energy costs to the vector and demands resources which cannot then be used for reproductive or other purposes. In most vector populations the rate of infection is commonly less than 10%; because of this, adoption of all-out refractoriness may not always be the best selective strategy as for most vectors it will be a waste of resources. It may be better to opt for a less radical solution. One such option has been demonstrated in the infection of the snail intermediate hosts of *Schistosoma mansoni*. Infection of the snail eventually leads to a reduction, or possibly complete inhibition, of egg-laying. Snails continuously exposed to infection partly overcome this by a significant increase in the level of egg-laying in the prepatent period (Minchella & Loverde 1983). This insurance against the possibility of being parasitized has the advantage that if the snail does not become infected it has not wasted any resources which could have been used in the reproductive effort.

An interesting example of the mechanisms determining how a particular parasite may come to be associated with a particular vector is beginning to be uncovered in the relationships of *Leishmania* and sandflies (Titus & Ribeiro 1990). When small numbers of *Leishmania major* promastigotes are artificially inoculated into a vertebrate host they fail to establish an infection. In contrast, the same number of parasites injected along with salivary extracts from sandflies successfully establish an infection and undergo an increase in numbers of two to three orders of magnitude. In the vertebrate host *Leishmania* are obligate intracellular parasites of macrophages. Consensus opinion suggests that the macrophage can kill *Leishmania* by the production of superoxide and other toxic metabolites. The macrophage is stimulated to do this by IFN-γ (gamma interferon) produced by activated parasite-specific T cells. The Leishmania Enhancing Factor (LEF) of the salivary glands of sandflies appears to inhibit this IFN-γ stimulated production of superoxide. In addition LEF inhibits the ability of the macrophages to present antigens to T cells thereby decreasing the specific production of IFN-γ. The saliva appears to inhibit the immune response of the vertebrate host for long enough to allow the parasite to become established. LEF is important and possibly essential to the parasite. LEF has not been found in blood-sucking insects other than sandflies. This is probably an important factor in the co-evolution of *Leishmania* and their sandfly vectors.

8.3 ORIGIN OF VECTOR–PARASITE RELATIONSHIPS

There has been considerable debate concerning the origin of vector-parasite associations. Because parasites leave few, if any, fossils, attempts to describe the evolutionary development of the associations is bound to be largely speculation; however, informed speculation has provided possible, and in some cases convincing, evolutionary routes for several of these combinations. Two conflicting evolutionary pathways have been proposed for many of the associations. In one route the parasites are seen as being originally associated with the invertebrate, the vertebrate becoming involved when a lasting interaction between vertebrate and invertebrate has developed. The second route proposes an origin of the parasite in the vertebrate with transmission to the invertebrate occurring when the vector developed the blood-sucking habit. Two such possible origins have been suggested for the family Trypanosomatidae, which contains the important genera *Leishmania* and *Trypanosoma* (Molyneux 1984, Lainson & Shaw 1987).

Monogenetic flagellates are common intestinal parasites of invertebrates. Majority opinion suggests that present-day parasitic forms had their origins in similar intestinal forms. Regular transmission can only have been established sometime after the evolution of a recurring, close association between the insect and vertebrate. Initially transmission was probably by contamination in the insect faeces; stercorarian (posterior station) trypanosomes, such as the causative agent of Chagas disease, *Trypanosoma cruzi*, are still found today. Transmission via the mouthparts is seen as a later development as the trypanosomes adapted to life in the anterior parts of the insect's intestine.

There is also a minority view that vector transmitted trypanosomes of vertebrates did not develop along this route, but instead developed from vertebrate gut-dwelling forms which invaded the bloodstream. Certainly if this occurred, and the levels of parasitaemia achieved in the blood were sufficiently great, then the chances of immediate transmission by blood-sucking insects are high (many present-day forms are mechanically transmitted by insects). In support of a vertebrate origin, it is pointed out that when flagellates parasitic in invertebrates are experimentally transferred to vertebrates they do not survive well. Conversely, vertebrate parasites successfully survive in a wide variety of insects, including species which are not vectors.

An often repeated tenet in parasitology is that there is an inverse relationship between the pathology caused to the host and the length of association of that parasite and host. This principle has often been invoked to claim an invertebrate origin for most insect-transmitted parasites, including the Trypanosomatidae. It is unlikely that this inverse relationship, which forms a useful rule of thumb, is always true (Ball 1943, 1960, Anderson & May 1982, May & Anderson 1983). Also, given the precarious life-style of blood-sucking insects, pathologies are likely to be a far more efficient selective agent in the insect than in the vertebrate, with the consequence that strains of insect showing minimum pathologies will be rapidly selected (Mattingley 1965). Also, given the much shorter generation time and increased fecundity of the insect compared to the vertebrate host, selection in the insect relationship may well be expected to occur over a shorter calendar period. Consequently it seems unwise to use the degree of pathology caused by the parasite as a guide to the origins of parasite–vector associations.

The occurrence of sexual reproduction in the insect host has occasionally been suggested as powerful evidence for an invertebrate origin (Lainson & Shaw 1987), but this is not a strong argument. As we shall now see the majority opinion is that malaria parasites originated in vertebrates even though sexual reproduction occurs in the insect.

It is generally accepted that malaria parasites evolved from gut-dwelling parasitic apicomplexans (Baker 1965). There are no gut-dwelling monogenetic, monoxenous parasitic apicomplexans in present-day Diptera (although they are represented, if rather poorly, in other insect orders). There are many such forms in vertebrates, many with a tendency to leave the gut and to become tissue dwelling, pointing to a vertebrate origin for malaria parasites. The strongest argument in favour of a vertebrate beginning is the observation that if malaria parasites were derived from invertebrate species, then it would be expected that the life cycle would be the inverse of that actually found. Exocoelomic schizogony would be expected to occur in the insect gut and schizogony and gametogony in the haemocoel, with sporogony occurring in the vertebrate host (Bray 1963). It would seem hard to refute this analysis, but arguments have been put in favour of an invertebrate origin. Greater apparent pathogenicity of malaria parasites in their vertebrate hosts and sexual reproduction occurring in the insect have both been cited in favour of an invertebrate origin. As argued above, neither of these is a strong thesis. More convincingly it has been pointed out that malaria parasites are only found in one order of insects, the Diptera, while they are found in three classes of vertebrates. This might suggest an origin in invertebrates, but equally may reflect a later radiation into various vertebrate groups once the highly efficient vectorial link had been forged. For example, if *Plasmodium* first parasitized early reptiles, then it may well have co-evolved into birds and mammals.

It is generally agreed that the filarial worms evolved from gut-dwelling vertebrate parasites (Anderson 1957). These parasites probably reached the orbit of the eye by migration via the oesophagus, nasal cavities and nasolacrimal duct. Once in the eye the larval stages could be picked up by eye-feeding insects. Present-day, orbit-dwelling spirurids, such as the ruminant parasite *Thelazii gulosa*, are transmitted in this way by non-blood-sucking flies such as *Fannia* and *Musca*. The next stage was probably the invasion of other tissues, either because the infective stage larva were returned to a wound away from the eye or, more likely, because of migration by the adult nematode. These early tissue-dwelling forms are likely to have caused wounds in the skin which attracted insects which then became the vectors of their larvae; modern examples of this sort of life cycle exist. The bovine-infesting filarial worm *Parafilaria bovicola* produces subcutaneous nodules which burst and bleed attracting vectors such as *Musca lusoria* and *M. xanthomelas*, insects which are not normally blood feeding. Both *Parafilaria multipapillosa* and *Stephanofilaria stilesi* have a similar life cycle, but this time the vector may be a blood-sucking insect such as *Haematobia atripalpis* or *Stomoxys calcitrans*. The next step on the evolutionary path is the abandonment of lesion formation as the larvae come to depend on insects which can readily

Table 8.2 Blood-sucking insects commonly take meals which are only a small proportion of the total blood present in the host animal (the ratio between total blood in the host and size of the insect's blood meal is given). This minimizes the chances of the insect ingesting any individual parasite during feeding. One strategy adopted by insect-borne parasites to overcome this problem is to produce large numbers of infective stages which circulate in the blood of the host.

Host	Approximate total blood volume in host adult (ml)	Ratio	Vector	Approximate blood meal volume (ml)
man	4000 (47.9 ml kg^{-1})	2×10^6	*Anopheles*	0.002
cow	34 200 (57 ml kg^{-1})	1.14×10^6	*Glossina*	0.03
chicken	140 (95.5 ml kg^{-1})	4.6×10^4	*Culex*	0.003

break the skin to obtain their blood meal. Initially the microfilariae are likely to have stayed in the dermal tissues and to have relied on pool feeders for their transmission. This is still seen in current-day *Onchocerca* spp. with their reliance on simuliid and ceratopogonid flies as vectors. Finally, it is suggested, microfilariae moved to the peripheral blood-stream and became dependent on capillary-feeding insects such as the mosquitoes, a situation seen in the important parasites of humans *Wuchereria bancrofti* and *Brugia malayi*.

8.4 PARASITE STRATEGIES FOR CONTACTING A VECTOR

Vector-borne parasites adopt strategies which enhance their chances of encountering a vector. There are several examples. Blood-sucking insects take small blood meals compared to the total volume of blood in the host animal (Table 8.2). This means that the chances of any one infective stage encountering a vector insect are slim – as high as two million to one against for one example given in Table 8.2. To overcome this problem, vector-borne parasites produce very large numbers of offspring; malaria gametocytes, microfilariae, trypanosomes and arboviruses are all good examples. Failure of some parasites to produce sufficiently high parasi-taemias may help explain why more of them are not transmitted by insects. For example, human immunodeficiency virus, the causative agent of AIDS, produces a level of viraemia in patients which is estimated to be about six orders of magnitude too low for it to be a candidate for regular mechanical transmission by arthropods (Piot & Schofield 1986).

In addition to being present in large numbers, the offspring of many vector-borne parasites concentrate at sites which gives them the optimum chance of encountering a suitable insect. For example, *Plasmodium inui* and *P. yoelii* gametocytes are not all equally infective to the vector. The younger, slightly larger forms are more infective than the older, smaller forms. It is the younger gametocytes which are preferentially ingested by blood-feeding mosquitoes because they are concentrated in the peripheral capillary beds of the skin from which the mosquitoes feed (Dei Cas *et al.* 1980a, b).

Many microfilariae are also found in precise locations in the vertebrate host in order to increase their chances of encountering a suitable vector. *Onchocerca volvulus* microfilariae are limited to the dermis, consequently they are not ingested by vessel-feeding insects such as mosquitoes, in which they cannot develop. Instead they are ingested by pool-feeding insects such as their blackfly vectors. In another filarial worm, *Parafilaria multipapillosa*, the adults produce nodules beneath the skin in which microfilariae congregate. These nodules burst to produce a bloody wound which is attractive to blood-sucking flies such as *Haematobia atripalpis*. Other microfilariae are found in the blood circulation from which they are picked up by vessel-feeding vectors. There is growing evidence that these vectors pick up more microfilariae than is to be expected from the microfilarial density measured in host blood and the size of the blood meal ingested. For example, *Culex quinquefasciatus* feeding on hosts showing the very low microfilarial density of three microfilaria per millilitre of blood ingest 30 times the expected number of parasites. The probable explanation for the apparent anomaly is that microfilaria concentrate in those parts of the peripheral circulation most likely to be lanced by the vector. Filarial worms also display microfilarial periodicity which results in the periodic concentration of parasites in the peripheral circulation in order to optimize their chances of encountering a suitable vector insect. There are two distinct strains of the filarial parasite of man, *Wuchereria bancrofti*. In the periodic strain the microfilariae are found in the lung during the day and only at night do they appear in numbers in the peripheral circulation (Fig. 8.3). The appearance of these microfilariae in the peripheral blood coincides with the peak biting time of the major vector (which is *Culex quinquefasciatus* over a large part of the parasite's range). Over a limited part of the parasite's range, particularly in the South Pacific, it is transmitted by day biting mosquitoes such as *Aedes polynesiensis*. In these areas the parasite is diurnally subperiodic with microfilariae present in the blood throughout the 24-h period but showing a clear peak in the afternoon. Similar synchronization between the appearance of microfilariae in the peripheral blood and the peak biting time of the major vector is seen in many other filarial worm–vector associations (Table 8.3). Clearly the appearance of the

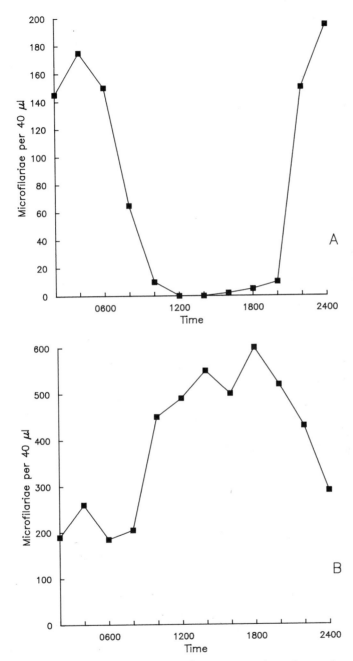

Figure 8.3 Microfilariae of *Wuchereria bancrofti* appear in peak numbers at the peak biting times of their vectors. The periodic strain (A) is carried by night biting vectors such as *Culex quinquefasciatus* and the subperiodic strain (B) by day biting mosquitoes such as *Aedes polynesiensis*. (Redrawn from Hawking 1962.)

165

Table 8.3 The microfilariae of many filarial worms display a pronounced periodicity, with microfilarial numbers in the peripheral blood coinciding with the peak biting time of locally abundant vector species.

Periodicity	Parasite	Major host	Major vector
nocturnal	*Wuchereria bancrofti* (periodic form)	man	*Culex quinquefasciatus*
	Brugia malayi	man	*Mansonia* spp.
	B. pahangi	cat	*Mansonia* spp., *Armigeres* spp.
	B. patei	cat	*Mansonia* spp., *Aedes* spp.
	Dirofilaria corynodes	monkey	*Aedes* spp.
	D. repens	dog	*Mansonia* spp., *Aedes* spp.
	Loa loa	monkey	*Chrysops* spp. (night biting)
diurnal			
	Loa loa	man	*Chrysops* spp. (day biting)
	W. bancrofti (subperiodic form)	man	*A. polynesiensis*
	Dipetalonema reconditum	dog	*Ctenocephalides canis*
	Ornithofilaria fallisensis	duck	*Simulium* spp.

microfilariae in the blood at particular times of the day is a means of maximizing the chances that the microfilariae will be ingested by a vector insect. But why not stay in the peripheral blood throughout the day? There must be a selective advantage in moving from the peripheral blood. The most likely explanation is that a microfilaria showing periodicity greatly reduces its chances of being ingested by a refractory vector in which it will die. It is also possible that outside the peak biting times of their vectors, microfilariae move into a more favourable part of the body, a move which will increase their longevity and, consequently, their chances of transmission.

Although the rhythmicity of trypanosomes has not been extensively investigated, 24-h periodicities are also known from this group. Indeed, it has been suggested that periodicity may be widespread among these protozoans (Worms 1972). In the frog *Rana clamitans*, individuals of the *Trypanosoma rotatorium* complex congregate in the kidney and appear as a flush in the peripheral blood around midday (Fig. 8.4) (Southworth *et al.* 1968). *Trypanosoma minasense* shows a mid-afternoon peak in the peripheral blood of the Brazilian marmoset. *Trypanosoma lewisi* and *T. duttoni* also show circadian periodicity in rats and mice, respectively (Cornford *et al.* 1976) with peak parasitaemia occurring soon after dusk. *Trypanosoma congolense* will also display periodicity under laboratory conditions (Bungener & Muller 1976, Hawking 1976, Allan & Lehane in prep.), although other workers have been unable to show such clear rhythmicity in cattle under field conditions.

Circadian variation in the infectivity of gametocytes for mosquitoes has been reported for several malaria parasites of birds and mammals (Hawking *et al.* 1966, 1968, 1971, 1972, Coatney *et al.* 1971). However, work on *Plasmodium falciparum* has failed to confirm these observations (Bray *et al.* 1976) and it remains to be seen if it is a real phenomenon under field conditions. The following hypothesis has been proposed to account for the periodicity observed in the laboratory. The period of ripeness of the gametocytes is short (much less than 24 h). The gameto-cytes arise at schizogony from the same merozoites as the asexual forms, with the timing of peak gametocyte maturity regulated by the timing of schizogony. The timing of schizogony in turn is almost constant for particular species or strains of parasite; in most Plasmodiidae it occurs either at 24-h intervals or at multiples of 24 h. This would enable the parasite to produce maximum numbers of mature gametocytes at precise times of the day. Another possible explanation of the circadian variation in infectivity may be the differential availability of gametocytes at various times of the day. This has been reported in a related species, *Leucocyto-zoon smithi*, in the turkey (Gore & Pittman-Noblet 1978) where the parasite changes the peripheral distribution of its gametocytes in parallel

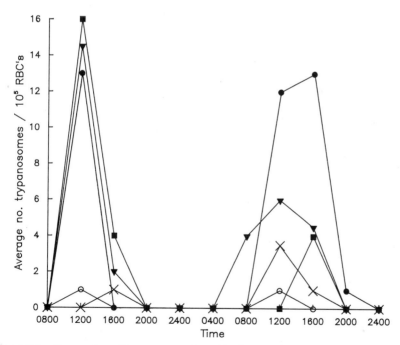

Figure 8.4 The number of trypanosomes found in the peripheral circulation of the frog, *Rana clamitans*, shows a circadian rhythm. The fluctuations for five separate frogs are shown. (Redrawn from Southworth *et al.* 1968.)

167

with diurnal changes in host body temperature (under the influence of hormones and the day–night cycle).

Parasites are also capable of manipulating the behaviour of the vertebrate host to facilitate their transmission to vector insects. The mouse, *Mus musculus*, in common with many other rodent species, shows very efficient anti-insect behavioural defences. In consequence, it is only rarely fed on by temporary ectoparasites. However, if the mouse is infected with malaria, there is a period during and just after peak parasitaemia when the mouse is disabled and unable to defend itself, during which time mosquitoes can feed readily from it (Day & Edman 1983). Under experimental conditions most mosquitoes feed on these mice about two days after peak parasitaemia and this coincides with a peak in gametocyte numbers (see Fig. 7.9). So the parasite has modulated the host's behaviour in such a way, and at a precise time, that will maximize the chances of parasite transfer to a vector insect.

Parasites can also affect haemostatic mechanisms in the vertebrate host in ways which enhance their chances of transmission. The longer an insect takes to obtain its meal the more likely it is to be swatted or disturbed by the host. To minimize host contact time insects produce various anti-haemostatic factors in the saliva (see Section 5.2). Parasites within the vertebrate host animal can also interfere with host haemostasis. Mosquitoes feeding on mice or hamsters infected with *Plasmodium chabaudi* or Rift Valley Fever virus, respectively, had their probing times reduced by at least one minute (Fig 8.5) (Rossignol *et al.* 1985). These and other pathologies in the vertebrate host blood system caused by vector-borne parasites (Aikawa *et al.* 1980, Wilson *et al.* 1982, Halstead 1984) may well be a parasite strategy to increase their chances of being successfully transmitted by a vector insect. The selective advantage to the insect, from the reduced dangers in feeding rapidly, might lead to vectors choosing to feed on infected hosts. Whether this short-term advantage would outweigh the longer-term disadvantage of becoming parasitized is not clear (see below).

Another intellectually attractive proposition is that the fevers so typical of many vector-borne illnesses are caused by the parasite as a means of attracting vectors to the infected host. This possibility has been proposed on many occasions (for example Gillett & Connor 1976), but experimental work on malaria-infected mice has shown no significant increase in numbers of feeding mosquitoes when the host is hyperthermic (Day & Edman 1984). While the extra heat generated by febrile hosts may not differentially attract vectors, experiments on Sinbis virus-infected chickens have shown there are other attractive factors associated with infected hosts which have not yet been identified. In these experiments, infected chickens held inside cages to which mosquitoes have no access attract more mosquitoes than uninfected controls, despite there being no

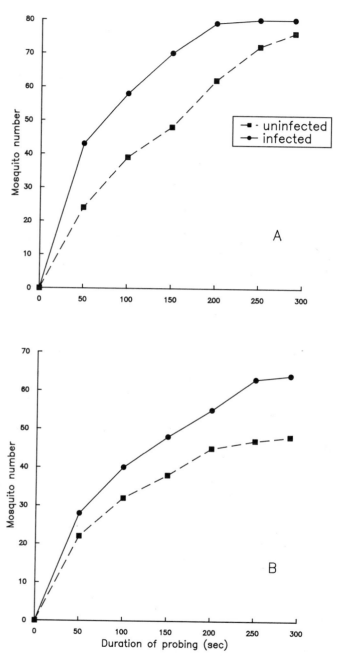

Figure 8.5 Mosquitoes feed more quickly on infected hosts. This can be seen in the trends in probing times observed in groups of mosquitoes variously fed on uninfected hosts (A ■ and B ■), mice infected with *Plasmodium chabaudi* (A ●), or hamsters infected with Rift Valley Fever virus (B ●). (Redrawn from Rossignol *et al.* 1985.)

consistent differences in temperature or carbon dioxide output from the traps (Mahon & Gibbs 1982). It is possible that these attractive factors are indirectly generated by the parasite to enhance its chances of trans- mission.

8.5 PARASITE STRATEGIES FOR CONTACTING A HOST

Parasites may alter the behaviour, particularly the feeding behaviour, of the vector insect to enhance their chances of entering the vertebrate host (Molyneux & Jefferies 1986). It has been reported that the tsetse flies *Glossina morsitans morsitans* and *G. austeni* will probe the host more frequently and feed more voraciously when they are infected with the salivarian trypanosome *Trypanosoma brucei* (Table 8.4) (Jenni *et al.* 1980). Similar results have been reported in tsetse infected with *Trypanosoma congolense* (Roberts 1981). The colonization of the foregut by the trypano- somes may possibly interfere with the feeding process, in particular, with the function of those labral mechanoreceptors which are responsible for detecting blood flow rate in the food canal. Also, colonies of trypanosomes in the gut reduce the cross-sectional area of the food canal which significantly interferes with the feeding process (see Section 5.6). To ingest the same size of meal the fly probably has to feed longer, or to take more, smaller meals, both of which may help in parasite trans- mission. Finally, colonies of trypanosomes occurring in the hypopharynx mean that considerably higher pressures have to be applied by the insect to secrete its saliva. This again is likely to increase the chances of trypanosome infection of the vertebrate host; however, the situation is not clear-cut because other workers have been unable to repeat these results (Moloo 1983, Jefferies 1984, Moloo & Dar 1985).

Similar explanations possibly account for the feeding difficulties of infected sandflies (Killick-Kendrick *et al.* 1977, Killick-Kendrick & Molyneux 1981) and their increased probing times compared to uninfec- ted sandflies (Adler & Ber 1941, Chung *et al.* 1951, Strangways-Dixon & Lainson 1966, Beach *et al.* 1985). Reduced feeding success and increased probing times have also been seen in some, but not all, female *Aedes triseriatus* mosquitoes infected with La Crosse virus (Grimstad *et al.* 1980).

The plague bacillus *Yersinia pestis* also interferes with the feeding of its host the plague flea, *Xenopsylla cheopis*. Blockage of the alimentary canal by the parasite leads to the regurgitation of the bacillus into the wound and its consequent transmission (Bacot & Martin 1914). To achieve this blockage the bacterium must be held in the flea's proventriculus by a fibrin clot formed in the blood meal. The fibrin clot is induced by a 'coagulase' produced by the bacterium and a trypsin produced in the

Table 8.4 Tsetse flies infected with trypanosomes feed more readily and probe more often than uninfected flies, thereby increasing the chances of parasite transmission. (Redrawn from Jenni et al. 1980.)

	Starving for 24 h			Starving for 48 h		
	No. feeding	No. feeding on first probe	Mean no. of probes ± S.E. before feeding	No. feeding	No. feeding on first probe	Mean no. of probes ± S.E. before feeding
40 infected flies	40	3	5.08 ± 0.40	40	4	4.68 ± 0.37
40 uninfected flies	28	13	1.89 ± 0.21	38	24	1.53 ± 0.13

flea's midgut. The clot forms successfully at temperatures below 25°C. Above 27°C the clot is rapidly destroyed by two temperature-dependent fibrinolytic factors, one from the flea and one from the bacterium, and under these conditions the bacterium is eliminated from the flea's gut (Cavanaugh 1971). These observations explain the physiological mechanism underlying the blockage of the flea's gut which leads to regurgitation and the transmission of the parasite. They also neatly explain the pattern of bubonic plague epidemics, which are always abruptly terminated by the arrival of the hot season (mean monthly ambient temperature >27.5°C). They do not explain why the parasite should produce a fibrinolytic agent which contributes to its own downfall at higher temperatures.

Malaria parasites have also been reported to interfere with the feeding process in vectors (Rossignol et al. 1984). This possibly occurs as a result of the sporozoite-induced pathology in the salivary gland and the attendant four- to five-fold decrease in the levels of apyrase produced in the saliva. Without sufficient apyrase the insect has trouble locating blood rapidly (see Section 5.2). This leads to an increase in feeding time and/or number of hosts bitten, with an associated increase in the number of parasites transmitted. It is quite possible that other parasites which invade the salivary glands, such as Trypanosoma brucei or the arboviruses, may also impair the function of the glands causing a similar increase in the duration of vector–host contact.

It is interesting to speculate on the fine evolutionary balances which exist in systems, such as those described above, where the parasite has modified the behaviour of the vector to its own advantage. In doing so the parasite has decreased the fitness of the insect (in these cases, decreased feeding efficiency) and the insect may reproduce less successfully as a result. This is a selection pressure which, if it occurred in isolation, could lead to the appearance of a refractory vector strain. In contrast, by its impact on the haemostatic factors in the vertebrate host, the parasite has increased the efficiency with which the insect obtains the infected blood meal. This is probably selectively advantageous to the insect as in most cases eggs which develop as a result of this meal will be deposited before the parasitic infection has matured and before the parasite-induced pathologies in blood-feeding occur. Mathematical models of these interrelated and counterbalancing effects are beginning to appear which deal with the impact either on the vector population (Dobson 1988) or on disease transmission (Kingsolver 1987, Rossignol & Rossignol 1988). Modelling studies are at an early stage. At present they only consider the impact of some aspects of parasite-induced changes in feeding efficiency. Clearly many other factors are of importance including the powerful impact other parasite-induced pathologies in the vector may have on both the make-up of the vector population and the epidemiology of disease.

8.6 VECTOR PATHOLOGY CAUSED BY PARASITES

Given the brief association of parasite and vector during mechanical transmission, it is not surprising that little, if any, pathological effect is seen. However, considerable damage can be caused during the more extensive associations seen during biological transmission. In the most extreme cases the ingested parasite can lead to the death of the vector insect; for example, *Rickettsia prowazekii* and *R. typhi* both cause the death of the louse, *Pediculus humanus* (Weyer 1960, Jenkins 1964). Normally, increased mortality occurs when the insect has ingested either an unusually large number of parasites, or an uncommonly large number of parasites have survived entry and are developing successfully in the insect. This is well illustrated in the infection of *Aedes trivittatus* with the filarial worm *Dirofilaria immitis*. These mosquitoes can tolerate low numbers of microfilariae, but when they feed on dog blood showing microfilaraemia as high as 347 microfilariae per 20 mm^3 a significantly increased rate of mortality occurs in the vector population (Fig. 8.6) (Christensen 1978). Reduced longevity of the vector can also be seen in malaria infections of mosquitoes (Gad *et al.* 1979, Klein *et al.* 1982). Increased mortality in the vector population, particularly if it occurs in the early phase of the infection, is not only harmful for the insect, but is also harmful for the parasite, reducing its chances of transmission. Also, if this increased mortality significantly reduces the reproductive success of the vector population then it may lead to the selection of more resistant strains of the vector species. For both reasons the parasite is likely to try to avoid reducing vector longevity. In one instance (tsetse and salivarian trypanosomes) it has been suggested that infection may actually increase the vector's longevity (Baker & Robertson 1957).

Parasites also have sublethal effects on vector insects. They may impair the reproductive potential of the vector, as has been shown in laboratory experiments where infected mosquitoes produce reduced numbers of eggs (Maier & Omer 1973, Hacker & Kilama 1974, Freier & Friedman 1976). While this may happen by decreasing the longevity of the vector, as mentioned above, it is likely that more subtle mechanisms are also operating. For example, the invasion of the salivary glands of *Aedes aegypti* by sporozoites of *Plasmodium gallinaceum* reduces salivary apyrase by as much as two thirds (Rossignol *et al.* 1984). Apyrase is a major component of the salivary anti-clotting mechanisms (see Section 5.2). This salivary pathology increases probing times in mosquitoes, thereby increasing the chances of the insect being disturbed (or killed) while feeding (Gillett 1967). These mosquitoes are also more ready to desist from feeding, which is also likely to reduce feeding success. Both effects will militate towards decreased egg production (Rossignol *et al.*

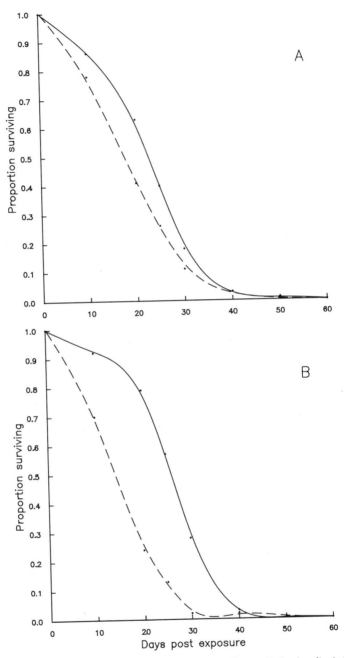

Figure 8.6 Mosquitoes ingesting low numbers of microfilariae (A, broken line) show little increased mortality over uninfected controls (B, solid line), but the ingestion of abnormally high numbers of microfilariae (B, broken line) leads to significant changes. (Redrawn from Christensen 1978.)

1986). Impaired reproductive success probably favours the selection of more refractory strains of the vector population.

It is also possible that reproductive success may be impaired because of competition with the parasites for available nutrients. Unambiguous evidence is lacking but there is evidence of metabolic competition affecting other vector activities. The flight capabilities of tsetse flies can be impaired because trypanosomes have used a considerable proportion of the reserves normally used for flight (Bursell 1981). In males this reduces flight potential by about 15% and by significantly more in females which, because of the reproductive effort, have less reserves to spare for other purposes. In mosquitoes, invasion of the flight muscles by filarial worms is followed by the loss of glycogen from the affected fibres (Lehane & Laurence 1977); this loss of reserves will almost certainly impair flight capability.

Impairment of flight activity (such as that discussed above) is a common result of parasite invasion of the vector. In *Anopheles stephensi* infected with *Plasmodium yoelii*, flight activity falls by about one third as the oocysts mature and rupture (Rowland & Boersma 1988). It has also been shown that *Plasmodium cynomolgi* infections in the same mosquito will reduce both speed and duration of flight (Schiefer *et al.* 1977). Similar effects are found in filarial infections of mosquitoes. *Aedes aegypti* parasitized by *Brugia pahangi* fly for significantly less time and will not show the characteristic increase in flight time with increasing hunger typical of uninfected mosquitoes (Fig. 8.7) (Hockmeyer *et al.* 1975). Filarial worms interfere with flight at two stages of development (Rowland & Lindsay 1986). Flight is temporarily diminished during the first two days after feeding but then recovers to normal levels. After this, insects harbouring small numbers of worms show normal flight activity. In mosquitoes harbouring 13 or more filarial worms, flight activity shows a second and dramatic fall seven to eight days after the infective blood meal, to about 10% of that seen in control insects. This coincides with the emergence of the infective stage larvae from the flight muscles and their migration to the mouthparts. Results on the effects of parasites on vector flight are not all in one direction; other authors have reported an increase, not a decrease, in the flight range of a parasitized vector (Alekseyev *et al.* 1984).

Common sense suggests that vectors with decreased flight capabilities will be less successful in reaching a host and transmitting the disease. If this is true it is another good reason for the parasite to try to minimize the damage that it causes to its vector, but it is still too early to come to such hard and fast conclusions about the interactions of parasites and vectors. Under some circumstances a limited amount of vector pathology might be to the parasite's advantage. Taking lymphatic filariasis as a speculative example, it is known that the number of worms in the vertebrate host can be modulated by the host's immune response. Hosts with elephantiasis

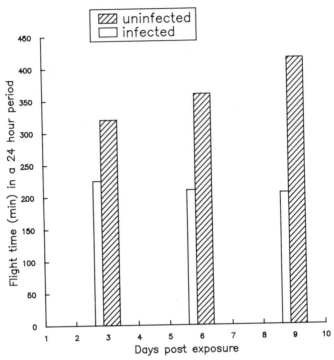

Figure 8.7 Infection of *Aedes aegypti* with the filarial worm *Brugia pahangi* reduces the insect's flight capacity. It also inhibits the gradual build-up in flight time after the blood meal, which is characteristic of uninfected flies. (Redrawn from Hockmeyer *et al.* 1975.)

show the greatest immune reaction and display the lowest number of parasites. The greatest number of parasites are seen in vertebrate hosts displaying little or no pathology. Children of infected mothers are exposed in the womb to filarial antigens and come to accept these as 'self'. Such children react little to filarial worms when they themselves become infected and so are excellent hosts for the worms. So lymphatic filariasis is a familial infection (Haque & Capron 1982) in which filarial worms prosper by being transferred to the offspring of infected mothers. In such circumstances there might be a selective advantage in causing limited damage to the flight capacity of the vector because this may localize transmission with the result that transfer is more likely to occur to a tolerant host (P. A. Rossignol, pers. comm.).

I have outlined some of the ways in which parasites can damage the insect and I have also pointed out that resistant strains of vectors may appear. At first glance, as with other parasitic diseases, it is hard to see why completely refractory populations of vector do not appear. The fact is they do not, and that parasites continue to exist; there must be reasons for this. The most obvious reason may be that insects cannot produce refractory strains against their parasites because as quickly as they adopt new defence mechanisms the parasites evolve strains capable of avoiding them. Another suggestion is that, under some circumstances, susceptible strains are more fit than refractory ones. For example, any physiologically based defence mechanism will have energy costs to the vector, demanding resources which cannot then be used for reproduction. In most vector populations the rate of infection is normally less than 10%. Because of the low probability of being infected in such populations, adoption of all-out refractoriness may not be the best selective strategy and a less radical solution may be more advantageous. One such option has been demonstrated in the infection of the snail intermediate hosts of *Schistosoma mansoni*. Infection of snails will eventually lead to a reduction, or possibly complete inhibition, of egg-laying. Snails continuously exposed to infection partly overcome this by a significant increase in the level of egg-laying in the prepatent period (Minchella & Loverde 1983, Minchella 1985). This insurance against the possibility of being parasitized has the advantage that if the snail does not become infected it has not wasted any resources on defensive mechanisms which could have been used in the reproductive effort. We might make a similar argument for the association between vector and malaria parasites discussed above. The parasites do not deleteriously affect the vector insect until the infective stages are produced. This takes between 8 and 14 days from the infective blood meal. During this time the mosquitoes are likely to have produced two or more egg batches. Given the heavy bias towards young insects in mosquito populations, the production of two to three egg batches may well represent a good reproductive effort. Resources put towards refractory mechanisms might be selectively disadvantageous by reducing this early reproductive effort.

A second way in which susceptible strains may be more fit than refractory ones is if a degree of mutualism has evolved between vector and 'parasite'. For example, if being parasitized causes only small losses to the vector then they may be outweighed by any of the following advantages: an incapacitated host is unlikely to damage or kill the vector (see Ch. 7); a parasitized host's haemostatic mechanisms may be impaired by the infection, so decreasing the time necessary for feeding and increasing the chances of successful feeding (see Section 8.5); the viscosity of the host's blood may be lowered by the parasite, this again may reduce the time required to complete feeding (see Section 5.6).

Disentangling and understanding the relationships between vectors and parasites presents a major challenge. Because these issues are central to disease transmission the rewards to be gained from such an understanding could be considerable.

8.7 INSECT DEFENCE MECHANISMS

Because of the bearing insect immunity has on the vectorial capacity of blood-sucking insects, I intend to give an outline of the topic here even though very little of the work has been done on blood-sucking insects.

Insects possess an immune system which protects them from the potentially damaging effects of biological invaders. Despite recent advances in our understanding of insect defences, we are still a long way from a clear view of their detailed operation. To give a broad view of the function of insect defence mechanisms let us compare them with the vertebrate immune system which has been studied in immense detail.

Perhaps the overriding feature of the vertebrate system is its ability to step up both the speed and intensity of the immune response on a second and subsequent exposure to a pathogen, a capacity termed acquired immunity. Vertebrate immune mechanisms displaying acquired immunity are characterized by their memory, the specificity of the response and its widespread dissemination and amplification as a result of challenge (Roitt 1985). Like vertebrates, insects clearly have the ability to distinguish self from non-self. They are also capable of amplification of both cellular and humoral responses to infection, and are able to disseminate these responses throughout their bodies. But insects appear to lack the specificity seen in vertebrates and there is debate as to whether insects possess the memory so characteristic of the vertebrate system. Although insect defence mechanisms may be less sophisticated than those of vertebrates, the tremendous success of insects as a group is proof that their immune mechanisms meet their survival needs. The difference in the degree of sophistication required in the two immune systems is possibly explained by the short generation time of most insects, where a memory component to immunity would add little to the insect's reproductive success (Anderson 1986).

To get inside the body of the insect, invading organisms must cross an epithelial barrier; there are several possible routes. The invader may directly attack the cuticle, perhaps simplifying the task by selecting the thinnest areas such as the intersegmental membranes or the trachea. But for insects, as for other highly organized Metazoa, the intestine is the main site of attack. There are two reasons for this. First, the intestine's role in absorption, and to lesser extent secretion, require that physically it is a relatively unprotected epithelial layer. Second, most insects ingest many potentially hostile organisms during the course of their feeding activities. Importantly, blood-sucking insects are in an unusual and privileged position here because blood taken directly from the host's circulatory system is largely a sterile food. For insects which feed exclusively on blood (e.g. tsetse flies, many lice and triatomine bugs) this means that the meal is only rarely a source of potential infection. This acts

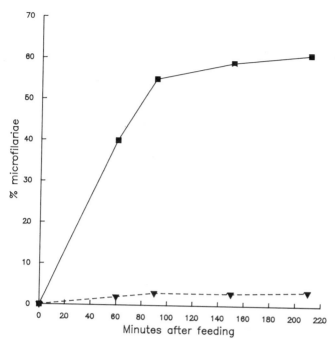

Figure 8.8 Several adult diptera possess a series of sclerotized spines and teeth in the foregut which are capable of fatally damaging invading microfilariae. These armatures are capable of limiting the numbers of microfilariae which successfully invade the insect. The percentage of *Wuchereria bancrofti* microfilariae which successfully migrate from the gut of *Aedes aegypti* (■, which lacks an armature) are compared with *Anopheles gambiae* species A (▼, which possesses a well-developed armature). (Redrawn from McGreevy *et al.* 1978.)

to lessen the selective pressure ensuring efficient intestinal defences in blood-sucking insects. This may well be an important factor in permitting invasion of the vector by parasitic organisms.

The insect fore- and hindgut are of ectodermal origin and are lined by cuticle. Occasionally this cuticle may be arranged into armatures which help protect the insect against invading organisms (Fig. 8.8). The midgut epithelium is of endodermal origin and does not have a cuticular lining. In most insects (but not in the Reduviidae, for example) the food (and any organisms ingested with it) is still separated from the midgut cells by an extracellular layer known as the peritrophic membrane. Peritrophic membranes may function as a barrier preventing invading organisms ingested with the food coming into contact with the midgut epithelium, although this is not always the case; to appreciate why we need to look at how peritrophic membranes are formed and at their structure.

Two types of peritrophic membrane are recognized based on their method of production. Type I peritrophic membranes are formed from

179

secretions of cells along the complete length of the midgut. This is the most common method for the production of peritrophic membrane and is widespread among insects and many other animal groups (Peters 1968). In haematophagous insects Type I peritrophic membranes are found, for example, in adult mosquitoes (Fig. 6.2), blackflies (Reid & Lehane 1984), sandflies and tabanids. Type I peritrophic membranes are not present in the hungry insect but are produced in response to the blood meal. They are secreted as a fluid which gradually develops into a hard peritrophic membrane. Even though relatively few studies have been performed it is clear that the thickness and rate of development of Type I peritrophic membranes is species-specific (Table 8.5). For example, if we compare peritrophic membrane formation in the three temperate simuliids, *Simulium equinum*, *S. ornatum* and *S. lineatum*, we see that at 24 h after the blood meal the mean peritrophic membrane thicknesses are 9.03 μm, 11.7 μm and 18.95 μm, respectively. In *S. ornatum* a third of the final thickness (3.91 μm) is achieved in the first two minutes following the blood meal and by one hour it is 80% (9.25 μm) of its final thickness. By comparison, in *S. equinum* the peritrophic membrane is only about 4.25 μm thick one hour after the meal and in *S. lineatum* only about 40% of the total thickness (8.07 μm) is achieved in one hour after feeding.

Type II peritrophic membranes are found in all larval Diptera and in adult tsetse flies, hippoboscids and biting muscids. They are produced by an organ known commonly as the proventriculus, but more correctly as the cardium, which is situated at the junction of the fore- and midgut. These peritrophic membranes pass backwards down the length of the midgut forming an unbroken cylinder which contains and separates the food from the midgut epithelium. Type II peritrophic membranes are

Table 8.5 Comparison of the rate of formation of the peritrophic membrane among various mosquito species.

Species	Formation (h)	
	First signs	Fully formed
Culex tarsalis	8–12	24
C. nigripalpus	6	24
C. p. pipiens	12	—
Aedes aegypti		4–6
	4–8	12
Ae. triseriatus	0.833	4
Anopheles stephensi	15–20	30
	12	48
An. atroparvus	—	24

Table 8.6 The correlation between the infection rate of tsetse flies with trypano-somes and the development of the peritrophic membrane. Whether this is cause and effect or coincidental is still unknown. (Information from Lehane & Msangi (forthcoming) and Wijers 1958.)

Age of adult tsetse (hours post-emergence)	Membrane length (% of midgut length)	Infection rate with *Trypanosoma gambiense* (%)
at emergence	0	
0–24		7.6
12–18	41.5	
24–48		1.9
36–42	73.1	
48–72		0.7
60–66	91	
>72		0
84–90	100	

continuously secreted and are fully formed on leaving the proventriculus (Fig. 6.3). Consequently, unlike the situation with Type I peritrophic membranes, there is usually no time during the blood meal when the peritrophic membrane is absent or only partially formed. An exception to this situation occurs in flies which feed soon after emergence from the pupal stage (see below and Table 8. 6).

Clearly, to permit digestion and subsequent absorption of the meal, while retaining the bulk of the meal within the endoperitrophic space, the peritrophic membrane must be a semi-permeable filter. The size of the 'pores' in the peritrophic membrane have been experimentally determined for relatively few insects. On the basis of experimental results on isolated peritrophic membrane preparations we estimate the pore size in the Type II peritrophic membrane of the adult tsetse, *Glossina morsitans morsitans*, to be about 9 nm making the peritrophic membrane permeable to globular molecules of up to a molecular weight of about 150 kDa (Miller & Lehane 1990). It has been suggested that 'pore' size in some Type I peritrophic membranes may be much greater (200 nm in *Locusta*; Peters *et al.* 1973).

So, there are two ways in which the peritrophic membrane may fail to be an effective physical barrier preventing invading organisms contac-ting midgut cells. The first is if the pores of the peritrophic membrane are larger than the invading organism. Clearly the 9 nm pores seen in the peritrophic membrane of the tsetse fly are too small to allow the free passage of any potential invaders, but the 200 nm pores reported in some Type I peritrophic membranes would allow the free passage of arbovi-ruses while denying entry to bacteria and eukaryote organisms.

Second, the peritrophic membrane may fail to be an effective physical barrier if the peritrophic membrane is not in position when the blood meal is taken. This is seen with Type I peritrophic membranes where there is a lag phase after feeding before an appreciable peritrophic membrane becomes visible between the food and the gut. The more rapidly the peritrophic membrane is produced the shorter this window of exposure will be; as noted above, the speed at which the peritrophic membrane is produced is species-specific and highly variable (Table 8.5). A potential window of exposure can also occur with Type II peritrophic membranes if a meal is taken soon after the emergence of the insect from the pupal stage. In the tsetse fly the peritrophic membrane takes about three days to form a complete sleeve down the full length of the midgut (Willett 1966, Msangi 1988) (Table 8.6). Apparently, the unfinished sleeve is closed at its posterior end during this period; the blood is retained in the endoperitrophic space and never directly contacts the midgut epithelium (Willett 1966). It is difficult to see how the organ producing Type II peritrophic membranes could engineer a closed sac and, with this in mind, it is notable that the period when the peritrophic membrane does not run the full length of the midgut is also the period when tsetse flies are particularly susceptible to trypanosome infection (Wijers 1958, Distelmans et al. 1982) (Table 8.6).

The developing peritrophic membrane is clearly not an absolute barrier preventing parasites entering the body of susceptible insects. If it were insects would rarely, if ever, become infected per os. Even so the peritrophic membrane can still be a factor limiting the intensity of infection. For example, microfilariae ingested along with the blood meal must penetrate the midgut of blackflies within about the first four hours, or they become trapped (Lewis 1953, Laurence 1966, Eichler 1973). Less than 50% of the microfilariae are normally successful in making this migration in simuliids and the rest die in the blood meal. Microfilariae ingested by mosquitoes are usually more successful and up to 90% escape from the gut (Ewert 1965). To achieve this they also leave the midgut quickly, beginning their migration within minutes of the blood meal being taken; usually over 50% have left within two hours (Wharton 1957, Laurence & Pester 1961). Clearly microfilariae trapping may be caused by several factors developing in the gut following feeding. One of these is the progressive clotting of the blood which reduces the ability of microfilariae to migrate (Kartman 1953, Sutherland et al. 1986). It is widely speculated, but not yet clearly proven, that it is the developing Type I peritrophic membrane which becomes the principal barrier.

Some interesting experiments suggest that the developing peritrophic membrane of susceptible mosquitoes may also be the factor limiting the numbers of successfully migrating malaria ookinetes. Malaria gameto cytes, ingested with the blood meal, have to undergo a maturation perio before they develop to ookinetes which can migrate through the midgu wall. During this maturation period the mosquito's digestive cycle begins, including the development and the thickening of the peritrophic membrane. If this maturation period is side-stepped by the feeding of *Plasmodium berghei* ookinetes rather than gametocytes to *Anop 'eles atroparvus*, then the number of parasites successfully invading the midgut increases greatly (Janse *et al.* 1985). The simplest explanation of these results is that more parasites penetrate the midgut because the peritrophic membrane encountered is thinner. This view is supported by a second series of experiments in which the peritrophic membrane of *Aedes aegypti* was artificially thickened. This significantly reduced the number of *Plasmodium gallinaceum* ookinetes successfully penetrating the gut (Ponnudurrai *et al.* 1988).

In addition to limiting the numbers of invading parasites, it has been speculated that the peritrophic membrane may sometimes be an absolute barrier to infection, so determining which species are refractory and which are susceptible to a particular infection. For example, *Plasmodium berghei* gametocytes require 18 h and *P. falciparum* gametocytes require 24 h to mature into ookinetes capable of penetrating the midgut wall. Based on this difference it has been suggested that *Anopheles stephensi* and *An. atroparvus* are both susceptible to *P. berghei* because their peritrophic membranes take longer than 18 h to form completely (Table 8.5), while only *An. stephensi* is susceptible to *P. falciparum* because its peritrophic membrane takes longer than 24 h to be fully formed (Ponnudurrai *et al.* 1988).

The peritrophic membrane is not the only gut barrier to invading organisms; some mosquitoes possess an intestinal barrier to the establishment of arboviruses (Chamberlain *et al.* 1954, Chamberlain & Sudia 1961). This can be demonstrated if the gut is bypassed by the direct injection of the virus into the haemocoel, when a full infection of otherwise refractory insects commonly occurs. This intestinal barrier can often be overcome if sufficient virus particles enter the gut. The threshold level for each viral infection varies among mosquito species with different viruses usually having different threshold levels in the same species. Given these facts, it seems possible that the barrier is a function of the number of surface receptors on gut epithelial cells and the probability of the virus encountering one; this in turn is dependent on several factors. First, different insect species and strains are likely to vary in the numbers of receptor sites present. Also, the gut environment is hostile to viruses and as digestion proceeds increasingly more virus particles are killed. So the speed with which digestion proceeds after feeding affects the time available for the virus to encounter a receptor site as does the timing of peritrophic membrane development, which masks the receptors on the cell's surface.

Organisms which are successful in crossing the outer epithelial barrier of an insect are then presented with the defensive mechanisms of the insect's blood system. Insect blood is contained in an open haemocoelic space which is not lined by endothelium, so the organs of the body are only separated from the blood by the basement membranes of their own cells. The blood consists of a fluid medium in which a variable number of cells are dispersed. As well as circulating blood cells, insects may also have accumulations of fixed cells which take part in the defence responses (Kaaya & Ratcliffe 1982). No clear-cut picture has yet emerged for the classification of insect blood cells, but essentially six categories of cell have been identified and these are outlined in Box 8.1 (Price & Ratcliffe 1974). Although this picture holds true for most insect orders, unfortunately the Diptera, which contain most blood-sucking species, vary considerably from this basic scheme (Ratcliffe 1986). Many adult flies (such as mosquitoes) tend to have very few blood cells; those cells which are present often do not fit easily into the classification outlined in Box 8.1 (Kaaya & Ratcliffe 1982). For example, cystocytes seem to be replaced by anucleate cellular fragments termed thrombocytoids (Zachary & Hoffman 1973). The cells of the insect blood system are capable of a range of defensive responses including encapsulation

Box 8.1   An outline of the properties of the six types of haemocyte commonly found in insect blood (Price & Ratcliffe 1974, Ratcliffe 1986)

Prohaemocyte   Small cells which are rarely more than a small percentage of the total blood cells. They are thought to be the stem cells from which the other blood cell types are formed.

Granular cells   Accounting for between 25 and 60% of the blood cells, they are thought to be involved in coagulation. They contain granules which are discharged when the cell contacts a foreign surface.

Plasmatocytes   Cells with a highly variable form accounting for between 25 and 60% of the cells in the blood. They are amoeboid in nature and are involved in phagocytosis and encapsulation responses.

Cystocytes   Small cells resembling prohaemocytes, but distinguishable from them by the possession of granules. They are particularly abundant, accounting for between 40 and 60% of the total blood cells. Like granular cells they are thought to be involved in blood coagulation but appear to be more easily triggered into degranulation than the granular cells.

Spherulocytes   Characterized by 1–3 μm inclusions in the cytoplasm, these cells account for less than 5% of the total blood cells and have an unknown function.

Oenocytoids   Large cells forming 1–2% of the total blood cells, which are believed to be involved in melanization because of the possession of components of the prophenoloxidase system.

---

coagulation and phagocytosis. In addition, it is becoming clear that insects possess inducible humoral defence mechanisms. The success of any of these agents relies firstly on the insect's ability to discriminate between self and non-self.

Self and non-self recognition is the basis upon which any defence system must be built. Work with transplants has shown that insects have a much less specific sense of self than vertebrates. There is still debate as to whether insects can recognize allogeneic grafts or even transplants between closely related species (Howcroft & Karp 1987, Lackie 1988). These weaknesses in insect defences are clearly of potential importance to invading parasites (Lackie 1986). Despite this relative insensitivity, insects are still able to recognize a wide variety of invading organisms. Three basic suggestions have been made concerning the recognition mechanisms involved. The first suggestion is that the insect is able to recognize differences in surface charge and hydrophobicity between its own tissues and those of the invader (Lackie 1988). The second possibility is that recognition is due to a direct interaction between molecules on the surface of the invader and receptors on the haemocyte. The third suggestion is that intermediary molecules (opsonins) may act as a bridge between the invader and haemocyte. The widespread occurrence of lectins in insect tissues (Renwrantz 1986), their specific binding properties and the fact that they are inducible (Takahashi et al. 1986, Minnick et al. 1986) has led to speculation that these are opsonins. This suggestion is strengthened because lectins are known to have such a role in other invertebrate systems (Renwrantz & Stahmer 1983) and evidence is beginning to appear for such a role in insects (Lackie & Vasta 1988). Components of the prophenoloxidase cascade are also candidates as opsonins in the insect immune system (Smith & Soderhall 1983a, b, Ratcliffe 1986) (see below). Whatever the mechanism, once the invader is recognized as foreign the insect can use a variety of defensive systems to try to kill or control it.

Phagocytosis is used by insects to control invasions by fungi, viruses, protozoa and bacteria. The main blood cell types involved are the plasmatocytes (Ratcliffe 1986). The details of the processes involved are poorly understood, particularly the interactions of protozoan parasites and their insect vectors. The information which is available mainly concerns the phagocytosis of bacteria in vitro. The phagocytic process in insects probably follows a similar pattern to that seen in vertebrates, with chemotaxis of the haemocyte to the assailant, followed by attachment, ingestion, and finally, the killing of the invader. However, the need for the chemotaxis stage in insects is questionable given the small haemocoelic space and the ample opportunity the cells within it have for chance contacts with an invading organism (Gotz & Boman 1985). If chemotaxis does occur it requires the recognition of the invader as foreign and the movement of the haemocyte along a chemical gradient to contact it. In theory the chemical attractant may originate from the invader itself or from insect cells affected by the invader (Renwrantz 1986), but no such attractant chemicals are known.

Humoral factors can enhance phagocytosis (Mohrig et al. 1979) possibly by opsonizing the invader. The factors involved may be elements of the prophenoloxidase cascade (see below) or may be haemolymph lectins (Ratcliffe 1986). Attachment itself is an energy-independent step and is presumably achieved by the interaction of surface receptors on both the haemocyte and invader, with or without the co-operation of other molecules. Once attachment is achieved the invader is internalized by phagocytosis. In vertebrates the vacuole containing the ingested organism fuses with a lysosome and killing is then achieved by oxygen metabolites such as hydrogen peroxide (H_2O_2) and superoxide (O_2^-). The killing mechanism in insects is unknown, but does not involve the oxygen metabolites so characteristic of the vertebrate system (Anderson et al. 1973). Insect haemocytes do contain enzymes which have bacteriolytic activity (Anderson & Cook 1979, Walters & Ratcliffe 1983) which may be involved in the killing mechanism.

A second defensive mechanism is encapsulation. Classic encapsulation occurs when the invading organism is too large to be phagocytosed and results in the segregation of the invader from the insect haemolymph. This is achieved in two stages. First, the invading organism encounters granular cells and coagulocytes which degranulate onto its surface (Ratcliffe & Gagen 1977, Schmit & Ratcliffe 1977) releasing the prophenoloxidase system. It has been suggested that this degranulation coats the invader in 'sticky' protein (probably a component of the prophenoloxidase cascade) in an opsonization process that clearly labels the invader as foreign and for the attention of other blood cells (Smith & Soderhall 1983a, b, Ratcliffe 1986). Sometime later the second part of the process begins and the invader becomes surrounded by other blood cells (mainly plasmatocytes) which progressively flatten over the surface forming an enveloping capsule many cells thick. Sometimes encapsulation stops here with the invader segregated from the haemolymph by a multilayered capsule of recognizable cells. In most systems the prophenoloxidase system becomes activated leading to a partly or wholly melanized capsule. Activation appears to be triggered by properties of the invader's surface. For example, $\beta 1,3$-glucans (such as laminarin) or bacterial peptidoglycans are both efficient triggers of the prophenoloxidase cascade. Serine proteases are activated by such triggers when they are presented in the presence of divalent cations. The serine proteases then promote the activation of prophenoloxidase to phenoloxidase by the cleavage of a 5 kDa peptide (Lackie 1988). Phenoloxidase causes the formation of melanin from phenolic precursors such as dihydroxyphenyl-

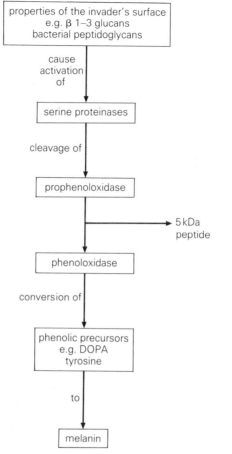

Figure 8.9 An outline of the prophenoloxidase cascade.

alanine (DOPA) and tyrosine (Fig. 8.9). This process usually begins at the invader's surface and proceeds outwards for a variable distance through the debris of necrotic blood cells (Fig. 8.10). It results in the formation of a capsule of melanin which eventually segregates the invader from the insect blood system. Encapsulation is not limited to parasites in the haemolymph, it can also be used as a defence against intracellular parasites. *Anopheles labranchiae atroparvus* are resistant to infection with the filarial nematode *Brugia pahangi*. Although the first stage larvae can become established in the flight muscles of the mosquito, they are soon enveloped in a melanizing capsule derived from material within the muscle cell (Fig. 8.11) (Lehane & Laurence 1977).

Encapsulation is not entirely limited to large organisms. Smaller

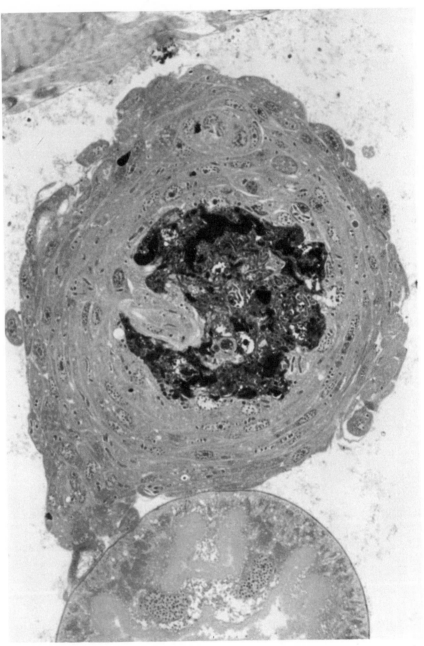

Figure 8.10 Encapsulation is a primary weapon in the defensive armoury of the insect. An encapsulated nodule in the haemolymph of *Pieris brassicae* formed in response to the injection of *Bacillus cereus* 24 h earlier is shown. (Photograph kindly supplied by N. Ratcliffe.)

particles such as bacteria, which the insect normally controls by phagocytosis, may be encapsulated if their numbers exceed a threshold. This class of encapsulation leads to the development of a nodule, and is a particularly rapid and efficient means of mopping up excessive numbers of small invaders (Ratcliffe & Walters 1983). The cellular details of nodule formation are similar to those of cellular encapsulation.

Humoral factors are also important in the defence mechanisms of blood-sucking insects. For example, many Diptera achieve encapsulation

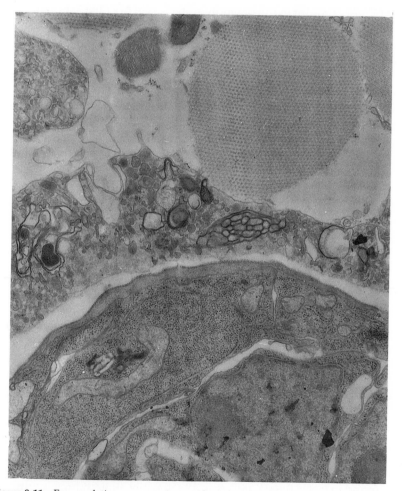

Figure 8.11 Encapsulation can occur in specific tissues as well as in the haemolymph. This micrograph shows microfilariae of *Brugia pahangi*, which are within the thoracic muscle fibres (i.e. are intracellular) of *Anopheles l. atroparvus*, beginning to be surrounded by vesicular material which will eventually melanize to form an enveloping capsule (see Lehane & Laurence 1977).

without the involvement of blood cells, the process being known as humoral encapsulation. Whether cellular or humoral encapsulation is the rule, in an insect it is a function of the number of circulating blood cells (Gotz et al. 1977). Humoral encapsulation, like cellular encapsulation, is a two-stage process. The invader, on entering the haemolymph, is first rapidly coated with droplets of material which make its surface sticky; this material may arise from the haemolymph and not from blood cells (Chen & Laurence 1985, Gotz 1986). The coat then gradually hardens and melanizes. But why does the whole of the haemolymph not melanize if the prophenoloxidase system is free in the haemolymph? We know there are inhibitors present in the haemolymph (Sasaki & Kobayashi 1984), but these alone cannot explain why melanization will occur in some haemolymph compartments and not others. The conundrum might be explained if the prophenoloxidase system can only be activated after attachment to a suitable surface, as has been suggested for Crustacea. Such a system would also explain the sequence of melanization seen in cellular encapsulation which proceeds outwards from the invader's surface (Fig. 8.10).

A primary difference between cellular and humoral encapsulation is the speed at which they occur. While cellular encapsulation may take anything from a few hours to several days to complete, humoral encapsulation may be achieved in 10 – 30 min (Gotz 1986). As well as its role in encapsulation and its possible role as an opsonin, the prophenoloxidase system may be important in blood coagulation. The system is always present in insect blood and is non-inducible.

Other humoral factors are also important in the insect immune system. Lectins are proteins or glycoproteins which recognize and bind specific carbohydrate moieties, and because of this capability are able to agglutinate cells or precipitate complex carbohydrates. Lectins, with a range of carbohydrate specificities, are widespread in insects and are inducible (Komano et al. 1980). It is unlikely, under most circumstances, that the number of invaders is sufficiently high for lectins to exert defensive control by agglutinating them. Instead, it seems more probable that lectins act as opsonins, binding invader and blood cell as a prelude to phagocytosis or encapsulation (Renwrantz 1986). Lectins are also found in the insect intestine in triatomine bugs (Pereira et al. 1981) and tsetse flies (Ibrahim et al. 1984). It has been suggested these gut-based lectins may also have a defensive function (Peters et al. 1983, Welburn et al. 1987). Specific antibacterial agents are probably also found in the haemolymph of blood-sucking insects, indeed cecropin and attacin-like molecules have been found in tsetse flies (Kaaya et al. 1987), while other inducible antibacterial factors have been found in the haemolymph of Rhodnius prolixus (de Azambuja et al. 1986). The cecropins and attacins act on bacterial outer membranes and lysozyme then attacks the bacterium's murein skeleton.

Other defence mechanisms are known to occur in the haemolymph although their biochemical basis is not understood. For example, the cell-free tsetse fly haemolymph contains a non-inducible, temperature-sensitive factor which is capable of inactivating some trypanosomes (Croft *et al*. 1982, East *et al*. 1983). This factor shows species specificity. It is active against trypanosomes which tsetse flies commonly encounter, such as *Trypanosoma brucei*, *T. vivax* and *T. congolense*, but it is not active against trypanosomes 'unusual' to tsetse flies such as *T. dionisii* from bats. The biochemical nature of this tsetse fly haemolymph factor is unknown.

When considering the blood defences of insects it should not be forgotten that the invader has crossed an epithelial barrier to enter the blood system. The wounding caused in this passage may itself may be an important part of the trigger for the specific defensive mechanisms discussed above (Ratcliffe 1986). In addition wounding will cause coagulation, which is a complex cascade involving prophenoloxidases (Hoffman *et al*. 1970), several serine proteases (Leonard *et al*. 1985), protease inhibitors (Eguchi *et al*. 1982) and coagulogens such as lipophorin (Gellissen 1983). The forming clot itself may entrap and kill invaders such as bacteria.

If a parasite is to develop successfully in a vector insect it must either avoid triggering the insect's immune response or be able to suppress or withstand its effects. Some work has now begun in this area and interesting data and ideas are beginning to come forward.

If an invader is to avoid triggering the insect's immune system, then it must escape recognition. It may do this by having inherent surface properties which mimic those of the host or by acquiring material from the host which enables it to masquerade as host tissue ('self'). *Aedes trivittatus* is refractory to infection with the filarial nematode *Brugia pahangi*. However, if the microfilariae are allowed to penetrate the midgut of a susceptible strain of *Ae. aegypti* before being intrathoracically injected into *A. trivittatus*, up to 31–43% of them will avoid encapsulation and melanization in the haemocoel (LaFond *et al*. 1985). Clearly something happens to change the properties of the parasite in its migration across the midgut of the susceptible mosquito. The first suggestion was that the parasite absorbs mosquito-derived material onto its surface and hides behind this coat, but attempts to demonstrate such adsorption, using indirect fluorescent antibody techniques, have failed. A second suggestion centres on the loss of the parasite's high electronegative surface charge as it migrates through the midgut of *Ae. aegypti* (Christensen *et al*. 1987). This change in charge may account for the reduction in immune response observed but conclusive evidence remains to be gathered.

Parasitoid wasps are known to be able to suppress the insect immune

191

Table 8.7 The melanization response to subsequent challenge of infected and uninfected *Aedes aegypti*, as shown by the intrathoracic injection of specific microfilariae (mff) which normally induce a strong melanization reaction. (From Christensen & LaFond 1986.)

	No.	Challenge	% mff melanized ± S.E.
Infected with *Brugia pahangi*	63	*B. pahangi* mff.	38.2 ± 6.3
	17	*Dirofilaria immitis* mff.	51.2 ± 6.6
Uninfected	68	*B. pahangi* mff.	64.5 ± 6.5
	20	*D. immitis* mff.	71.3 ± 3.9

system. Several effecter mechanisms have been suggested to account for this, including the reduction in numbers of blood cells (Rizki & Rizki 1984) or their function, the inhibition of the phenoloxidase system (Stolz & Cook 1983) or the diminution of substrate levels (Sroka & Vinson 1978). Suppression may be a sensible strategy for parasitoids which, in the short term, lead to the host's death; it is, however, a dangerous strategy for parasites to adopt. The parasite's successful transmission relies on the continued fitness of the vector and immunosuppression may well open the door to invaders which prevent this (Lackie 1986). Nevertheless reports are appearing which suggest that some parasites do depress the immune function of their vector insects. For example, developing micro-filariae actively suppress the immune response of the mosquito (Table 8.7) and suppression of the insect immune system is also suggested by the decreased xenograft rejection seen in *Triatoma infestans* infected with *Trypanosoma cruzi* (Bitkowska *et al.* 1982). From these data, it seems possible that immune suppression of the insect, over the short time period which parasites need for transmission, may be an acceptable risk for the parasite.

CHAPTER NINE

The blood-sucking insect groups

This section of the book gives an **outline** of the major groups of insect which feed on blood, concentrating on those groups which are habitual blood feeders. Detailed coverage of each blood-sucking group is not attempted, that would be far beyond the space available in a book such as this. Instead, this chapter has been written as a quick reference section for those new to medical and veterinary entomology and for those who need a quick outline of a particular group in order to make most use of other parts of the book. A basic outline of each assemblage is given covering the essential details of the group's medical and veterinary importance, morphology, life history and bionomics. For those who need more detailed information on particular blood-sucking insects the following two abstracting journals provide a useful way to scan the literature quickly and give a good coverage of papers produced: (1) *The Review of Applied Entomology, series B, Medical and Veterinary*, published monthly by the Commonwealth Agricultural Bureau. It is available as hard copy or for on-line computer searches in various ways including CD-ROM; (2) *Entomological Abstracts* covers insect biology in general but includes many papers on the physiology, ecology, cell biology and behaviour of blood-sucking insects which may have been missed by the more 'applied' abstracting services.

Insects are individuals and even within a single species insects often do things in subtly different ways to each other; this is, after all, a prerequisite for evolution. Consequently, generalizing about a genus, family and whole class of insects is an imprecise business, but bearing in mind that there are many exceptions to these generalizations, this approach can still be a good way of introducing typical group characters. In discussing each group, quantitative values are often given for fecundity, duration of each stage in the life cycle, longevity, etc. These figures are good guidelines to expected values under optimum conditions, but in the field one can expect a good deal of variation.

Before looking at each group separately we need to briefly look at how they fit into a general classification scheme and to explain some of the rules and conventions by which insects are named.

193

9.1 INSECT CLASSIFICATION

The Insecta are the dominant class in the extremely successful phylum Arthropoda. The class Insecta is divided into two subclasses, Apterygota and Pterygota. There are 28 orders of insects and most of these (including all blood-sucking species) fall into the Pterygota (Table 9.1). This subclass is further subdivided into two divisions, Exopterygota (Hemimetabola), in which the young (often called nymphs) resemble the parents except that they lack wings and are smaller, and Endopterygota (Holometabola), in which the young (usually called larvae and pupae) do not resemble the adults and in which the transformation from the larval configuration to the adult occurs in a quiescent stage, called the pupa.

Orders of insects may be subdivided into suborder, division, super-family, family, subfamily, tribe, genus, subgenus, species and sub-

Table 9.1 An outline classification of the class Insecta

Phylum	Class	Subclass	Division	Order	Common name
Arthropoda					
	Insecta				
		Apterygota		Thysanura	silverfish, fire brats, bristle-tails
				Diplura	bristle-tails
				Protura	
				Collembola	spring-tails
		Pterygota			
			Exopterygota	Ephemeroptera	mayflies
				Odonata	dragonflies
				Dictyoptera	cockroaches and mantids
				Isoptera	termites
				Grylloblattodea	
				Plecoptera	stoneflies
				Dermaptera	earwigs
				Phasmida	stick insects
				Orthoptera	grasshoppers and crickets
				Embioptera	web-spinners
				Zoraptera	
				Psocoptera	book-lice and psocids
				Thysanoptera	thrips
				Hemiptera	bugs, plant-lice, scale insects
				Phthiraptera	lice
			Endopterygota	Neuroptera	lacewings, ant-lions
				Mecoptera	scorpionflies
				Trichoptera	caddisflies
				Lepidoptera	butterflies and moths
				Siphonaptera	fleas
				Coleoptera	beetles
				Strepsiptera	
				Diptera	flies, mosquitoes, etc.
				Hymenoptera	bees, wasps, ants

194

species. In most everyday work order, family, genus and species are the most common and useful groupings. To help with the complexities of taxonomy the *International Code of Zoological Nomenclature* (1985) lays down certain articles and opinions about the naming of these taxonomic compartments. So most orders of insects end in the term '-ptera', all superfamilies in the term '-oidea', all families in the term '-idae', all subfamilies in the term '-inae' and tribes in '-ini'. Species are indicated by the generic followed by the specific name, e.g. *Rhodnius pallescens* (usually abbreviated to the form *R. pallescens* after one full citation). In the scientific literature, especially when naming a little known species, it is good practice on the first citation of a species to go further and to include the name of the author of the original taxonomic description, e.g. *Rhodnius pallescens* Barber. Rarely the date of the original description is also included, so we get *Anopheles labranchiae* Falleroni, 1927. The appearance of the author in brackets after the specific name, such as *Triatoma infestans* (Klug), indicates the generic name has been changed since the naming and original description of the species by that author. Subspecies are identified by a trinomial, such as *Glossina morsitans centralis*. In the case of species complexes (see Section 3.2) such as the *Simulium damnosum* complex, another convention has been adopted. Because *Simulium damnosum* refers to a species as well as the complex as a whole, it is often necessary to distinguish between the use of the term in the two instances. This is done by appending the terms *sensu lato*, usually abbreviated to *s.l.*, when the name refers to the complex and *sensu stricto*, usually abbreviated to *s.s.*, when the name refers to the species.

9.2 PHTHIRAPTERA

The order Phthiraptera contains the lice of which about 4000 species have been described. The traditional classification divides Phthiraptera into three suborders, Mallophaga, Anoplura and Rhynchophthirina, and I will use this convenient classification here. Many authorities now raise Mallophaga and Anoplura to the status of order and then separate Mallophaga into three suborders, Amblycera, Ischnocera and Rhynchophthirina. Others suggest that Anoplura, Amblycera, Ischnocera and Rhynchophthirina be considered as four suborders of the Phthiraptera.

The Anoplura are the smaller suborder and they contain the sucking lice, all of which are blood feeding parasites of eutherian mammals; most (67%) are found on rodents. Although anoplurans are haematophagous, only a few Mallophaga regularly feed on blood. To be more specific, the Ischnocera do not as a rule feed on blood, although it may be taken if readily available on the skin's surface, the Amblycera may take blood as a regular component of the diet and the two species in the third suborder, the Rhynchophthirina, are both blood feeders.

Three species of lice are regularly found on man, *Pediculus humanus* (=*P. humanus humanus* = *P. corporis*) the body louse, *P. capitis* (=*P. humanus capitis*) the head louse, and *Phthirus pubis* the crab or pubic louse. There is some confusion concerning the status of the first two because they interbreed under laboratory conditions, but there is little evidence of this occurring naturally.

As well as the psychological damage humans may suffer because of the stigma associated with louse infestation, they also run the risk of physical damage. Individuals heavily infested with lice commonly show allergic responses to saliva injection or to the inhalation of louse faeces. Such individuals literally feel 'lousy' as a result of their lice. There is evidence that such individuals can be educationally impaired (hence the term 'nitwit', with 'nits' being a colloquial term for louse eggs). The allergic responses to louse infestation may cause the skin of heavily infested individuals to become thickened and heavily pigmented. Itching and scratching may lead to secondary infection of scarified areas. Such itching and scratching may lead to economically noticeable damage to the wool or hides of domesticated animals and the value of wool may be further impaired by staining from louse faeces. Cattle, particularly calves, also respond to lice by increased licking and this may lead to the formation of hair balls in the gut. Lice populations reach peak numbers during winter when the stock has longer and thicker coats. The irritation caused by the lice may disturb eating or sleeping habits to the extent that egg or milk yields may be impaired. Heavy infestations of *Haematopinus eurysternus*, particularly on adult beef cattle, may cause anaemia, lowering of food conversion efficiency, abortion, sterility and possibly even death. The foot louse of sheep, *Linognathus pedalis*, may cause lameness.

Pediculus spp. are the vectors of *Rickettsia prowazekii* the causative agent of epidemic typhus; *Borrelia recurrentis*, the causative agent of epidemic relapsing fever; and less commonly, *Rochalimaea quintana*, the causative agent of trench fever. Louse-borne epidemic typhus has been a scourge of impoverished, overcrowded, undernourished communities throughout history, particularly in the temperate regions of the world. To take just one example, around the end of the First World War up to 30 million people are believed to have been afflicted by the disease in Europe with a fatality rate of more than 10%. Although the disease can be experimentally transmitted by all three human lice it appears that in epidemics the disease is transmitted by the body louse. Typhus is not transmitted by the bit￢ of the louse but rather in the louse's faeces or when the infected louse ￯ crushed. The infective material may be scratched through the skin, ￯ it can gain access to the body across mucous membranes. The rickettsiae are fatal to the louse whose gut is damaged or ruptured by the

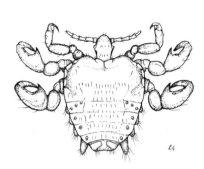
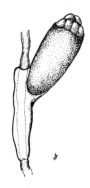

Figure 9.1   An adult crab louse, *Phthirus pubis*, and an egg of the same species firmly glued to a host hair. (Redrawn from Smith 1973.)

invading and replicating organisms. There are two defined ways in which the organism persists between epidemics. In the short term the rickettsiae remain viable in dry faeces for over two months. In the longer term infected, but relatively unaffected, humans are the reservoir. As many as 16% of these asymptomatic carriers, who have recovered from the primary attack of the disease, may subsequently fall ill with Brill-Zinsser's disease and, if louse-infested, form the focus of another epidemic. There is some evidence that *R. prowazekii* is also found in domesticated and other animals and that transmission among these may occur through ticks, lice or fleas, but the relationship and importance of this to human disease is unknown.

Epidemic relapsing fever, caused by the spirochaete *Borrelia recurrentis*, occurs throughout the world. Mortalities as high as 50% were associated with this much-feared disease before the advent of modern antibacterial agents. Transmission does not occur by the bite of the louse or in the faeces, instead the infected louse must be crushed to release the infective organisms which enter the body through skin abrasions or across mucous membranes. Lice, unlike ticks, do not transovarially transmit the organisms to the next generation of potential vectors.

Adult lice vary in size from about 0.5 mm to about 8 mm depending on species. They are dorsoventrally flattened, wingless, brownish-grey insects with three to five segmented antennae, ocelli are absent and eyes are reduced or absent (Fig. 9.1). In Amblycera the antennae are partly protected in grooves on the head. The Anoplura, which are blood feeders, have mouthparts which are highly modified for sucking. The mouthparts are housed in the stylet sac inside the head with only the opening of the sac visible at the anterior tip of the head. The Mallophaga, only some species of which are blood feeders, have chewing mouthparts which are used to cut their various foods. Incidentally in some cases,

197

certainly by design in others (such as the Rhynchophthirina), this may lead to bleeding of the host and the mallophagan may then feed on the released blood. In fact the Rhynchophthirina hold their mouthparts on an extended rostrum which is almost certainly an adaptation to facilitate blood-feeding through the thick skin of their hosts (see Fig. 2.1). Lice have a leathery tough cuticle capable of considerable expansion after a blood meal because of the provision of concertina-like infoldings.

In the Anoplura the three thoracic segments are fused. In the Mallophaga the prothorax is free, although the meso- and metathorax may be fused. Lice tend to have short, stout legs each bearing one (most mammal-infesting forms) or two (bird-infesting forms) pairs of terminal claws which are used for holding onto the host covering. Louse species differ considerably in their speed of movement in the pelage. The crab louse, *Phthirus pubis*, can only move 15 cm a day and often remains stationary for days attached to the skin by its mouthparts. In contrast, the body louse, *P. humanus*, can move much more rapidly, at rates of $0.5\,\text{cm s}^{-1}$. In *P. humanus* the foreleg is modified for holding the hind leg of the female during copulation.

Lice are ectoparasitic at all stages of their lives, taking blood meals of up to a third of their own body weight every few hours. Lice will die within one or two days if deprived of blood. The newly emerged female usually feeds before copulation. The pregnant female louse seeks out a particular temperature zone within the host's covering in which to lay her eggs, gluing them to the host's covering using material from special cement-producing glands. *Pediculus capitis* glues eggs to the base of human hair and in the sheep louse, *Damalinia ovis*, most of the eggs are laid within 6 mm of the host's skin. Egg-laying sites are also determined by other factors particularly the diameter of the host hair. So, for example, *Haematopinus asini* lays its eggs on the coarser hairs of the horse's tail, mane, forelock and that zone above the hooves of some horses known as the feathers. Lice produce relatively few eggs in a batch. Human lice produce up to ten eggs per day; the cattle louse *Haematopinus eurysternus* produces about two eggs per day. Because they feed several times a day, lice produce eggs every day. Adult female *P. humanus* live for a month or so and in that time produce about 300 eggs. *Pediculus humanus*, which has adapted to life in the artificial body covering presented by man's clothes, attaches most of its eggs to the clothes. Hatching of these eggs is dependent on temperature, but eggs will not hatch beyond about a month after laying and so clothes not worn for this period are unlikely to be a source of infestation. The nymphs feed on blood and pass through their three larval instars to the adult stage in about two weeks.

In common with other obligate haematophages lice have symbiotic micro-organisms which provide nutritional supplements. In the Anoplura the mycetome housing the symbionts is situated on the ventral surface of the ventricular region of the midgut and in adult females symbionts are also associated with the ovary from which they are transovarially transmitted to the offspring. In the Mallophaga the myceto-cytes are distributed among the cells of the fat body and, in females, symbionts are also associated with the ovary.

9.3 HEMIPTERA

The order Hemiptera contains those insects usually known as bugs. Most hemipterans are either entomophagous or plant feeders but three families contain several species of small to very large blood-sucking insects. The first of these is the Cimicidae, all members of which are blood-feeding. Cimicids are found throughout the world but are particularly well represented in the Northern Hemisphere. The majority are parasites on bats and birds, although a minority of species feed on mammals. These include the two species normally known as bed bugs, *Cimex lectularius* and *C. hemipterus*, which normally feed on man. In addition the primarily bat-feeding form, *Leptocimex boueti*, also regularly feeds on humans. *C. lectularius* has been carried to all corners of the world in man's belongings but is encountered most often in temperate regions. *C. hemipterus* is largely a tropical species while *L. boueti* is only found in tropical West Africa. Bed bugs bite at night, heavy infestations can disturb the sleeping human and some individuals develop marked responses to their bites which may include oedema, inflammation, and an indurated ring at the site of the bite. Any role for cimicids in the transmission of disease has yet to be proven, although in the laboratory bed bugs have been shown to excrete hepatitis-B surface antigen for up to six weeks after an infected meal (Ogston & London 1980). Cimicids are also important economic pests of poultry; prominent species include

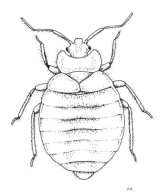

Figure 9.2   An adult bed bug. (Redrawn from Smith 1973.)

Haematosiphon inodorus in Central America and *Ornithocoris toledo*, formerly important in Brazil.

Adult bed bugs are typical cimicids, being wingless, dorsoventrally flattened, brownish insects about 4–7 mm in length. Viewed from above they have a rather oval shape (Fig. 9.2); the short rudimentary wings, or hemielytra, are clearly seen and the broad head and well-developed eyes are clearly visible. When these insects are not feeding the mouthparts are folded beneath the head and thorax. During feeding they are swung forwards in front of the head. The adult male carries a curved aedeagus at the tip of the 11-segmented abdomen. On the ventral surface of the female's fourth abdominal segment is the opening of the organ of Berlese used to store sperm. Mating in bed bugs is very unusual. The male penetrates the cuticle of the female and deposits his sperm into the organ of Berlese. The sperm eventually reach the ovaries by migrating through the haemolymph to the base of the oviducts which they ascend to reach the unfertilized egg. The female lays about eight eggs a week and may produce 100 or more by the end of her long life. The eggs are cemented into cracks and crevices in which the adults congregate and hide between feeding forays. The eggs hatch in about seven to ten days. There are five nymphal instars each of which requires one or two blood meals before moulting to the next stage. Under good conditions the egg-to-adult period may be as short as five weeks, but bed bugs are characterized by a marked ability to withstand starvation – in the laboratory adult bugs may go as long as 18 months between meals! Consequently adult lifespans and the length of time in each nymphal instar is particularly variable in these insects.

Nymphs and adults of both sexes feed on blood. As is normal in insects living entirely on blood, cimicids have symbiotic micro-organisms which provide supplementary nutrition. In beg bugs these are housed in a mycetome sited in the abdomen. Bed bugs are unusual in having two different symbiotic oganisms in their mycetome, one pleiomorphic and one rod-shaped. Bed bugs feed mainly at night, with a feeding peak just before dawn. During the ten-minute-or-so feeding period they take two to five times their own body weight in blood from their hosts, who are usually asleep. Once fed, bugs return to the cracks and crevices in which they hide. Bug congregation in such places is influenced by an aggregation pheromone and then by thigmotaxis. The bugs also produce an alarm pheromone which can cause the rapid dispersal of these congregations. Because they are unable to fly, bed bugs are heavily reliant on passive dispersal from one home to another.

The second hemipteran family containing blood-feeding bugs is the Reduviidae, which contains the subfamily Triatominae, all 155 species of which are obligate blood feeders. These bugs feed on a variety of vertebrates including man and many are intimately associated with the

Table 9.2 The geographical distribution of triatomine species which have become highly adapted to the domestic–peridomestic environment of man and so represent a particular threat as vectors of Chagas' disease. (Information largely from Schofield 1988.)

Vector species	Geographical distribution
Triatoma infestans	widespread from S. Argentina to N.E. Brazil
Triatoma brasiliensis	arid regions of N.E. Brazil
Triatoma dimidiata	humid coastal areas from C. America to Brazil
Panstrongylus megistus	humid coastal regions of E. and S.E. Brazil
Rhodnius prolixus	drier savannah areas from S. Mexico to northern S. America

habitual resting sites or nests of birds, mammals and other animals. Most triatomines (102 species) are confined to the Americas with one pantropical species (*Triatoma rubrofasciata*), five species in the Indian subcontinent and seven species in South-east Asia. Triatomines that are important from a human perspective are largely confined to the Americas, from 40°N to 46°S. These bugs have many local names, including cone nose bugs, kissing bugs or vinchucas. Some species, *Triatoma infestans* in particular and to a lesser extent *Rhodnius prolixus* and other species have become closely associated with man's domestic and peridomestic environment. Sylvatic populations of *Triatoma infestans* are limited to the Cochabamba valley of Bolivia where they live among rocks feeding largely on guinea-pig colonies. Throughout the rest of its range, which extends through parts of Argentina, Bolivia, Brazil, Chile, Paraguay, Peru and Uruguay, the insect is an entirely domestic and peridomestic species. The transition from a sylvatic to a domestic lifestyle is believed to have occurred in pre-Columbian times. The bugs readily adapted to the rough dwellings that are still common in rural and unplanned urban areas of South America; they are so common that the species is still extending its geographical range (Schofield 1988). These bugs will feed on humans and other available vertebrates such as chickens, dogs, goats or guinea-pigs. Houses only become infested if they have unplastered or cracked walls, thatch roofs or other places where bugs can easily hide. This species and others (Table 9.2) are the vectors of *Trypanosoma cruzi*, the causative agent of South American sleeping sickness or Chagas disease. This is a zoonosis which, in its sylvatic cycle, is transmitted between a variety of animals, mainly rodents and marsupials, by sylvatic Triatominae. This is perhaps the classic parasitic disease of poverty because of the sharp economically-based distinction in vector insect distribution. Poor-quality houses in which infected people live can often be found next door to uninfested, higher quality houses in which no one is infected. The high standard of living in California means that although

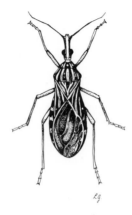

Figure 9.3   An adult triatomine bug viewed from above. (Redrawn from Faust *et al.* 1977.)

25% of *Triatoma protracta* in that state are infected with *Trypanosoma cruzi* (Wood 1942) and will bite man, cases of Chagas disease are thankfully very rare indeed with only three reported instances.

The smallest triatomine species, *Alberprosenia goyovargasi*, has adults which are only about 0.5 cm long, and the largest species, such as *Dipetalogaster maximus*, may have adults over 4.5 cm long! Triatomines are often distinctly marked with yellow or orange patches on the fringes of the abdomen, pronotum and bases of the forewings, which contrast against brown to black bodies. They all possess a characteristically elongated head bearing two prominent eyes and four-segmented antennae attached to the dorsal surface of the head (Fig. 9.3). As in the bed bugs, the non-feeding insect folds the straight, three-segmented elongated rostrum (proboscis) under the head. These mouthparts are swung forwards in front of the head for feeding. When viewed from above the characteristic triangular shape of the pronotum is obvious. The proximal part of the forewings is relatively hard and rigid, gradually leading into the more membranous distal part. In the non-flying insects these forewings, or hemielytra, hide the completely membranous hind wings and cover most of the dorsal surface of the abdomen. In the unfed bug the edges of the flattened abdomen may form a raised rim around the wing-covered area.

The feeding adult bug takes up to three times its unfed body weight in blood and the abdomen swells to look like a ripe berry. In a normal house infestation, a person can expect to receive about 25 bites per night and so the bug's attentions can contribute to chronic anaemia in the human population. They normally feed every four to nine days, but bugs can withstand starvation for months, particularly the later nymphal instars. Adult female *T. infestans* produce small batches of eggs (10–30) subsequent to each blood meal; in her lifespan of six months to a year she may typically produce 100–300 eggs. The eggs are oval and about 2 mm long, white, yellow or pink, depending on species and stage of development. Eggs hatch in 10–30 days and there are five nymphal instars. Triatomine bugs are obligate haematophages; most feed at night, are stealthy, and have an almost painless bite as is evinced by *T. infestans* which commonly feeds on the facial mucous membranes of their sleeping human hosts. Early instar nymphs may take up to 12 times their unfed body weight in blood while adult bugs rarely take three times their unfed body weight. Each nymphal instar requires only one full blood meal to stimulate the moult into the next instar. Even so, most nymphs commonly take more, smaller meals at each instar. Full development from egg to adult takes eight months to a year to complete. In common with other insects feeding exclusively on blood, triatomine bugs have symbiotic micro-organisms. These are housed in the lumen of the anterior midgut and provide additional nutrients without which nymphs do not successfully develop into adulthood.

The domesticated species, *T. infestans*, probably relies heavily on passive transfer in hand luggage, furniture, etc., for movements into previously uninfested regions. The adult insects are capable of strong flight, which is probably an important dispersal agent at a local level. Sylvatic species almost certainly use flight for dispersal, but it is also thought that the early stage nymphs may disperse phoretically on their hosts.

The third family of blood-sucking bugs is the Polyctenidae (Fig. 9.4). All members of this group are permanently ectoparasitic, obligate haematophages living on microchiropteran bats of both the New and Old Worlds.

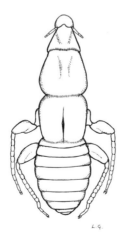

Figure 9.4   An adult polyctenid, *Eoctenes spasmae*. (Redrawn from Marshall 1981.)

They are of no economic importance. These eyeless insects are oviparous and there are only two postnatal nymphal instars.

9.4 SIPHONAPTERA

The order Siphonaptera contains the fleas of which about 2500 species and subspecies have been described. All adult fleas are ectoparasitic on warm-blooded hosts; about 94% infest mammals (74% on rodents alone) and 6% birds. Fleas are found throughout the world with a concentration in temperate regions. There are about 20 flea species which will feed on man. In the past *Pulex irritans* had been a serious nuisance insect on man, but it is now becoming rare in most industrialized countries. Meanwhile the cat flea, *Ctenocephalides felis*, originally restricted to North Africa and the Middle East, has emerged to take its place, and this insect is now a serious household nuisance throughout Europe and North America.

Because of the stigma attached to flea infestation, affected humans may often suffer more mentally than physically from the presence of fleas. However, fleas can present a very serious disease risk. Heavy infestations of fleas are usually restricted to animals in poor condition, the health of which may be directly affected by infestation. Hosts show varying degrees of sensitivity to flea bites; some display severe reactions which may seriously impair their own condition. For example, the dog and cat fleas (*Ctenocephalides canis* and *C. felis*, respectively) may cause moderate to severe pruritic reactions appearing as areas of moist dermatitis on their hosts. Secondary infection of these areas can occur, and local alopecia may result from excessive grooming by the affected animal.

As well as fleas which move freely in the home and pelage of the host there are those, like the tungids (Tungidae) and alakurts (Vermipsyllidae), which burrow into its skin and become neosomic (Fig. 9.5). There are also the sticktights which attach to the host for long periods by their mouthparts, and in some species display neosomy. Humans are afflicted by the chigoe, sand or jigger flea, *Tunga penetrans*, which causes a painful infestation by burrowing into the feet, often beneath the toenails. Originally a New World species, it has spread throughout the Neotropical and Afrotropical regions over the last 150 years. Heavy infestations can be crippling and secondary infection of the ulcer caused by the body's response to dead jiggers can lead to other complications. The 'sticktight' flea, *Echidnophaga gallinacea*, is an important poultry pest in the tropics and subtropics. Young birds infested with this flea quickly die; older birds suffer anaemia, reduced egg output, and may also perish due to heavy infestations.

Fleas are also the transmitters of disease, being most notorious for their involvement in the Black Death. This was the pandemic of plague which

204

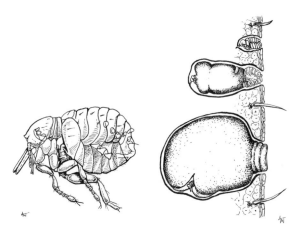

Figure 9.5 The embedding of the neosomic flea *Tunga calcigena*. (After Jordan 1962.)

killed a quarter of the population of Europe (25 million people) in the 14th century. We know of two other plague pandemics. The first, in the sixth century A.D., spread from Central Asia through the Middle East and into Africa. The last began in China in the 1890s and spread rapidly along trade routes to India, the Americas, Australia, South Africa, and on to many other parts of the world. It caused over 10 million deaths in India at the turn of this century. This third pandemic is still lingering on although the annual number of cases is now small. Plague is caused by the bacterium *Yersinia pestis*; it is a zoonotic disease usually cycled between rodents by their fleas. Man becomes involved in the cycle when infected domestic or peridomestic rodents such as the black or brown rat die. Their infected fleas then leave and look for another host which may be man. The bacterium blocks the gut of infected fleas and is regurgitated into the wound when the flea attempts to feed. Mechanical transmission may also be of some importance during an epidemic of the disease. The two flea species most commonly involved in carrying the disease to humans are *Xenopsylla cheopis*, which is the major transmitter in urban areas of the tropics and subtropics, and *X. brasiliensis*, which is an important rural vector in Africa and India. In its pneumonic form, plague is directly transmitted between humans without the mediation of an insect vector.

Another bacterium, *Francisella tularensis*, is also transmitted by fleas and causes a plague-like disease in mammals called tularaemia. The bacterium occurs in domesticated animals such as cattle, horses and sheep, in which the disease is particularly severe. There are many wild reservoir hosts including rabbits, rodents and birds. This disease occurs mainly in the Northern Hemisphere, possibly more commonly in the Old rather than the New World. The rodent populations are thought to be the major natural reservoir of the bacterium with transmission among them by fleas and lice. The bacterium, which can infect man through unbroken skin, can spread through a number of additional agencies. Mechanical transmission by insects, particularly tabanids, is well documented; it can also be airborne and can spread in contaminated food and water.

Fleas also transmit the causative agent of murine typhus, *Rickettsia typhi (mooseri)*. The disease is a mild infection of wild rodents throughout the world and may cause significant mortality in affected human populations. It is transmitted between wild rodents by their fleas, lice and possibly their mites. Like plague, human infections are associated with domestic and peridomestic infestations of rats. The major vector is *X. cheopis* and, although other flea vectors may be involved, the evidence is inconclusive. The rickettsia is either released in the flea's faeces, or when the flea is crushed on the skin, gaining entry to the body through skin abrasions or cuts across mucous membranes or after inhalation. The faeces can remain infective for several years. Rat salmonellosis, caused by *Salmonella enteritidis*, is also transmitted by flea bite and in flea faeces.

Although in most of its range myxomatosis is mechanically transmitted to rabbits largely by mosquitoes, in Great Britain the flea *Spilopsyllus cuniculi* is the major vector. This flea is a much more efficient vector of the virus than mosquitoes and has been artificially introduced into Australia as part of that country's long standing campaign against rabbits. The flea *Echidnophaga myrmecobii* may also play a minor role in transmission of the disease in Australia.

Fleas are also the intermediate hosts of some helminths. The common dog and cat tapeworm, *Dipylidium caninum*, develops in the dog and cat fleas, *C. canis* and *C. felis*. The eggs are ingested by the flea larvae and the cysticercoid stage of the parasite develops in the haemocoel of the adult

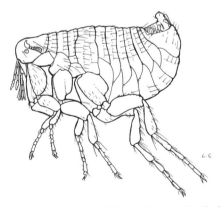

Figure 9.6 External anatomy of adult flea (*Xenopsylla cheopis*). (Redrawn from Marshall 1981.)

flea. The mammal acquires the tapeworm when it ingests the infected flea (normally while grooming). This parasite also occasionally occurs in children. The tapeworm of rodents (and also man), *Hymenolepis diminuta*, uses fleas as well as flour beetles and other insects as intermediate hosts, as may another tapeworm of rodents (and man) *H. nana*. Fleas also transmit a filiarial worm of dogs, *Dipetalonema reconditum*.

There is some evidence for the involvement of fleas in the transmission of many other bacterial, rickettsial and viral diseases. So, although the diseases outlined above are undoubtedly the most important transmitted by fleas, some surprises may still be in store.

Adult fleas are small (< 5 mm), brown, wingless, flattened insects (Fig. 9.6). Their characteristic laterally flattened shape is a feature they share only with some streblids. The flea's body is generally well covered with backwardly projecting spines, and combs. The rather arrow shaped, immobile head bears the mouthparts which project downwards, and the three-segmented antennae which are protected in grooves on the head. The size of the mouthparts depends on the life-style of the flea with sticktight fleas having much longer mouthparts than more mobile forms. Fleas lack compound eyes but most have lateral ocelli which are best developed in fleas infesting diurnally active hosts. The three-segmented thorax bears three pairs of well-developed legs. The hind pair are specialized for jumping and can propel the flea 20–30 cm in one jump! The abdomen of male fleas has an upturned appearance which distinguishes it from the female, whose abdomen is more rounded.

Adult fleas show a spectrum of associations with the host. At one extreme are fleas which are more or less permanently attached to the host; examples are the jigger, *Tunga penetrans*, which is buried beneath the host's skin, or the sticktight fleas such as *Echidnophaga* spp., which remain attached to the host by their mouthparts. In the middle of the spectrum we have those active forms which remain largely on the host, such as the bat fleas. At the other extreme we have active forms such as the bird fleas which reside mostly in the nest, only visiting the host to feed. Many species fall between the last two categories, the adults are not permanent ectoparasites but spend a considerable part of their lives away from the host animal, commonly in the host's nest. Adult fleas of both sexes feed on blood and will also feed on watery solutions. The frequency of feeding varies considerably among species and in the same species at different times in the reproductive cycle. In general, male fleas feed more frequently than females, but take less blood in the process. Fleas are capable of withstanding starvation periods of several months while their host animal is absent. Although they have distinct preferences for hosts, fleas will feed on other animals when the major host is absent. This serves to keep the insect alive, but normally reduces its fecundity, sometimes drastically. This capacity to utilize animals other than their usual host is the basis of their importance as the vectors of the zoonoses described above.

Fleas are anautogenous and so require a blood meal to produce eggs. Although the female only produces small numbers of eggs a day (<20), during her lifetime, which may last a year or more, she will produce several hundred. The large, whitish eggs are normally deposited in the host animal's nest, but some fleas, like the vermipsyllids, deposit eggs which fall indiscriminately to the ground. The eggs of some species have a sticky coat which may stick them to, or coat them in, debris. The delay period to hatching, normally two to six days, is affected by humidity as well as temperature. The whitish, eruciform, apodous larva which hatches from the egg is very active and negatively phototropic and geotropic so that it moves into the substrate. Like the eggs, the larval stages are susceptible to low humidities and move to zones of higher humidity. Most species have three larval instars, but some species have only two. The mature larva of some species is up to a centimetre long. Larvae feed on organic material which, because so many are nest dwellers, is usually plentiful. Much of the organic debris in the nest is derived from the host, but in addition adult fleas of some species do not completely digest their blood meals so that their faeces are commonly a food source for the larval stages. In the rabbit flea, S. cuniculi, the frequency of feeding and defecation increase dramatically during the reproductive period. As a consequence the burrow becomes well supplied with faeces which the forthcoming larval states utilize. In species such as Nosopsyllus fasciatus, where the larval as well as the adult diet consists largely of vertebrate blood, the fleas possibly harbour intestinal symbionts to supplement the diet. Species with a more catholic larval diet probably do not harbour such symbionts.

The larval stages are normally complete in under a month, although the larval span can be prolonged over many months allowing the flea to survive through a difficult period. At pupation the larva empties its intestine and produces a silken cocoon in which it pupates over a period of several days. The pupa is sensitive to low humidity and low temperature. Under ideal conditions the pupal stage is completed within a week, but it can extend for a year or more and is used by the flea as a means of spanning periods of adverse conditions. After emergence, most adult fleas will commence feeding before mating; the exceptions are mostly bird fleas. Mating is influenced by species specific, contact pheromones; more than one mating may be required to fertilize all the available eggs.

9.5 DIPTERA

The order Diptera contains many of the most important and familiar blood-sucking insects. They are of tremendous significance to man, from both an economic and health point of view, because of their role as parasite vectors. Because of the significance of this assemblage of insects I will briefly outline the characteristics and taxonomy of the order before discussing the particular details of each group separately.

As their name implies, the obvious character separating adult Diptera from all other insects is that they possess only a single pair of wings. The second pair of wings (the hindmost) have been modified into short, knob-like structures called halteres, which are used as balance organs. Wing tracheation is reduced and is so remarkably constant that wing venation can often be used for taxonomic purposes. The wings are carried on a thorax in which the mesothorax is greatly enlarged and the prothorax and metathorax correspondingly reduced.

Most adult Diptera have large and highly mobile heads bearing well-developed eyes which are often larger in the male than the female, and mouthparts which lack mandibles and are suctorial. Many of the blood-sucking species show considerable morphological specialization of these sucking mouthparts to allow for penetration of the host's skin.

The Diptera are holometabolous (endopterygotes) and have four life stages: egg, larva, pupa and adult. Depending on species the adult female may produce eggs or live young. The larvae, which are apodous and eruciform, prefer humid to wet habitats; in some species larvae are entirely aquatic and many terrestrial larval forms live in the humid surroundings of rotting or fermenting organic matter. Other larval forms are carnivorous, some living in the flesh or internal organs of vertebrate hosts and some, like the Congo floor maggot, feeding on blood. Some larval forms feed on live plant tissue. In the 'higher' forms the pupal stage is normally immobile, but in the 'lower' forms the pupal stage may exceptionally display various degrees of mobility as is seen in mosquitoes. In 'higher' forms the pupa is commonly enclosed in the retained and hardened last larval skin and the structure arising is known as a puparium. The cyclorrhaphan flies (see below) have a special expandable bag called a ptilinum which is associated with the head. They use the ptilinum to force a cap from the end of the pupa at the time of eclosion. Adult Diptera are usually highly active and mobile.

The order Diptera is subdivided into three suborders. The first of these, the Nematocera, contains the soft-bodied flies such as the mosquitoes, sandflies, blackflies, etc. (Fig. 9.7). The second suborder, the Brachycera, contains the tabanids and the blood-sucking rhagionids (Fig. 9.7). The third suborder, the Cyclorrhapha, contains the insects most commonly termed 'flies' (Fig. 9.7). The taxonomy of the suborder Cyclorrhapha is complex. It is subdivided into two series, the Ashiza and the Shizophora. The Shizophora is subdivided once more into three sections, Acalypterae, Calypterae and Pupipara. The families of insects which fit into each of these divisions are shown in Table 9.3.

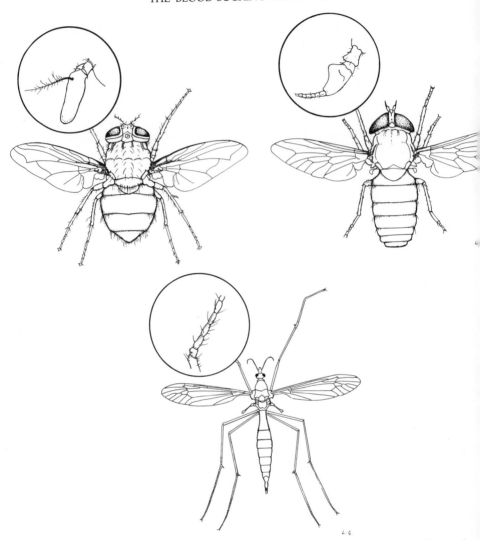

Figure 9.7 A typical example of a nematoceran (bottom), a brachyceran (right), and a cyclorrhaphan (left), emphasizing the antennal differences between the three groups. (Redrawn from Colyer & Hammond 1968.)

9.5.1 Culicidae

Mosquitoes are perhaps the most familiar of all blood-sucking insects. They fall into the nematoceran family, the Culicidae, which is divided into three subfamilies and 37 genera. The Anophelinae and the Culicinae are blood feeders but the third subfamily, the Toxorhynchitinae, do not feed on blood and so do not concern us here. Anophelinae contain about 430 species in three genera, *Bironella*, *Chagasia* and, containing by far the

Table 9.3 The divisions of the order Diptera and the major families in each division. Families containing blood-sucking species are in bold type.

Suborder	Series	Section	Family
Nematocera			Tipulidae
			Psychodidae
			Culicidae
			Chironomidae
			Ceratopogonidae
			Simuliidae
			Anisopodidae
			Bibionidae
			Mycetophilidae
			Sciaridae
Brachycera			Stratiomyidae
			Rhagionidae
			Tabanidae
			Asilidae
			Bombyliidae
			Empididae
			Dolichopodidae
Cyclorrhapha			
	Aschiza		Phoridae
			Syrphidae
	Schizophora		
		Acalypterae	Drosophilidae
			Chloropidae
			Sepsidae
			Gasterophilidae
			Piophilidae
		Calypterae	**Muscidae**
			Anthomyiidae
			Glossinidae
			Calliphoridae
			Sarcophagidae
			Oestridae
			Tachinidae
		Pupipara	**Hippoboscidae**
			Streblidae
			Nycteribiidae

largest number of species (416), *Anopheles*. Culicinae, containing some 2908 species, are a taxonomically more complex group. The most commonly encountered genera are *Culex, Aedes, Sabethes, Mansonia* (= *Taeniorhynchus*), *Culiseta* (= *Theobaldia*), *Psorophora, Wyeomyia, Haemagogus* and *Armigeres*.

Mosquitoes are found throughout the world except the Antarctic. In many parts of their distribution, most particularly in tundra areas of the

211

Northern Hemisphere, mosquito populations reach pest, and sometimes plague, proportions. However, their importance as pests is insignificant compared with their role as vectors, particularly of human diseases such as malaria. The World Health Organisation estimated in 1985 that 220 million people were suffering from malaria and that 200 million of these were in sub-Sahelian Africa where an estimated one million children a year die from the disease (W.H.O. 1985). They also estimated that over 2.5 billion people are living in areas which are endemic for malaria. Each of the four species of human malaria parasites are transmitted exclusively by *Anopheles* spp. Anopheline mosquitoes also transmit malaria parasites to other animals, for example *P. knowlesi* and *P. cynomolgi* to monkeys and *P. berghei* to rodents. Culicine mosquitoes transmit malaria parasites such as *P. gallinaceum* to wild and domesticated birds.

Mosquitoes also transmit filarial worms such as *Wuchereria bancrofti*, *Brugia malayi* and *B. timori*, which are parasitic in man and cause elephantiasis; there are as many as 250 million infected people in the tropics. These debilitating parasites are transmitted by many different mosquito species in various parts of their geographical distribution. To generalize, *Anopheles* and *Mansonia* spp. are important vectors of bancroftian and brugian filariasis, respectively, in many rural parts of the tropics and subtropics. In urban areas *Culex quinquefasciatus*, which thrives in the polluted waters common in urban slum areas, is often the major vector of bancroftian filariasis. The filarial worms of the dog, *Dirofilaria immitis* and *D. repens*, are also transmitted by mosquitoes, causing the disease known as dog heartworm.

Mosquitoes also transmit more than 200 arboviruses to man and other animals. Culicine mosquitoes are most commonly involved as vectors but anophelines do transmit a few viruses. The most important of the arboviruses transmitted are again infections of man. These include dengue, which is largely an urban disease endemic in many parts of South-east Asia and the Western Pacific, also occurring in the Caribbean. The major vector is *Aedes aegypti* with *Ae. albopictus* playing a subsidiary role in Asia. Yellow fever is a zoonotic disease of forest monkeys found in Africa, Central and South America. It is transmitted between monkeys by tree hole breeding, forest-dwelling mosquitoes such as *Haemagogus* spp. and *Sabethes chloropterus* in Latin America and *Aedes africanus* in Africa. These insects bite their hosts high in the forest canopy. *Haemogogus* and *Sabethes* bite man when he enters the forest transferring the disease from monkey to man; in Africa *Aedes bromeliae* (= *simpsoni*) transfers the virus from monkey to man. Once the virus is in man domesticated vectors such as *Ae. aegypti* are very efficient at spreading the disease among the urban population. Mosquitoes also transmit arboviruses causing encephalitis. In the Americas, Venezuelan equine encephalitis, Eastern equine encephalitis and Western equine encephali-

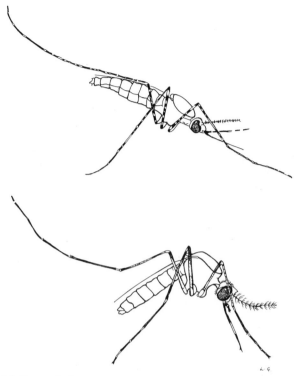

Figure 9.8 Adult anophelines (above) and culicines (below) commonly have these characteristic resting postures which is a useful identification feature in the field. (Redrawn from Kettle 1984.)

tis are disseminated by species of *Aedes* and *Culex* which are primarily bird feeders. In Japan, South-east Asia and India the major vector species transmitting Japanese encephalitis to man is the rice-field-breeding *Culex tritaeniorhynchus*.

Adult mosquitoes are small (usually about 5 mm long) rather delicate insects, with slender bodies, long legs and elongated, forwardly projecting mouthparts (Fig. 9.8). When resting, mosquitoes hold their single pair of wings over the abdomen like a pair of closed scissors. In most anophelines the wings have a dappled appearance because of alternating blocks of dark and light scales on the wings. In contrast most culicines have wings which lack distinct markings. Characteristically all mosquitoes have scales on their wing veins and the trailing edges of the wings. Males of important vector species are easily distinguished by their conspicuous, plumose antennae which contrast with the pilose antennae of the female (Fig. 9.9). Adult members of the blood-sucking subfamilies can be distinguished in the field by their resting postures. Adult anophelines usually rest with the three divisions of the body in a straight

213

line with the abdomen tilted up from the resting surface (Fig. 9.8). In contrast the body of culicine adults forms an angle about the thorax and they tend to stand more or less in parallel with the resting surface (Fig. 9.8).

Adult mosquitoes of both sexes feed on sugary solutions, but only females take blood. Mosquitoes as a group feed on a tremendous range of vertebrates from fish and reptiles to birds and mammals, but each species typically has a narrow range of preferred hosts from which it normally feeds. Man is a major host for most, but not all, important mosquito vectors of human disease. Many mosquitoes feed in a particular place; the canopy rather than the forest floor, or the reeds at the edge of the river rather than on birds resting on open water. This is important to man because although most mosquitoes bite in the open, some of the more important vector species feed largely within dwellings (see below). Each species usually has a characteristic peak biting time or times, so *Anopheles gambiae* bites in the early hours of the morning and *Ae. aegypti* shows two

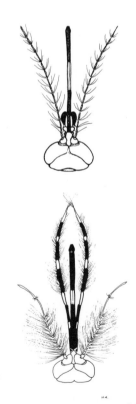

Figure 9.9 Female mosquitoes have pilose antennae while the antennae of male mosqui-toes are plumose. The examples shown here are *Culex annulirostris*. (Redrawn from Kettle 1984.)

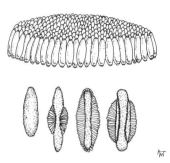

Figure 9.10 Mosquito eggs come in a variety of forms. *Culex* eggs are commonly produced as a floating raft, a typical aedine egg is shown bottom left and eggs of three species of anophelines are shown to illustrate the egg floats typical of this genus. (Redrawn from Faust *et al.* 1977.)

biting peaks, one at dawn and another at dusk. To generalize, most anopheline species are night biters while the Culicinae contains both night and day biting species. In the tropics mosquitoes take a meal every two to four days. Once the meal has been taken they find a quiet spot to digest it. For some important vector species this is inside man's dwellings, a behaviour which provides an opportunity for effective vector control of disease by spraying insecticides within these relatively accessible and limited spaces.

Although some mosquitoes need two blood meals to mature the first batch of eggs, most mosquitoes display gonotrophic concordance through most of their reproductive lives, each blood meal normally leading to the development of a batch of about 50–200 dark brown to black ovoid eggs. A few species, such as *Culex molestus*, can develop the first egg batch autogenously, but production of subsequent egg batches relies on blood-feeding. *Anopheles*, most *Culex* and some *Mansonia* (subgenus *Coquillettidia*) species lay their eggs directly onto the water surface. Other *Mansonia* (subgenus Mansonoides) species lay their eggs in groups on the undersurface of the leaves of aquatic plants. Anopheline eggs are laid singly and usually possess floats (Fig. 9.10) which keep them on the water's surface. *Culex, Culiseta, Coquillettidia* and some *Uranotaenia* species lay their eggs in clusters arranged together into floating rafts (Fig. 9.10). In tropical regions these eggs hatch within three days of deposition. *Haemagogus, Psorophora* and *Aedes* species lay their eggs on damp substrates just above the water's surface. These eggs, unlike those laid directly onto the water's surface, soon become resistant to desiccation and can remain viable for many months delaying their hatch until they are immersed in water. While immersion is commonly the hatching trigger, these eggs are also capable of entering diapause so that hatching may be delayed beyond the first immersion. Egg diapause is commonly used as an overwintering device by aedine mosquitoes in temperate

215

regions, while others overwinter as larvae or as hibernating, inseminated female adults. In tropical areas mosquitoes may survive the dry season as eggs or as aestivating adults, but in most areas mosquitoes breed continuously throughout the dry season, albeit at a greatly reduced rate.

Various mosquito species have specific requirements for larval habitats. Because mosquito control commonly involves offensives against the larvae, an understanding of their habitats is often central to the successful control of vector or nuisance populations. Almost all bodies of fresh or brackish water can be utilized as a larval habitat by one species or another. The exceptions are fast-flowing water (although backwaters or marshy areas along their edges may be used, especially by anophelines) and expanses of open water such as lakes, particularly if well stocked with larvivorous fish. Many species, particularly culicines, make use of natural container habitats such as broken coconut shells, rot holes in trees, bamboo stumps, leaf axils, and pitcher plants. Some of these species have been able to make use of the tremendous number of man-made containers available. The classic example is *Ae. aegypti*, which has spread to urban and particularly to slum areas throughout the tropics and subtropics. Here it finds discarded tins and tyres, water storage vessels, flower pots in cemeteries and a host of other man-made containers suited to its breeding. Other species prefer water contaminated with organic matter. Slum areas throughout the tropics and subtropics, with their inadequate sanitation and rubbish disposal facilities, have provided polluted water which is an ideal breeding ground for one such mosquito, *Culex quinquefasciatus*. This mosquito has soared in numbers with the rapid, unplanned urbanization in such areas. Other

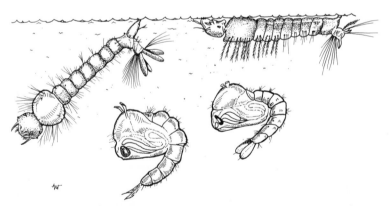

Figure 9.11 Culicine larvae hang head down from the surface film and are easily distinguished from anopheline larvae, which lie with their bodies parallel to the water's surface. The comma-shaped pupae of culicines can be distinguished (with experience) from the very similar pupa of anophelines by the shape of the respiratory horns. (Redrawn from Faust *et al.* 1977.)

216

mosquitoes, such as the important malaria vector *An. gambiae*, need temporary sunlit pools and man's farming and forestry activities in Africa have provided extensive additional breeding sites for these mosquitoes.

Each of the four larval stages of mosquitoes are aquatic, legless and have a thorax wider than the other body regions. A few larval mosquitoes are carnivorous (and in some cases cannibalistic), but the majority feed on micro-organisms using their mouth brushes to gather food particles. Most anopheline species are surface feeders, many other mosquito larvae feed on the bottom. Most culicine and anopheline larvae and pupae come to the surface to breathe atmospheric oxygen. Culicines bear a respiratory siphon on the penultimate, eighth abdominal segment and they use this to attach themselves to the water's surface film from which they hang head down (Fig. 9.11). Anopheline larvae are easily distinguished from culicines because they do not have a respiratory siphon although they are still air breathers and have paired spiracles on abdominal segment eight. When at the surface they lie with their bodies parallel to the water's surface (Fig. 9.11) suspended by a pair of thoracic palmate hairs and palmate hairs present on most of the abdominal segments. *Mansonia* (including subgenus *Coquillettidia*) spp. larvae breathe by plugging into the air vessels of plants using the modified valves of their conical respiratory siphons to cut their way into the plant tissues, to which they remain attached throughout their development.

In the hot tropics the larval stages can be completed within seven to ten days. The larvae then moult to becomed pupae, which are comma-shaped, non-feeding, air-breathing, aquatic stages in which the head and thorax are fused to form a cephalothorax. On the dorsal surface of the cephalothorax are paired breathing trumpets which attach the pupa to the surface film where it spends most of its time. If disturbed the pupa uses the paired paddles on the last abdominal segment to swim rapidly away from the surface. *Mansonia* and *Coquillettidia* spp. are different in that, like the larvae, the pupae remain attached to aquatic vegetation by their respiratory siphons. Anopheline and culicine pupae are structurally similar and distinguishing between them requires experience. In the tropics the pupal stage is completed in two to three days.

Males emerge before the females; they undergo a post-eclosion maturation period of about 24 h in which the terminalia must rotate through 180° before they are ready to mate. In some species mating of the female occurs at emergence, but in most species it occurs sometime after emergence but before the female takes her first blood meal. In some species mating involves crepuscular swarming of the males above marker points (such as trees or prominent rocks) of species-specific height and setting. Females respond to the wing beat sounds of males and enter the swarm; coupling takes place and they leave the swarm and insemination occurs. Not all species need to swarm to mate. For example, *Aedes aegypti* use the host animal as a meeting point for the sexes with male *Ae. aegypti* being attracted to host animals during the same period of the day as the females.

Male mosquitoes do not usually travel more than about 100 m from the larval site. Females travel further to find both blood meals and new larval sites to colonize. There are considerable inter- and intraspecific differences in the distances females travel. Probably few actually fly more than one or two kilometres, but it is well known that individuals of some species may be carried considerable distances on the wind. For example, in the USA *Aedes vexans* has been observed to be displaced by over 200 miles on weather fronts and the saltmarsh mosquito, *Ae. taeniorhynchus*, has often caused a nuisance in built-up areas many miles downwind of its breeding sites. The malaria vector *Anopheles pharoensis* has been recorded many tens of miles downwind of breeding sites in North Africa. It is likely that under tropical field conditions male mosquitoes live for about seven to ten days and a few females for up to a month. In temperate regions females live two or more months, and hibernating adults survive for up to eight months.

9.5.2 Simuliidae

The nematoceran family Simuliidae contains some 1500 species in 19 genera. Four genera, *Austrosimulium*, *Cnephia*, *Prosimulium* and *Simulium*, are of economic importance. These flies are known locally by a variety of names but most commonly as blackflies or buffalo gnats. In temperate and particularly sub-Arctic regions of the world, the flies can appear in huge swarms which can make life a misery for man, his livestock and wildlife. Indeed, both man and on some occasions thousands of livestock have died from the attacks of these huge swarms (16 474 domestic animals in Southern Rumania in 1923; Baranov 1935, Bradley 1935). Death probably occurs from a combination of effects, mainly toxaemia (or anaphylactic shock) from the insect's bite, blood loss and breathing problems from inhalation of large numbers of insects. Lower biting levels can cause economic losses in livestock from the worry they cause to the animal (Hunter & Moorhouse 1976).

Blackflies are also important vectors of disease organisms; they are mechanical transmitters of myxomatosis, vectors of *Leucocytozoon* and avian trypanosomes of domestic and wild birds, and occasional vectors of the arbovirus, Venezuelan equine encephalitis. But most importantly blackflies are vectors of filarial worms, the most important being *Onchocerca volvulus*, which causes the human disease river blindness in Africa and Central and South America. Blindness can be so common in areas of intense transmission, such as the Volta and Niger river basins of West Africa, that it seriously interrupts the social organization of the population and thus impairs the development of the country (Senghor & Samba 1988). Onchocerciasis in Africa is mainly transmitted by members of the *Simulium damnosum* complex which contains at least 30 separate species or cytotypes (Service 1988). The Onchocerciasis Control Program of the World Health Organisation was instituted in 1974 to attempt to control the disease. This immense programme, which is based on the control of the vectors, demands that up to 50 000 km of river are monitored or treated with insecticide each week. This successful programme has reduced prevalance from an original level of 25-30% to below 5%. Because adult *Onchocerca volvulus* live in man for 15 years the campaign must continue for at least this period to achieve any long-term success. The programme may be usefully supplemented in the future with drug-based control using ivermectin, which was experimentally introduced into the programme in 1987 (Bradshaw 1989).

Adult buffalo gnats are small flies 1–5 mm long. They have a single pair of clear, scaleless, broad wings between 1 and 6 mm long with large anal lobes and anterior veins that are usually prominent. When the fly is at rest the wings are folded over the body like a closed pair of scissors. The flies are usually blackish or greyish but often have silvery, orange or golden hairs patterning the thorax and/or legs. The eyes in the female fly are dichoptic; in the male the eyes are holoptic and the upper ommatidia are larger than the lower (Fig. 9.12). The male eye specializations are used for detecting flying females in the mating swarm. There are no ocelli. Antennae are cigar shaped, commonly 11-segmented and lack prominent hairs (Fig. 9.13). The thoracic scutum is well developed and gives the flies their characteristic hump-backed appearance (Fig. 9.13). Compared to other Nematocera the legs of blackflies are distinctly short and stout (Fig. 9.13). Adults are diurnal and show an essentially bimodal behaviour pattern with maximum activity in the early morning and late afternoon.

Blackflies breed in water, commonly in rapid-flowing 'white' water.

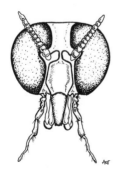

Figure 9.12   Male blackflies are holoptic (eyes meeting on top of the head) and females are dichoptic (two separate eyes). (Redrawn from Smith 1973.)

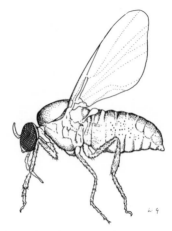

Figure 9.13 Side view of adult female *Simulium damnosum s.l.* (Redrawn from Smith 1973.)

Each species has a preference for particular breeding sites. Thus, the African species *Simulium damnosum s.l.* breeds in the rapids of small to very large rivers, *S. ochraceum* in very small streams of South America and the East African species *S. neavei* in small streams and rivers attached in a phoretic association with freshwater crabs of *Potamonautes* spp. Most species lay eggs in the evening, in clusters containing 150–600 eggs on submerged vegetation or rocks, etc., but in some species (e.g. *S. ochraceum*) the female flies over the water broadcasting eggs onto the surface. In many temperate species the egg may diapause; in others the larval stage may overwinter. In the tropics the egg and larval stages may be completed in about a week. The larvae are commonly encountered in water bodies and are easily recognized from their characteristic shape (Fig. 9.14). The larva spins a small pad of silk on the substrate from its salivary glands and attaches firmly to this by a ring of hooklets on its

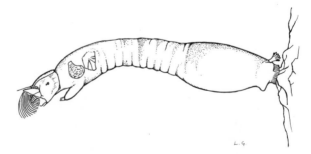

Figure 9.14 Blackfly larva are easily recognized from their cigar shape, mid-ventral proleg just behind the well-formed head and the circle of hooklets on the end of the body used to attach the larva to the substrate. When viewed in water the two large cephalic fans may also be seen. (Redrawn from Smith 1973.)

posterior end. The larva is sedentary throughout most of its life, but it has another, smaller circlet of hooks on the proleg (Fig. 9.14) and it can detach each circlet in turn and move using looping movements of the body. The larva can also produce silk from its salivary glands and can use these threads to relocate downstream and to move back upstream if the thread has been anchored. The larvae are filter feeders using their pair of cephalic fans to abstract fine particulate matter (10–100 μm) from the passing water flow. Larvae grow to about 12 mm long undergoing up to eight larval moults to achieve this. The mature larva, or gill spot-larva (which is actually a prepupa or pharate pupa), pupates in a cocoon spun from salivary gland silk. During daylight, two to six days after pupation, the adult emerges rapidly from the pupa. It reaches the water surface by climbing up an available surface, or by floating up in a gas bubble, and is capable of immediate flight.

Males form mating swarms near the emergence site or occasionally near the female's vertebrate host. Male flies grasp females in flight and insemination occurs once the pair have landed or fallen to the ground. In some species mating may occur in the absence of swarming with males seeking out females on, or close to, the female's host animal. Insemination may take only seconds or nearly an hour depending on species.

Blackflies are strong fliers and are capable of considerable dispersion and migration from the emergence site. This has created problems for the Onchocerciasis Control Program of re-invasion of treated areas in West Africa, where flies have been reported to travel up to 400 km on the prevailing winds of the Intertropical Convergence Zone (Garms et al. 1979). Both male and female blackflies feed on nectar which is stored in a gut diverticulum called the crop. It is thought that nectar-feeding is an important energy source for flight. It has been calculated that following a sugar meal, Simulium venustum is capable of flying well over 100 km in still air (Hocking 1953).

Only female flies take blood meals; they are exophilic and exophagic and take blood meals during the day. Blackflies are pool feeders; the mouthparts are used to cut the skin and blood is taken from the pool that forms in the wound. The female completes the meal in a few minutes. Blackflies display host preferences; most species such as Simulium euryadminiculum and S. rugglesi are ornithophilic, but there are many mammophilic species such as S. venustum, and others such as S. damnosum s.l. feed on both birds and mammals. Although some species of blackfly are autogenous, a blood meal is required by most females for egg batch maturation. The ovarian cycle takes four to five days in S. damnosum and about 500 eggs are matured in each cycle. The female has a maximum life expectancy of about 30 days and so can complete about six ovarian cycles. Other species, such as Austrosimulium pestilens, may complete the gonotrophic cycle within one day (Hunter & Moorhouse

1976) and other species have field life expectancies of at least 85 days (Dalmat 1952). Fly numbers usually show seasonal peaks; in the tropics most flies appear during and immediately after the rainy season, in the more temperate regions of the world they appear in the summer months.

9.5.3 Ceratopogonidae

The nematoceran family Ceratopogonidae contains the biting midges which are known locally as punkies, no-see-ums or even sandflies (not to be confused with the phlebotomines – see Section 9.5.4). The blood-sucking midges are contained within four genera, *Forcipomyia, Leptoconops, Austroconops* (a single species limited to Western Australia) and, by far the largest genus, *Culicoides*, with about 1000 described species. These tiny flies can make life a misery when they are present in large numbers. The reaction to their bite is often intense. They are a serious threat to the tourist and leisure industries in several parts of the world including parts of Scotland, Florida and the Caribbean. They are known vectors of several arboviruses of veterinary importance including bluetongue, bovine ephemeral fever, African horsesickness and Akabane virus. It is possible that the role of biting midges as vectors of arboviruses is considerably underestimated at present. Biting midges also transmit parasitic protozoa including *Hepatocystis* spp. to mammals, some *Leucocytozoon* spp. and *Akiba* spp. to birds, and *Parahaemoproteus* spp. mainly to birds. They also transmit some filarial worms such as *Mansonella ozzardi* to man and *Onchocerca gibsoni* to livestock.

The adult flies are very small, usually less than 1.5 mm long (Fig. 9.15). The black and white patterning typical of the single pair of wings of most species is the first and easiest clue for field identification. At rest the short, broad wings are folded scissor-like over the body. There is no distinct wing venation in the genus *Culicoides*, but it is more pronounced

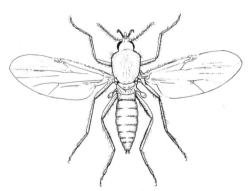

Figure 9.15 Adult *Culicoides nubeculosus*. (Redrawn from Smith 1973.)

in the other genera. The mouthparts, which are used to cut the skin rather than pierce it, are held beneath the head. The long, usually 15-segmented antennae (12–14 in *Leptoconops* spp.) project well clear of the head and are plumose in the male and non-plumose in the female. The most important genus of biting midges, *Culicoides*, is distinguished by the presence of two depressions, known as humeral pits, placed anteriorly on the dorsal surface of the thorax.

In temperate parts of the world biting midges are largely a summer problem, in the tropics they may be present all year round. Both sexes feed on sugary solutions but only the females feed on blood. Ceratopogonids are exophilic and exophagic and mammophilic and ornithophilic species occur. Peak feeding time varies among species. Many are crepuscular, but perhaps the majority are active in the evening and first half of the night. Most activity takes place in still, warm, humid conditions; feeding is inhibited in winds of $1-2 \text{ m s}^{-1}$ probably by the inability of the fly to find hosts efficiently. Any exposed surface may be bitten, but when biting man the scalp is often paid particular attention. Although autogeny is common in biting midges most species require a blood meal for egg maturation to occur. Mating usually occurs during swarming. The ovarian cycle take two to four days and females show gonotrophic concordance, maturing from 40–50 eggs in each cycle. These are laid in wet substrates, commonly marshy or sandy areas on the edge of water bodies (fresh water or salt). Some species use decaying organic matter as larval sites, breeding in rot holes in trees or in rotting banana stumps, etc. Larval sites are very species specific and adult flies do not move far from their breeding sites.

We know little of the egg-laying behaviour in the biting midges, but females of *Leptoconops spinosifrons* burrow 3–6 cm into the sandy substrate to deposit their eggs (Duval *et al.* 1974). In temperate areas biting midges may overwinter in the egg state whereas in the tropics eggs usually hatch in under ten days. There are four larval instars. The majority of ceratopogonid larvae are typically nematoceran having 11 body segments lacking appendages or conspicuous setae, a dark, sclerotized head and two retractable anal papillae probably used in salt absorption (Fig. 9.16); the final instar larva may be up to 6 mm long. These larvae develop within their particular substrate and only rarely appear on the surface or swimming, in their characteristic serpentine fashion, in overlying water. It is possible that migration does occur at the wet fringe of expanding or contracting water bodies. (Larvae of the 50 or so species in the Forcipomyiinae do have anterior and posterior prolegs and have a more active, crawling life-style than most vermiform ceratopogonid larvae.) Although larval development of some species in the tropics may be complete in two weeks, larval development in biting midges is usually an extended process which, in the extreme case of some Arctic forms, may take two

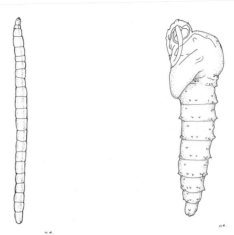

Figure 9.16 Larva and pupa of *Leptoconops spinosifrons*. (Redrawn from Smith 1973.)

years. Larvae of some species are predatory, feeding on soil-dwelling metazoa such as nematodes; larvae of other species feed on micro-organisms. The short-lived pupae (< 10 days), which are culicid-like (Fig. 9.16), live in the surface layers of the substrate.

9.5.4 Psychodidae

The 700 or so species of phlebotomine sandflies are contained within the nematoceran subfamily the Phlebotominae, of the family Psychodidae. Two genera, *Lutzomyia* and *Phlebotomus*, contain the major vector species. Most phlebotomine sandfly species occur in the tropics, but some important disease vectors are found in temperate areas such as the Mediterranean. Two broad patterns of vector distribution are seen: in the Old World vector species are largely confined to the drier regions of the southern part of the temperate zone but in the New World vector species are mainly found in wetter forested areas.

Phlebotomine sandflies transmit the protozoan parasites causing leish-maniasis in man and, in some cases, other animals. In many, if not most, areas leishmaniasis is a zoonotic disease. Cutaneous leishmaniasis (also called dermal leishmaniasis or oriental sore) is an Old World disease occurring in the south of the Palaearctic region and transmitted prin-cipally by *Phlebotomus* spp. American dermal leishmaniasis (also called chiclero's ear, espundia and uta, and which includes mucocutaneous leishmaniasis) occurs largely in wet, forested areas of Central and South America from Mexico to Chile, where its vectors are *Lutzomyia* spp. Visceral leishmaniasis (also called kala-azar) is widely distributed through the drier areas of the southern Palaearctic region of the Old World, where its vectors are *Phlebotomus* spp.; it is also found in the drier

224

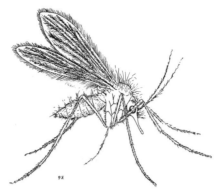

Figure 9.17 Female of *Phlebotomus papatasi*. Note the erect wings. (Redrawn from Smith 1973.)

parts of North-east Brazil, where the vectors are *Lutzomyia* spp. In the Andean region of South America phlebotomine sandflies also transmit *Bartonella bacilliformis*, the mysterious rod-like micro-organism which is the causative agent of Carrion's disease of man (also known as Bartonell-osis or Oroya fever). Sandflies also transmit viral diseases including vesicular stomatitis virus to cattle and horses, and sandfly fever (also known as three-day fever or pappataci fever) to man. Although not important medically or economically, sandflies can also transmit trypa-nosomes (Anderson & Ayala 1968), and probably reptilian malaria parasites (Ayala & Lee 1970). Although their importance is primarily as vectors of disease, sandflies may occasionally be present in sufficient numbers to reach pest status because of the severe irritation caused by their bites.

Adult sandflies are small (1.5–4 mm long) slender-bodied, hairy, brownish insects with long stilt-like legs. These blood-sucking flies are readily recognized at rest because they hold their single pair of narrow, lanceolate wings almost erect over the body (Fig. 9.17) whereas resting, non-biting psychodids fold their wings roof-like over the body. Closer examination shows that vein two of the wings of phlebotomines branches twice towards the middle or tip of the wing, whereas in non-biting psychodids branching is nearer the wing base (Fig. 9.18). The

Figure 9.18 Phlebotomine wing to show that vein two (arrow) branches twice. (Redrawn from Smith 1973.)

225

Figure 9.19   Phlebotomine larva showing matchstick hairs and pupa with last larval skin and caudal hairs. (Redrawn from Kettle 1984.)

non-blood-feeding male sandfly is easily distinguished from the female by the pair of prominent claspers at the abdomen tip.

In temperate areas sandflies are only present in the summer months; in the tropics they may be present throughout the year. Mating probably does not occur through swarming, but rather occurs among insects dancing on surfaces at twilight. Both sandfly sexes feed on sugar solutions, but only the female feeds on blood. Cold-blooded vertebrates and mammals are common hosts for various species, fewer species seem to be ornithophilic. Flies bite exposed parts of the skin either at twilight or after dark during periods of settled calm weather. Most are exophagic but some will rest and feed indoors; not surprisingly these endophilic, endophagic forms contain some of the most important vector species. Autogenous forms are known (e.g. *Phlebotomus papatasi*), but most species require a blood meal for egg production to occur. Less than 100 tiny (<0.5 mm), ovoid eggs are laid singly at each oviposition and darken to a brown or black colour within hours of being laid. Eggs, even of dry savanna dwellers, cannot withstand desiccation, and so oviposition sites need to be moist or wet. Oviposition sites are species-specific and range from termite mounds to cracked masonry, buttress roots of forest trees to leaf litter. In warm climates eggs hatch in under three weeks and the 12-segmented larvae feed on organic debris. There are four larval instars. The larvae are grey with rather darker heads and are very difficult to locate, but if found are easily identifiable as phlebotomine larvae by the presence of so-called matchstick hairs on each segment (Fig. 9.19). In temperate areas sandflies probably overwinter as mature larvae, in the tropics the larval period is usually less than two months. Pupae retain the skin of the final larval stage with its characteristic double pair of caudal bristles and can be recognized for this reason (Fig. 9.19). The pupal stage is completed in under two weeks. Like the egg stage, both larvae and pupae are very sensitive to drying out and require a humidity of more than 75%. Because of their sensitivity to environmental conditions adult sandflies are often restricted to particular microhabitats (such as termitaria or mammal burrows) in an otherwise hostile landscape. Phlebotomine sandflies are weak fliers and rarely disperse more than about a kilometre from the larval site so that biting and transmission areas may be very localized.

9.5.5   Tabanidae

The Brachyceran family Tabanidae consists of about 3500 species of medium to very large biting flies; indeed they contain some of the largest blood-sucking insects. The three most important genera are *Chrysops*, *Haematopota* and *Tabanus*. The sheer size of tabanids and the pain their bite can cause mean they are not easily ignored. In consequence they have gained many local names including clegs, mainly for *Haematopota* spp., deerflies, mainly for *Chrysops* spp., and horseflies and hippo-flies, mainly for *Tabanus* spp. Large accumulations of flies do occur in some areas and with their painful bite tabanids can be pests of both stock and man. Stock worry can reach a level where economic losses occur. Tabanids also transmit disease to both man and livestock. Their painful bite results in the flies often being disturbed while feeding but they are persistent and will commonly move directly to another animal to recommence feeding. Because of this, and the rather 'spongy' nature of the labella which can hold up to a nanolitre of uncongealed blood, they are excellent mechanical vectors of several parasites.

Tabanids are a major mechanical transmitter of the flagellate *Trypanosoma evansi* which causes severe disease (surra) in horses, camels and dogs, and less severe illness in several other mammals including cattle; surra occurs over large areas of the tropics and subtropics. Tabanids are also the mechanical transmitters of other trypanosomes, including *T. vivax viennei* of cattle and sheep, *T. equinum* the causative agent of mal de Caderas of equines, *T. simiae* of pigs, and the African trypanosomes normally transmitted by tsetse flies. Tabanids are also the vectors of the posterior station (stercorarian) trypanosome *T. theileri* of cattle. In addition, tabanids are mechanical vectors of the bacterium *Francisella* (=*Pasteurella*) *tularensis*, the causative agent of tularaemia in man. This zoonotic disease is widespread throughout the holarctic; the primary hosts are wild rodents. It is transmitted by a variety of agents but in the USA the tabanid *Chrysops discalis* is a major vector. Tabanids also mechanically transmit other infectious agents including anthrax, equine infectious anaemia virus, California encephalitis, Western equine encephalitis, rinderpest and anaplasmosis.

Tabanids also transmit filarial worms including *Loa loa* of man, the

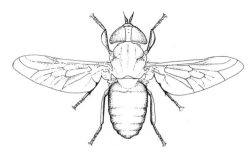

Figure 9.20 Adult female *Ancala africana*. (Redrawn from Smith 1973.)

closely related *L. loa papionis* of monkeys, the arterial worm of sheep, *Elaeophora schneideri*, and *Dirofilaria roemeri* of macropodid marsupials. Transmission of filarial worms is not mechanical, the tabanid is an obligatory stage in transmission with the parasites undergoing development in the insect. The major vectors involved in the transmission of the human parasite *Loa loa* are *Chrysops dimidiata* and *C. silacea* in West Africa and *C. distinctipennis* in Central Africa.

Adult tabanids are medium to very large (up to 25 mm), stocky insects (Fig. 9.20) variously coloured from black through brown to greens and yellows. The large, semi-lunar head bears two prominent eyes which, in the live insects, are often patterned with iridescent reds and greens which usually fade after death. The eyes are commonly spotted in *Chrysops* spp., have zig-zag bands in *Haematopota* spp. and horizontal stripes and no patterning in *Tabanus* spp. In the male the eyes are holoptic; in the female, dichoptic (Fig. 9.20). This forms an easy means of distinguishing the sexes. The thickset, porrect antennae consist of three components, the scape, pedicel and flagellum, and antennal shape is used to distinguish the three major genera, *Tabanus*, *Chrysops* and *Haematopota* (Fig. 9.21). Only the females suck blood. Their mouthparts are adapted for cutting rather than piercing the skin and project below the head, not in front of it.

The thorax and abdomen are commonly patterned in various colours which often provides a good guide for field identification. The single pair of wings may be very large (a span of up to 65 mm in the largest species),

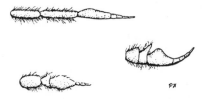

Figure 9.21 Antennal shape can help to distinguish the three major genera of tabanids: *Chrysops* (top), *Haematopota* (bottom) and *Tabanus* (middle). (Redrawn from Service 1986.)

228

they may be clear (many *Tabanus* spp.) or may bear coloured patterns (usually mottled in *Haematopota* spp. and with a distinct stripe in *Chrysops*). At rest tabanids either hold the wings flat over the body like a pair of partially open scissors (*Chrysops* and *Tabanus* spp.), or they may hold the wings roof-like over the body (*Haematopota* spp.).

In temperate countries adult tabanids appear in summer. In the tropics they also commonly show a seasonal pattern with the smallest numbers in the dry season. Both male and female flies feed on sugary solutions which are important for their energy and water budgets. Adult flies survive three or four weeks in the field. The blood-feeding female utilizes a wide variety of mammals, particularly ungulates; some species also feed on reptiles and amphibians, but relatively few tabanids feed on birds. Most tabanids are restricted to certain habitats and host choice is influenced by availability within the habitat. Most tabanids are woodland or forest dwellers, including *Chrysops* vectors of loiasis; man only becomes infested if he enters the forest habitat of these insects. Most species feed during full daylight and are often most in evidence on the hottest, sunniest days. Most species are exophilic and exophagic, an exception being the loiasis vector, *C. silacea*, which will enter houses to feed.

Mating occurs after virgin females enter male swarms. In the very many canopy-inhabiting, forest species, swarming happens above the canopy normally in the early morning or late afternoon. A few species are autogenous, but most require a blood meal for egg maturation. Different species produce between 100 and 1000 eggs which are laid beneath a waterproof layer on plants or other surfaces adjacent to muddy or wet sites, surfaces chosen for egg laying are often species-specific. The eggs hatch in one or two weeks and the larvae drop to the ground. In general, larval *Haematopota* spp. live in relatively dry soil, *Tabanus* spp. in wet soil close to water bodies, and the larvae of *Chrysops* spp. in wet mud often in semi-submerged situations. There are between 4 and 11 larval instars, larval development is prolonged and may take three years in some species (e.g. *Tabanus calens*). The larvae are greyish and have typically brachyceran reduced heads. Tabanid larvae may be identified by the raised tyre-like rings between body segments, by the six pseudopods which can often be found on segments 4–10, and by the presence of Graber's organ (a black pyriform sensory structure) on the dorsal surface in front of the spiracle-bearing terminal siphon. In general, *Chrysops* larvae are detritovores, whereas *Tabanus* and *Haematopota* larvae are carnivores (and cannibalistic). Pupation occurs in the soil at the drier fringes of the larval habitat and the pupae are capable of limited movement. Adults emerge in under three weeks. Although adult tabanids are strong fliers capable of considerable movement away from the breeding site, they do not normally become widely dispersed.

Figure 9.22 Adult rhagionid, *Spaniopsis longicornis*. (Redrawn from Smith 1973.)

9.5.6 Rhagionidae

The Rhagionidae (= Leptidae) are a relatively little-studied family of brachyceran flies commonly known as snipe flies. Most are predatory upon other insects, but some species feed on blood. They are not known to be the vectors of any parasitic organisms. Because their bite is very painful and some humans react badly to it, these flies can be of nuisance value. The best known snipe flies occur in the genus *Symphoromyia*; these are found in the Holarctic region where species such as *S. atripes* and *S. sackeni* are troublesome to both man and other animals. *Spaniopsis* and *Austroleptis* spp. are nuisance flies in Australia and *Atherix* spp. are found in the Nearctic and Neotropical regions.

Rhagionids are medium to large, elongated, rather bare, sombrely coloured flies (Fig. 9.22). The mouthparts are heavily sclerotized and are used for cutting and piercing the host's skin, blood is then sucked up from the wound. Man, horses, dogs, deer, cattle and frogs have all been reported to be attacked by snipe flies. The adults also feed on sugary solutions. Little is known about the larvae, but larval sites include wet soil, leaf mould and rotting wood. Larvae of some species are predatory upon various insect larvae, earthworms and other soil fauna, but other species feed on mosses.

9.5.7 Muscidae

The cyclorrhaphan family Muscidae contains some 3900 species, only a few of which are haematophagous as adults. Some, like the stablefly

Stomoxys calcitrans, have mouthparts which are well developed for penetrating vertebrate skin. Others, such as *Musca planiceps*, have strong, rasping prestomal teeth which dislodge scabs or scrape through thin skin. Other flies, such as the headfly *Hydrotaea irritans*, are facultative haematophages feeding from open wounds or sores. From an economic and health point of view the most important genera are *Stomoxys* and *Haematobia* and these two genera will be used as examples of biting flies.

The stablefly, *S. calcitrans*, is a pest species of world-wide distribution. It is particularly important as a worry to stock and causes considerable annual losses to the agricultural industry. In some areas it may also be a direct nuisance to man. It can mechanically transmit trypanosomes and is especially important in the transmission of *Trypanosoma evansi*, which causes severe disease (Surra) in horses, camels and dogs, and less severe illness in several other mammals including cattle. It also transmits the Neotropical *T. equinum*, which causes mal de Caderas in equines, sheep, cattle and goats. Stableflies also play a minor role in the transmission of the African trypanosomes associated with tsetse fly. It is also the vector of the nematode *Habronema majus*, a stomach worm of equines. *Stomoxys calcitrans* has also been implicated in the transmission of other organisms such as polio virus, equine infectious anaemia, anthrax and fowl pox, although the regularity and hence importance of such events is unclear. *Stomoxys nigra, S. omega* and *S. inornata* are more localized pest and vector species found in tropical Africa and Asia.

The horn fly, *Haematobia irritans irritans*, is a major agricultural pest species throughout the Northern Hemisphere. Like the stablefly, it causes substantial economic losses in the agricultural industry through stock worry. It is also a vector of *Stephanofilaria stilesi*, a skin parasite of cattle. The buffalo fly, *Haematobia irritans exigua*, is particularly important as an agricultural pest in the Australian region.

Other muscids, such as the head fly *Hydrotaea irritans*, are facultative haematophages. Unable to penetrate the skin themselves, they gain access to blood by disturbing other insects which have penetrated the skin for them. Some *Hydrotaea* spp. may crowd round feeding insects such as tabanids, feeding concurrently with them or even crowding them off the wound. The headfly is a woodland species and in Europe causes damage to sheep, particularly lambs, both by animal annoyance, and through secondary bacterial infection of wounds and sores. It is also a vector of summer mastitis.

In temperate regions adult stableflies are commonly encountered throughout the summer and autumn basking on sunlit surfaces around stock pens but, despite the name, surprisingly rarely around stables. They are housefly-sized, but are easily distinguished from them by the forwardly projecting proboscis (Fig. 9.23). Both male and female flies feed on blood as well as sugary solutions. A wide range of mammalian

231

Figure 9.23 The characteristic forwardly projecting proboscis of the stablefly *Stomoxys calcitrans*. (Redrawn from Smith 1973.)

hosts are used. On sunny summer days they may feed twice but one blood meal a day is probably normal. The bite is painful and often disturbs the host, leading to interrupted feeding and movement to another host – ideal conditions for the mechanical transmission of disease. Males congregate in sunlit patches which serve as mating leks.

The hornfly is much smaller than the stablefly, about half the size of a housefly. Its forwardly projecting proboscis is squatter and more heavily built than the stablefly's. Both males and females take blood meals, mainly from bovids. The flies are almost permanently associated with their hosts. They commonly feed several times a day and defecate partially digested blood. Mating occurs on the host animal.

The female hornfly leaves the host animal to lay her brownish eggs. Up to 24 eggs are laid during any one cycle, in, under or close to the freshly deposited dung of the host animal where they hatch in less than a day. The larvae feed in the dung and after three to eight days migrate to drier areas at the edge of the dung pat or in adjacent soil and pupariate. In temperate regions the flies overwinter as puparia although in ideal conditions adults emerge from the puparium after three to eight days.

The anautogenous female stablefly lays her white eggs in rotting or fermenting vegetation, urine-soaked straw, etc., but rarely in dung alone. She lays up to 50 eggs during each ovarian cycle. These hatch one to four days after laying and the three larval stages are complete in 10–20 days. Pupation occurs in the drier parts of the medium and the adult emerges in under ten days. Development time of the immature stages of these flies is often not readily correlated with ambient temperature because rotting and fermenting vegetation is warmer than its surroundings. In temperate regions the flies overwinter in the larval or pupal stage.

Both adult stableflies and hornflies are strong fliers and both species may disperse considerable distances from their emergence sites. Stableflies are largely inactive after dark, but it is thought that hornflies actively disperse at night.

9.5.8 Calliphoridae

No adult calliphorids are blood-sucking but in some species the larvae feed on blood. The best-known example is the African species *Auchmer-omyia senegalensis*, the Congo floor maggot. The adult fly is mainly coprophagous. The female lays batches of up to 60 eggs in dry sandy soil on the floor of huts, caves, etc. The three larval stages are extremely resistant to dry conditions and to starvation. The non-climbing larvae emerge from cracks and crevices to take nightly blood meals from hosts sitting or sleeping on the floor of the shelter. Recorded hosts are humans, suids, hyena and aardvark. The larva penetrates the skin using its mouth hooks and maxillary plates and completes the blood meal in about 20 min. The larval period can be completed in under two weeks given freely available food, but may take up to three months if food is scarce. Pupation occurs in protected situations and the pupal period is about two weeks. Several genera of calliphorids including *Protocalliphora* have larval stages which attack nesting birds.

9.5.9 Glossinidae

The cyclorrhaphan family Glossinidae contain medium to large insects commonly known as tsetse flies. Tsetse flies are important as vectors of trypanosomes to man and to his domesticated animals. They are rarely, if ever, present in sufficient numbers to present a fly worry problem to stock. As vectors of animal trypanosomiasis, or nagana, they preclude the use of up to ten million square kilometres of Africa for cattle-rearing (Steelman 1976) and have also restricted development on that continent by limiting the use of draught and pack animals. The counter argument has also been put that the tsetse has prevented desertification of large areas of land from overgrazing and has been the saviour of Africa's game animals. These arguments have been clearly outlined by Jordan (1986). It is clear that nagana also slowed the conquest and colonization of Africa by preventing the free passage of the horse. The transmission of human sleeping sickness occurs on a more restricted scale with only about 10 000 new cases reported each year, but the World Health Organisation estimates that 45 million people are at risk. The danger from the disease was shown in the great epidemics of the colonial era. More than half a million people died of the disease in the Congo basin between 1896 and 1906 and as many as a quarter of a million in Uganda between 1900 and 1906. The circumstances of these disasters are discussed by Duggan (1970).

Adult tsetse flies are brownish, rather slim, and elongated; the smallest are about 6 mm long and the largest are about 14 mm. They have a characteristic resting attitude in which the single pair of wings are folded

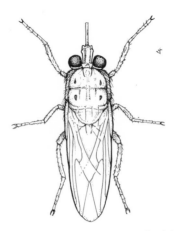

Figure 9.24 Plan view of a resting tsetse fly showing the folded wings. (Redrawn from Smith 1973.)

like closed scissors over the fly's dorsal surface, their tips extending for a short distance beyond the rear of the abdomen (Fig. 9.24). The proboscis, which has a distinct basal bulb, projects in front of the head and is flanked by palpi of almost equal length. Two characteristic features of tsetse flies are the branched arista of the antennae (Fig. 9.25) and the 'hatchet' shape of the discal cell of the wing (Fig. 9.26). Another distinctive feature of tsetse flies is the high-pitched buzzing sounds which they produce when heating themselves endothermically (Howe & Lehane 1986), and after which they have gained their onomatopoeic name. Tsetse flies are exophilic and exophagic and live at low-population densities estimated at an average of about ten per hectare. They are diurnal insects which essentially show a bimodal pattern of activity (morning and evening), although there are species-specific differences. The pattern is modulated by environmental conditions and by the fly's physiological status (Brady 1975). The tsetse is a viviparous insect: the

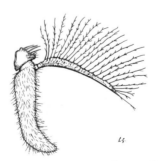

Figure 9.25 The branched hairs of the arista, a characteristic feature of tsetse flies. (Redrawn from Smith 1973.)

234

Figure 9.26 The hatchet shape of the discal cell of the wing, a characteristic feature of tsetse flies. (Redrawn from Kettle 1984.)

female gives birth to a single, preposterously large, third instar larva about every nine days.

There are 22 living species and a further 14 subspecies of tsetse fly and these are commonly arranged into three groups (Table 9.4): *fusca*, *palpalis* and *morsitans*. These groups are sometimes accorded the status of subgenera, when they are termed *Austenina*, *Nemorhina* and *Glossina*, respectively. Tsetse flies are now restricted to sub-Sahelian Africa (*c.* 4°N to 29°S) although fossil specimens have been found in the Oligocene shales of Florissant, Colorado, North America (Cockerell 1918) suggesting that in the past they had a more extensive range into the Nearctic. Tsetse flies currently compass more than 11 000 km^2 of Africa, being restricted in the north by the aridity of the deserts and in the south by low seasonal temperatures, although they are not evenly distributed within this range (Fig. 9.27).

The three tsetse groups are each closely associated with particular vegetation types. Eleven of the twelve species in the *fusca* group are forest species; some members (e.g. *Glossina nigrofusca*) are restricted to lowland rain forest, others (e.g. *G. fusca* and *G. brevipalpis*) are restricted to the edge of the rain forest and/or isolated areas of forest in the savanna, while one member, *G. longipennis*, is found in arid habitats. The *palpalis* group occurs in lowland forest, but many (e.g. *G. palpalis*) also extend into the drier savannahs along the woodlands of river banks and lake shores. Members of the *morsitans* group are restricted to woodlands, scrub and thicket in the savannah. Because of the relative intolerance of tsetse flies to low temperature (adults are inactive below *c.* 16°C), they are not found in highland areas (above *c.* 1300–1800 m depending on latitude).

Male flies complete spermatogenesis before emergence from the puparium, but need several blood meals before they become fully fertile at about seven days post-emergence (Langley 1977). In contrast, the females become sexually receptive within a day of emergence. Because of the low density of tsetse fly populations mating is most likely to occur when flies congregate around vertebrates to feed. The non-feeding, mature males which constitute the 'following swarm' found around moving hosts (Bursell 1961) are probably there to mate with incoming, hungry, particularly virgin females; the male locates the female visually.

235

Table 9.4 Characteristics of the three tsetse fly (*Glossina*) groups: *fusca, palpalis* and *morsitans*. (Information from several sources, notably Jordan 1986 and Weitz 1963.)

Group	Species	Subspecies	Habitat
fusca	G. *haningtoni* G. *nashi* G. *tabaniformis* G. *van hoofi* G. *severini* G. *nigrofusca*		lowland rainforest species
		nigrofusca *hopkinsi*	
	G. *fusca*	*fusca* *congolensis*	edge of the rainforest, forest areas outside the lowland forest, forest areas along water courses
	G. *fuscipleuris* G. *medicorum* G. *schwetzi*		
	G. *brevipalpis* G. *longipennis*		forest islands, often along watercourses arid habitats
morsitans	G. *morsitans*		
		morsitans *submorsitans* *centralis*	savannah woodlands, in the drier areas restricted to mesophytic vegetation around waterways
	G. *swynnertoni* G. *longipalpis*		
	G. *pallidipes*		
	G. *austeni*		
palpalis	G. *palpalis*		
		palpalis *gambiensis*	lowland rain forest often associated with its waterways. Extending its range into savannah-following waterways
	G. *fuscipes*	*fuscipes*	
		martinii *quanzensis*	
	G. *pallicera*	*pallicera* *newsteadi*	rain forest
	G. *caliginea*		coastal mangrove swamp and rain forest
	G. *tachinoides*		along water courses in savannahs

236

Table 9.4 (continued)

Geographical distribution	Vector status	Major hosts
W. and C. Africa		river hog, porcupine, ox, bushbuck
W. and C. Africa and isolated areas of E. Africa		bushbuck, river hog, aardvark
E. Africa E. Africa	locally important	bushpig, hippo, buffalo, bushbuck, rhino, elephant, buffalo, giraffe, ostrich
extensively in E. and Southern Africa	economically the most important vector	warthog, kudu, buffalo
Mozambique/Zimbabwe up to Tanzania		warthog, ox, man, buffalo, kudu
Ethiopia across to Senegal Botswana/Angola up to Southern Uganda		warthog, man, bushbuck, buffalo
N. Tanzania, S. Kenya Guinea to Cameroon	locally important	warthog, buffalo, giraffe, rhino bushbuck, buffalo
Mozambique up to Ethiopia	major economic importance	bushbuck, warthog, bushpig, buffalo
East Coast from Mozambique to Somalia	locally important	bushpig, ox, duiker
extensively in W. and C. Africa Benin down to Angola	important medically	
Sierra Leone to N. Benin		man, reptiles, bushbuck, ox
extensively in W. and C. Africa extensively in north zone of lowland forest	important medically	
in south-east zone of lowland forest		
in south-west zone of lowland forest		
forest areas of W. Africa		
W. African mangrove and adjoining forest		
W. Africa and pockets as far east as Ethiopia	important medically	man, ox, porcupine

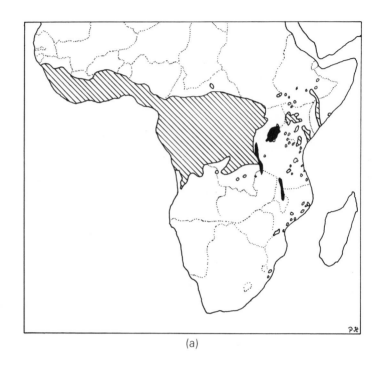

(a)

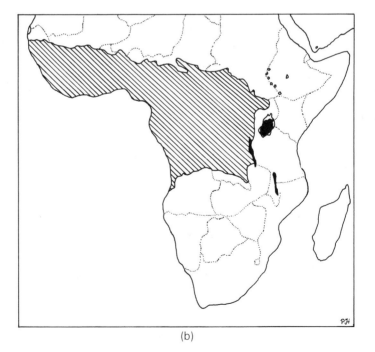

(b)

238

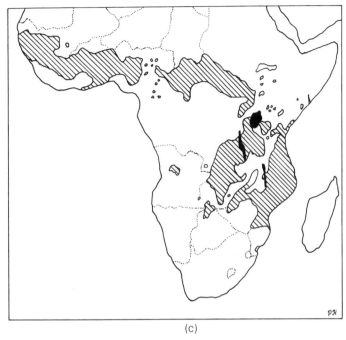

(c)

Figure 9.27 Distribution maps for the three major groups of tsetse in Africa: (a) *Glossina fusca* group, (b) *Glossina palpalis* group, (c) *Glossina morsitans* group. (Redrawn from Jordan 1986.)

Mating requires the presence of a contact sex pheromone which is present in the cuticular waxes of the female (Langley *et al*. 1975). After mating the sperm migrates into the female's spermathecae where it remains viable for the rest of her life. Despite laboratory evidence that females are willing to mate more than once (Jordan 1958), it seems that this rarely occurs in the field (Jordan 1986). Mating for more than 60 min is required to trigger the generation of hormones in the female, which stimulate ovulation (Chaudhury & Dhadialla 1976). The first ovulation occurs about nine days post-emergence and subsequently at nine- to ten-day intervals. Because female tsetse only produce a single offspring during each reproductive cycle the female must be relatively long-lived to produce sufficient offspring for the perpetuation of the species. Many studies have determined the age of flies in the field from the structure of the ovary. Every female has two ovaries, each with two polytrophic ovarioles which ovulate in turn in a predictable order. At each ovulation structural changes occur in the ovarioles which can be recognized and quantified on dissection, allowing the estimation of the physiological age of field-caught female flies (Saunders 1962, Challier 1965). Using such techniques it is estimated that females live in the field for an average of

239

100 days and occasionally up to 200 days. Until recently, no accurate technique existed for determining the age of males. The best estimates suggested they live only a few weeks in the wild. The recent introduction of a new biochemical technique for determining fly age will permit easier and more accurate estimates of the lifespan (Lehane & Mail 1985, Lehane & Hargrove 1988, Langley *et al.* 1988).

About four days after fertilization the egg hatches in the female's uterus, where it is nourished by secretions (milk) from the mother's uterine (milk) gland. The larva moults about one, two and a half, and five days after hatching from the egg and the resulting third stage larva is deposited by the female about nine days after ovulation. The larva has been so well nourished *in utero* that at birth it weighs about the same as the unencumbered female fly. It is stubby in shape, white, with two prominent, black polypneustic respiratory lobes on its posterior end. Larval deposition occurs in daylight, probably in the afternoon. Larvae are generally deposited in dense shade on a substrate which is both dry enough and loose enough for larval penetration (Buxton 1955). The free larva does not feed, is negatively phototactic and positively thigmotactic, and rapidly buries itself in the substrate. Within five hours of deposition it pupariates and turns black. The final larval ecdysis and pupation occurs within the puparial case about four days later, with no visible change in the appearance of the puparium. About a month after deposition the developed fly emerges from the puparium. Females normally emerge one or two days before the males. The emerging fly forces the cap from the puparium using its ptilinum and burrows up through the soil. It hardens off and is capable of flight about an hour after emergence. Over about the next nine days the fly's cuticle continues to develop, thicken and harden and the fly increases the mass of its thoracic flight musculature. This interval of post-emergence development is called the teneral period.

Both male and female flies feed exclusively on blood. Like other insects feeding in this way, tsetse flies maintain symbiotic micro-organisms to supplement their nutrition. These symbionts are held intracellularly in a specialized portion of the midgut called the mycetome. Because there is no feeding in the larval stage the adult fly emerges from the puparium with few remaining reserves and it is critical that an early blood meal is obtained (Bursell 1960). Tsetse flies then feed every two to three days and take blood meals which average two to three times their unfed body weight although, due to the size of the growing larva, the female takes progressively less blood as pregnancy proceeds. Tsetse flies are attracted to their food source by a mixture of olfactory and visual stimuli. There are five main patterns of host selection seen in *Glossina* spp. (Weitz 1963):

(1) Species with catholic tastes, such as *G. tachinoides, G. palpalis* and *G. fuscipes,* which feed on most available hosts.

(2) Species such as *G. austeni*, *G. tabaniformis* and *G. swynnertoni*, which feed largely on suids.
(3) Species such as *G. morsitans*, which feed on suids and bovids.
(4) Species such as *G. pallidipes*, *G. fusca* and *G. longipalpis*, which feed largely on bovids.
(5) Species such as *G. longipennis* and *G. brevipalpis*, which feed on other hosts such as rhinoceros and hippopotamus, respectively.

Disease problems only arise when the fly's host choice includes man or his domesticated animals; one of the reasons why the vector status of the 22 species varies considerably (Table 9.4) is because the degree of overlap varies among fly species.

Tsetse flies are unusual in using proline as an energy source for flight (see Section 6.6). One consequence is they spend less then an hour a day flying (Bursell & Taylor 1980). Resting sites are species-specific and an understanding of them is important in the planning of insecticide-based control campaigns (Davies 1964). During the dry season flies tend to be restricted to certain primary foci; appreciation of this fact permits the use of reduced quantities of insecticide in their control. Similarly, while *G. morsitans* may rest at heights of up to 12 m, *G. pallidipes* in the same habitat is rarely found resting more than 3 m above the ground. Clearly, appreciation of this difference means great savings can be made in the insecticide quantities used for treatments solely for *G. pallidipes* control. Day resting sites are usually the woody parts of the vegetation and the higher the temperature the nearer the ground the fly tends to be. At night the fly moves to the upper surfaces of leaves (Jordan 1986).

9.5.10 Hippoboscidae

The Hippoboscidae are cyclorrhaphous flies which are exclusively ecto-parasitic in the adult stage. They are often included in the artificial grouping, the Pupipara. At present about 200 species are recognized, but our knowledge of the group is increasing rapidly. The flies are locally known as keds or forest flies. The best known insect in the group is the sheep ked, *Melophagus ovinus*, which is found on sheep throughout the world (Fig. 9.28). They are irritating to the sheep whose scratching, rubbing and biting damages the wool which also becomes badly marked with the ked's faeces. The feeding activities of the keds also cause damage to the sheep's skin known as 'cockle' which may make the hide unsaleable. Heavy infestations cause loss of condition in the sheep and can cause anaemia. Various species of hippoboscid will attack man and the bite of the pigeon fly, *Pseudolynchia canariensis*, is said to be painful. Apart from the sheep ked, hippoboscids are of little importance to man.

About 75% of hippoboscids infest birds (the pigeon being the only

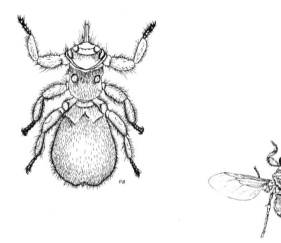

Figure 9.28 Adult of the sheep ked, *Melophagus ovinus* (left), and the winged hippoboscid, *Hippobosca equina* (right).

domesticated species afflicted), the rest being found on mammals, particularly bovids and cervids. Hippoboscids transmit several relatively non-pathogenic parasites to these hosts. For example, *Trypanosoma melophagium* is transmitted to sheep by the sheep ked *T. theileri*, to cattle by *Hippobosca* spp., and the filarial worm *Dipetalonema dracunculoides* to cats and dogs by *H. longipennis*.

Adult keds are highly adapted for life on the surface of the host animal. They are about 5–10 mm long leathery insects; most have a single pair of rather tough wing membranes, all of which act as protection against abrasion by the host covering. Further protection is afforded by the partial sinking of the head into the thorax, the placing of the antennae in deep grooves, and having partially retractable mouthparts which are further protected by the rigid palpi. Like many other ectoparasites keds are dorsoventrally flattened. Flattening may well allow increased freedom of movement among hair or feathers and allows adpression against the host or its covering. Both help the insect to evade the host's grooming activities. The legs are stout and bear paired claws in mammal-infesting forms and toothed claws in bird-infesting forms, which are used to cling to the covering of the host. Adult hippoboscids present a spectrum of forms from permanently winged to wingless. About three quarters of all Hippoboscidae are, like *Hippobosca* spp., winged throughout life (Fig. 9.28). *Allobosca* spp. lose only the wing tips on landing on a host, but in *Lipoptena* spp. the wings are lost at their base by abrasion, breakage or shedding once the host is located. *Melophagus* spp. do not have functional wings.

Female hippoboscids are viviparous and produce a single larva which is retained and nourished in the common oviduct (uterus) of the female until it is ready to pupate. The mature larvae are deposited in a variety of species-specific sites. The sheep ked, *M. ovinus*, is unique in that the deposited larva and pupa are attached to the fleece of the sheep. In *Lipoptena* spp. the pupa fall at random from the host. In *H. equina* the larva is commonly deposited in humus under bracken. Maturation of the larva in the female takes about a week, hardening and darkening of the deposited larva about six hours. Pupa produced in the summer will take about a month to develop. In temperate countries flies overwinter in the pupal stage and peak numbers of flies occur in the summer months. Sheep keds are unusual in that peak numbers occur in the winter and it is thought this is due to the physical removal of keds at shearing, the development of resistance among ewes, and possibly the adverse effects of temperature increases in the fleece in the summer months. Adult flies emerging at a distance from the host, fly immediately to the host; once on the host hippoboscids are usually difficult to dislodge. Although in some species mating can occur on the wing, most species mate on the host. The first mating usually occurs soon after the female's emergence, but repeated matings are common with subsequent matings occurring after larval deposition. Both sexes feed exclusively on blood. Once on the host *Hippobosca* spp. commonly feed several times a day whereas the sheep ked only feeds once every 36 h. In common with other insects in which all the life stages are dependent solely on blood as the nutrient source, hippoboscids have symbionts. These are housed in a mycetome on the intestine and are transferred to the offspring in the nutrients provided by the mother for her intra-uterine larva. As might be expected from the low fecundity of these flies, the adults are relatively long-lived, adult sheep keds living for about five months and *Hippobosca* spp. living for six to ten weeks.

9.5.11. Streblidae

The Streblidae are a relatively little-studied family of cyclorrhaphous flies in which the adults are exclusively ectoparasitic on bats. They are mainly found on colonial bats and many are strongly host-specific. Along with the Nycteribiidae, which are also exclusively ectoparasitic on bats, they are known as bat flies. They are often grouped with the Hippoboscidae and Nycteribiidae into the artificial grouping, the Pupipara. Streblids are largely a New World group found in the tropics and subtropics where winter temperatures do not fall below 10°C. Although they have occasionally been recorded as biting man they are of no medical or economic importance.

Streblids vary in length from about 1 mm up to about 6 mm (Fig. 9.29).

Like all ectoparasites streblids are exposed to the abrasive outer covering of the host. As a protective measure they are leathery and hold the appendages, such as the antennae, in grooves or pits. In common with many other ectoparasites their bodies are flattened, allowing increased freedom of movement through the host's covering and allowing adpression against the host, both of which help the fly avoid the host's grooming activities. Although most streblids are dorsoventrally flattened, interestingly the New World genus *Nycterophilia* are, like fleas, laterally compressed. Streblids show a range of forms from fully winged during adult life (about 80%), through caducous, to apterous. Some streblids display a further modification in being able to fold away their wings beneath protective setae on the abdomen, which minimizes wing abrasion when the fly is on the host. Wing form correlates with leg form; fully winged species have well-developed, rather long legs; species with reduced or no wings have shorter legs which are modified in various ways for clinging to, or parting, the host's hair.

Both sexes of streblid feed on blood. It is probable that the pelage-dwelling forms such as *Megistopoda aranea*, which tend to remain more or less permanently on the host, feed at least once a day. Volant forms such as *Trichobius yunkeri*, which often leave the host, probably feed less often. Newly emerged females may mate on the host or, in forms like *T. major* which commonly leave the host, it may occur elsewhere in the roost. In common with other Pupipara, multiple matings are usual with subsequent matings occurring after larval deposition. Unlike other Pupipara, it is believed that each mating is necessary for the production of further offspring. Streblids are viviparous and there is no free-feeding larval stage, so that all life stages are dependent upon blood for their nutrition. In consequence, it is virtually certain that streblids, like the tsetse flies, triatomine bugs and other insects in this position, have symbiotic micro-organisms providing nutritional supplements; however the site of the mycetome is unknown. The mature larvae are deposited by the female and immediately pupate. In *Trichobius* spp. the larva pupates within the mother, and it is the pupa not the larva which is deposited.

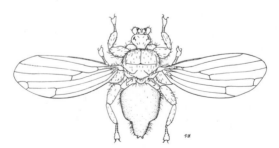

Figure 9.29 Adult streblid, *Trichobius lonchophylla*. (Redrawn from Marshall 1981.)

Most streblid offspring are deposited in bat roosts; more active streblids may leave the host to deposit their offspring on particular surfaces or within cracks or crevices in the walls of the roost, in other less active species the larva is broadcast. Newly emerged flies seek out a host which is most commonly a female or juvenile bat as these are more colonial than the mature males.

Female *Ascodipteron* are neosomic (Audy *et al.* 1972); on reaching a suitable host they migrate to a species-specific site, shed their wings and legs, and using their very well-developed mouthparts cut their way into the subdermal tissues of the bat so that only the end of the abdomen is protruding. They become anchored in this position by their mouthparts. The abdomen swells greatly overgrowing the rest of the body so the head and thorax become embedded within it. In this position the fly feeds and matures its offspring which, as third stage larvae, fall from the bat to the floor of the roost.

9.5.12 Nycteribiidae

The Nycteribiidae are a small family of relatively little-studied cyclorrhaphous flies in which the adults are exclusively ectoparasitic on bats. Along with the streblids, which are also exclusively ectoparasitic on bats, they are known as bat flies. They are often grouped with the Hippoboscidae and Streblidae into the artificial grouping, the Pupipara. The various species of bat fly show a high degree of host specificity. They are largely an Old World group of insects, found in temperate as well as tropical areas and are of no medical or economic importance.

Nycteribiids are highly modified for their ectoparasitic existence, being rather leathery, dorsoventrally flattened, wingless insects in which the head is held protectively in a groove on the dorsal surface of the thorax (Fig. 9.30). Nycteribiids have body combs which possibly protect delicate areas of the body from abrasion and/or which prevent removal of the insect during host grooming (see Section 7.2). Nycteribiids can be divided into two main types. The first group are small, like many *Nycteribii* spp., and tend to remain in the fur of the host's trunk, in which they are capable of 'swimming'. The second type is exemplified by several species of *Basilia* and *Penicillidia*; they tend to be larger and to live on the surface of their host's hair into which they can escape if disturbed. Perhaps not surprisingly these larger forms tend to congregate in areas protected from the host's grooming activities such as between the shoulder blades, under the chin or in the axilla of roosting hosts or, in flying hosts, in the tail region. The legs of nycteribiids are elongate and bear claws which allow the large forms to move rapidly over the host's surface. These large forms will often leave the host.

Adults of both sexes feed exclusively on blood. Nycteribiids are

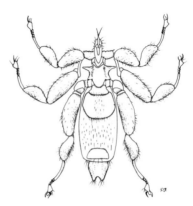

Figure 9.30 An adult nycteribiid. (Redrawn from Marshall 1981.)

viviparous and there is no free-feeding larval stage, so that all life stages are dependent upon blood for their nutrition. In common with other insects in a similar position (e.g. tsetse flies and triatomine bugs) nycteribiids have symbiotic micro-organisms providing nutritional supplements. The mycetome housing these symbionts is sited in the dorsal part of the abdomen. Newly emerged females may mate before taking a blood meal. Multiple matings are common with subsequent matings occurring at the time of larval deposition. From egg production to larval deposition takes about nine days in *Basilia hispida*. Nycteribiids leave the host to deposit their offspring in a suitable crack or crevice, often on a vertical surface in the bat roost. The pupal stage lasts about a month. Newly emerged flies seek out a new host, most commonly a female or a juvenile bat as these are more colonial in their habits than the mature males. Nycteribiids produce between three and 16 offspring in a lifetime. Nycteribiids are associated with hibernating bats on which, although they continue to feed, reproduction is greatly slowed or ceases altogether, so population size falls to its lowest levels in the winter months.

9.6 OTHER GROUPS

In addition to the insects already discussed there are a few instances of blood-sucking in other orders. In the main, these are probably reports of rare occurrences such as the account of blood feeding by a vespid wasp (Phipps 1974). In other cases, however, it is clear that blood-feeding is a regular occurrence.

In the order Lepidoptera there is one noctuid, *Calpe* (= *Calyptra*) *eustrigata*, which is known to feed on blood which it has obtained by piercing the skin of its forest-dwelling mammalian hosts (Banziger 1971).

This moth, which is found in South-east Asia, pierces the host's skin using its straight, lance-like proboscis. This insect probably evolved from species specialized for piercing fruits. Other geometrid and pyralid moths will feed on blood oozing from wounds, or from drops of blood released from the anus of feeding mosquitoes.

Species in which adults are ectoparasitic on mammalian hosts are found in the coleopteran groups Leptininae, Quediini, Amblyopinini and Languriidae. The best-known examples are in the four genera *Leptinus, Leptinillus, Silphosyllus* and *Platypsyllus*. These insects appear to feed mainly on ectodermis and its products and its seems that blood is not a regular component of their diet. *Platypsyllus castoris* is unusual in that the larvae also live on their beaver hosts, and when present in sufficient numbers can cause superficial abrasions which permit blood feeding. These beetles appear to be progressing along the well-trodden evolutionary highway, from carnivore–detritovore nest inhabiters to phoretic epidermis feeders – which may eventually lead to fully fledged blood-feeding (see Section 2.1).

References

Aboul-Nasr, A. E. 1967. On the behaviour and sensory physiology of the bed bug (*Cimex lectularius*). I. Temperature reactions. *Bulletin Societe Entomologique d'Egypte* **51**, 43–54.

Adler, S. & M. Ber 1941. The transmission of *Leishmania tropica* by the bite of *Phlebotomus papatasi*. *Indian Journal of Medical Research* **29**, 803–9.

Ahmadi, A. & G. A. H. McClelland 1985. Mosquito-mediated attraction of female mosquitoes to a host. *Physiological Entomology* **10**, 251–5.

Aikawa, M., M. Suzuki & Y. Gutierez 1980. In *Malaria*, J. P. Kreier (ed.). Vol. II. New York: Academic Press.

Akov, S. 1965. Inhibition of blood digestion and oocyte growth in *Aedes aegypti* by 5-fluoruracil. *Biological Bulletin Marine Biological Laboratory (Woods Hole)* **129**, 439–53.

Albritton, A. 1952. *Standard values in blood*. Saunders.

Alekseev, A., S. Rasnitsyn & L. Vitlin 1977. On group behaviour of females of blood-sucking mosquitoes. Communication I. Discovery of the 'effect' of invitation. [In Russian]. *Meditsinskaya Parazitologiya i Parazitarnye Bolezni* **46**, 23–4.

Alekseyev, A. N., I. T. Abdullayev, S. P. Rasnitsyn & M. Martsinovskiy 1984. Comparison of flight capability of *Aedes aegypti* that are infected and not infected with plasmodia. *Meditsinskaya Parazitologiya i Parazitarnye Bolezni* **1**, 11–13.

Allan, S. A. & J. G. Stoffolano 1986a. Effects of background contrast on visual attraction and orientation of *Tabanus nigrovittatus* (Diptera: Tabanidae). *Environmental Entomology* **15**, 689–94.

Allan, S. A. & J. G. Stoffolano 1986b. The effects of hue and intensity on visual attraction of adult *Tabanus nigrovittatus* (Diptera: Tabanidae). *Journal of Medical Entomology* **23**, 83–91.

Allan, S. A., J. F. Day & J. D. Edman 1987. Visual ecology of biting flies. *Annual Review Entomology* **32**, 297–316.

Altman, P. L. & D. S. Dittmer 1971. Blood and other body fluids. In *Respiration and circulation*. Federation of American Societies of Experimental Biology, Bethesda, Ma.

Altner, H. & R. Loftus 1985. Ultrastructure and function of insect thermo and hygroreceptors. *Annual Review of Entomology* **30**, 273–95.

Amin, O. M. & M. E. Wagner 1983. Further notes on the function of pronotal combs in fleas (Siphonaptera). *Annals of the Entomological Society of America* **76**, 232–4.

Anderson, J. R. & S. C. Ayala 1968. Trypanosome transmitted by a *Phlebotomus*: first report from the Americas. *Science* **161**, 1023–5.

Anderson, J. R. & J. B. Hoy 1972. Relationship between host attack rates and CO_2 baited insect flight trap catches of certain *Symphoromyia* sp. *Journal of Medical Entomology* **9**, 373–92.

Anderson, R. C. 1957. The life cycles of dipetalonematid nematodes (Filarioidea,

Dipetalonematidae): the problem of their evolution. *Journal of Helminthology* **31**, 203–24

Anderson, R. M. 1986. Genetic variability in resistance to parasitic invasion: population implications for invertebrate host species. In *Immune mechanisms in invertebrate vectors*, A. M. Lackie, (ed.) Zoological Society of London Symposia, no. 56. Clarendon Press; Oxford University Press.

Anderson, R. M. & R. M. May 1982. Coevolution of hosts and parasites. *Parasitology* **85**, 411–26.

Anderson, R. S. & M. L. Cook 1979. Induction of lysozyme-like activity in the hemolymph and hemocytes of an insect, *Spodoptera eridania*. *Journal of Invertebrate Pathology* **33**, 197–203.

Anderson, R. S., B. Holmes & R. A. Good 1973. *In vitro* bactericidal capacity of *Blaberus craniifer* hemocytes. *Journal of Invertebrate Pathology* **22**, 127–35.

Aschner, M. 1932. Experimentelle unter suchungen über die Symbiose der Kleiderlaus. *Naturwissenschaften* **20**, 501–505.

Aschner, M. 1934. Studies on the symbiosis of the body louse. I. Elimination of the symbionts by centrifugation of the eggs. *Parasitology* **26**, 309–14.

Aschner, M. 1946. The symbiosis of *Eucampsipoda aegyptica* Mcq. *Bulletin Societe Entomologique d'Egypte* **30**, 1–6.

Audy, J. R., F. J. Radovsky & P. H. Vercammen-Grandjean 1972. Neosomy: radical intrastadial metamorphosis associated with arthropod symbioses. *Journal of Medical Entomology* **9**, 487–94.

Ayala, S. C. & D. Lee 1970. Saurian malaria: development of sporozoites in two species of phlebotomine sandflies. *Science* **167**, 891–2.

Bacot, A. M. & C. J. Martin 1914. Observations on the mechanism of the transmission of the plague by fleas. *Journal of Hygiene*, **13** (Plague supplement III), 423–39.

Bailey, L. 1952. The action of the proventriculus of the worker honeybee. *Journal of Experimental Biology* **29**, 310–27.

Baker, J. R. 1965. The evolution of parasitic protozoa. In *Symposium of the British Society for Parasitology*, no. 3. A. E. R. Taylor, (ed.). Oxford: Blackwell.

Baker, J. R. & D. H. H. Robertson 1957. An experiment on the infectivity to *Glossina morsitans* of a strain of *Trypanosoma rhodesiense* and of a strain of *Trypanosoma brucei*, with some observations on the longevity of infected flies. *Annals of Tropical Medicine and Parasitology* **51**, 121–35.

Baker, R. C. 1986. Pheromone-modulated movement of flying moths. In *Mechanisms in insect olfaction*, T. L. Payne, M. C. Birch & C. E. J. Kennedy (eds), Oxford: Clarendon Press.

Ball, G. H. 1943. Parasitism and evolution. *American Naturalist* **77**, 345–364.

Ball, G. H. 1960. Some considerations regarding the Sporozoa. *Journal of Protozoology* **7**, 1–6.

Banziger, H. 1971. Blood-sucking moths of Malaya. *Fauna* **1**, 5–16.

Baranov, N. 1935. New information on the Golubatz fly, *S. columbaczense*. *Review of Applied Entomology B* **23**, 275–6.

Barrera, A. 1966. Hallazgo de *Amblyopinus tiptoni* Barrera, 1966 en Costa Rica, A. C. (Col.: Staph.). *Acta Zoologica Mexicana* **8**, 1–3.

Barrera, A. & C. E. Machado-Allison 1965. Coleopteros ectoparasiticos de Mamiferos. *Ciencia Mexico* **23**, 201–8.

Bar-Zeev, M., H. I. Maibach & A. A. Khan 1977. Studies on the attraction of *Aedes aegypti* (Diptera: Culicidae) to man. *Journal of Medical Entomology* **14**, 113–20.

Bates, M. 1949. *The natural history of mosquitoes*. New York: Macmillan.

Baudisch, K. 1958. Beitrage zur zytologie und embryologie einiger Insektensymbiosen. *Zeitschrift für Morphologie und Oekologie der Tiere* **47**, 436–88.

Beach, R., G. Kiilu & J. Leeuwenburg 1985. Modification of sandfly biting behaviour by *Leishmania* leads to increased parasite transmission. *American Journal of Tropical Medicine and Hygiene* **34**, 279–83.

Beckett, E. B. & W. W. Macdonald 1971. The development and survival of subperiodic *Brugia malayi* and *B. pahangi* larvae in a selected strain of *Aedes aegypti*. *Transactions of the Royal Society for Tropical Medicine and Hygiene* **65**, 339–46.

Beklemishev, V. N. 1957. Some general questions on the biology of bloodsucking lower flies. *Meditsinskaya Parazitologiya i Parazitarnye Bolezni* **5**, 562–6.

Bell, J. F., S. J. Stewart & W. A. Nelson 1982. Transplant of acquired resistance to *Polyplax serrata* (Phthiraptera: Hoplopleuridae) in skin allografts to athymic mice. *Journal of Medical Entomology* **19**, 164–8.

Belzer, W. R. 1978a. Factors conducive to increased protein feeding by the blowfly *Phormia regina*. *Physiological Entomology* **3**, 251–7.

Belzer, W. R. 1978b. Patterns of selective protein ingestion by the blowfly *Phormia regina*. *Physiological Entomology* **3**, 169–257.

Belzer, W. R. 1978c. Recurrent nerve inhibition of protein feeding in the blowfly *Phormia regina*. *Physiological Entomology* **3**, 259–63.

Belzer, W. R. 1979. Abdominal stretch receptors in the regulation of protein ingestion by the black blowfly, *Phormia regina*. *Physiological Entomology* **4**, 7–14.

Bennet-Clark, H. C. 1962. Active control of the mechanical properties of insect cuticle. *Journal of Insect Physiology* **8**, 627–33.

Bennet-Clark, H. C. 1963. Negative pressures produced in the pharyngeal pump of the blood-sucking bug, *Rhodnius prolixus*. *Journal of Experimental Biology* **40**, 223–9.

Bennett, G. F., A. M. Fallis & A. G. Campbell 1972. The response of *Simulium* (*Eusimulium*) *euryadminiculum* Davies (Diptera: Simuliidae) to some olfactory and visual stimuli. *Canadian Journal of Zoology* **50**, 793–800.

Bequaert, J. C. 1953. The Hippoboscidae or house-flies (Diptera) of mammals and birds. Part 1. Structure, physiology and natural history. *Entomologia Americana* **32**, 1–209.

Bidlingmayer, W. L. & D. G. Hem 1979. Mosquito (Diptera: Culicidae) flight behaviour near conspicuous objects. *Bulletin of Entomological Research* **69**, 691–700.

Bidlingmayer, W. L. & D. G. Hem 1980. The range of visual attraction and the effect of competitive visual attractants upon mosquito (Diptera: Culicidae) flight. *Bulletin of Entomological Research* **70**, 321–42.

Billingsley, P. B. 1990. The midgut ultrastructure of haematophagous arthropods. *Annual Review of Entomology* **35**, 219–48.

Billingsley, P. F. & A. E. R. Downe 1983. Ultrastructural changes in posterior midgut cells associated with blood-feeding in adult female *Rhodnius prolixus* Stal (Hemiptera: Reduviidae). *Canadian Journal of Zoology* **61**, 1175–87.

Billingsley, P. F. & A. E. R. Downe 1985. Cellular localisation of aminopeptidase in the midgut of *Rhodnius prolixus* Stal (Hemiptera: Reduviidae) during blood digestion. *Cell and Tissue Research* **241**, 421–8.

Billingsley, P. F. & A. E. R. Downe 1986. The surface morphology of the midgut cells of *Rhodnius prolixus* Stal (Hemiptera: Reduviidae) during blood digestion. *Acta Tropica* **43**, 355–66.

Billingsley, P. F. & A. E. R. Downe 1988. Ultrastructural localisation of cathepsin B in the midgut of *Rhodnius prolixus* Stal (Hemiptera: Reduviidae) during blood digestion. *International Journal of Insect Morphology and Embryology* **17**, 295–302.

Billingsley, P. F. & A. E. R. Downe 1989. Changes in the anterior midgut cells of adult female *Rhodnius prolixus* Stal (Hemiptera: Reduviidae) after feeding. *Journal of Medical Entomology* **26**, 104–8.

Bitkowska, E., T. H. Dzbenski, M. Szadziewska & Z. Wegner 1982. Inhibition of xenograft rejection reaction in the bug *Triatoma infestans* during infection with a protozoan, *Trypanosoma cruzi*. *Journal of Invertebrate Pathology* **40**, 186–9.

Bos, H. J. & J. J. Laarman 1975. Guinea pig lysine, cadaverine, and estradiol as attractants for the malaria mosquito *Anopheles stephensi*. *Entomologia Experimentalis et Applicata* **18**, 161–72.

Bracken, G. K. & A. J. Thorsteinson 1965. The orientation behaviour of horse flies and deer flies (Tabanidae: Diptera). IV. The influence of some physical modifications of visual decoys on orientation of horse flies. *Entomologia Experimentalis et Applicata* **8**, 314–18.

Bracken, G. K., W. Hanec & A.J. Thorsteinson 1962. The orientation of horseflies and deerflies (Tabanidae: Diptera). II. The role of some visual factors in the attractiveness of decoy silhouettes. *Canadian Journal of Zoology* **40**, 685–95.

Bradbury, W. C. & G. F. Bennett 1974. Behaviour of adult Simuliidae (Diptera). I. Response to colour and shape. *Canadian Journal of Zoology* **52**, 251–9.

Bradley, G. H. 1935. Notes on the southern buffalo gnat *Eusimulium pecuarum* (Riley) (Diptera: Simuliidae). *Proceedings of the Entomological Society of Washington* **37**, 60–4.

Bradshaw, H. 1989. Onchocerciasis and the Mectizan donation programme. *Parasitology Today* **5**, 63–4.

Bradshaw, W. E. 1980. Blood-feeding and capacity for increase in the pitcher plant mosquito, *Wyeomyia smithii*. *Environmental Entomology* **9**, 86–9.

Bradshaw, W. E. 1983. Interaction between the mosquito, *Wyeomyia smithii*, the midge *Metriocnemus knabi* and their carnivorous host *Sarracenia purpurea*. In *Phytotelmata: terrestrial plants as hosts of aquatic insect communities*, J. H. Frank & L. P. Lounibos (eds). London: Plexus.

Bradshaw, W. E. & C. M. Holzapfel 1983. Life cycles strategies in *Wyeomyia smithii*: seasonal and geographic adaptations. In *Diapause and life cycle strategies in insects*, V. K. Brown, & I. Hodek (eds) The Hague: Junk.

Brady, J. 1970. Characteristics of spontaneous activity in tsetse flies. *Nature* **228**, 286–7.

Brady, J. 1972. The visual responsiveness of the tsetse fly *Glossina morsitans* Westw. (Glossinidae) to moving objects: the effects of hunger, sex, host odour and stimulus characteristics. *Bulletin of Entomological Research* **62**, 257–79.

Brady, J. 1973. Changes in the probing responsiveness of starving tsetse flies (*Glossina morsitans* Westw.) (Diptera, Glossinidae). *Bulletin of Entomological Research* **63**, 247–55.

Brady, J. 1975. Circadian changes in central excitability – the origin of behavioural rhythms in tsetse flies and other animals? *Journal of Entomology (A)* **50** (2), 79–95.

Brady, J. & A. Shereni 1988. Landing responses of the tsetse fly *Glossina morsitans morsitans* Weidemann and the stablefly *Stomoxys calcitrans* (L.) (Diptera: Glossinidae & Muscidae) to black and white patterns: a laboratory study. *Bulletin of Entomological Research* **78**, 301–11.

Bray, R. S. 1963. The exo-erythrocytic phase of malaria parasites. *International Review of Tropical Medicine* **2**, 41.

Bray, R. S., A. W. R. McCrae & M. E. Smalley 1976. Lack of a circadian rhythm in the ability of the gametocytes of *Plasmodium falciparum* to infect *Anopheles gambiae*. *International Journal of Parasitology* **6**, 399–401.

Brecher, G. & V. B. Wigglesworth 1944. The transmission of *Actinomyces rhodnii*

251

Erikson in *Rhodnius prolixus* Stal. (Hemiptera) as its influence on the growth of the host. *Parasitology* **35**, 220–4.

Breev, K. V. 1950. The behaviour of blood-sucking Diptera and warble flies when attacking reindeer and the responsive reactions of reindeer. [In Russian]. *Parasitologicheskii Sbornik* **12**, 167–89.

Brooke, M. de L. 1985. The effect of allopreening on tick burdens of moulting eudyptid penguins. *Auk* **102**, 893.

Browne, S. M. & G. F. Bennett 1980. Colour and shape as mediators of host-seeking responses of simuliids and tabanids (Diptera) in the Tantramar marshes, New Brunswick, Canada. *Journal of Medical Entomology* **17**, 58–62.

Browne, S. M. & G. F. Bennett, G. F. 1981. Response of mosquitoes (Diptera: Culicidae) to visual stimuli. *Journal of Medical Entomology* **18**, 505–21.

Bruce, D. 1895. *Preliminary report on the tsetse fly disease or nagana, in Zululand.* Durban: Bennett and Davis.

Buchner, P. 1965. *Endosymbiosis of animals with plant microorganisms.* New York: Wiley.

Bungener, W. & G. Muller 1976. Adharenz-phanomene bei *Trypanosoma congolense*. *Tropenmedizin und Parasitologie* **27**, 307–71.

Burkot, T. R. 1988. Non-random host selection by anopheline mosquitoes. *Parasitology Today* **4**, 156–62.

Bursell, E. 1960. The effect of temperature on the consumption of fat during pupal development in *Glossina*. *Bulletin of Entomological Research* **51**, 583–98.

Bursell, E. 1961. The behaviour of tsetse flies (*Glossina swynnertoni* Austen) in relation to problems of sampling. *Proceedings of the Royal Entomological Society of London A* **36**, 9–20.

Bursell, E. 1975. Substrates of oxidative metabolism in dipteran flight muscle. *Comparative Physiology and Biochemistry* **52B**, 235–8.

Bursell, E. 1977. Synthesis of proline by fat body of the tsetse fly (*Glossina morsitans*). Metabolic pathways. *Insect Biochemistry* **7**, 427–34.

Bursell, E. 1981. The role of proline in energy metabolism. In *Energy metabolism in insects*, R. G. H. Downer, (ed.). New York: Plenum Press.

Bursell, E. 1984. Effects of host odour on the behaviour of tsetse. *Insect Science and its Applications* **5**, 345–9.

Bursell, E. 1987. The effect of wind-borne odours on the direction of flight in tsetse flies, *Glossina* spp. *Physiological Entomology* **12**, 149–56.

Bursell, E. & P. Taylor 1980. An energy budget for *Glossina* (Diptera: Glossinidae). *Bulletin of Entomological Research* **70**, 187–96.

Burton, R. 1860. *The lake regions of Central Africa.* Vols. 1 & 2. London: Longman.

Buxton, P. A. 1930. The biology of a blood-sucking bug, *Rhodnius prolixus*. *Transactions of the Royal Society for Tropical Medicine and Hygiene* **78**, 227–36.

Buxton, P. A. 1947. *The louse.* London: Edward Arnold.

Buxton, P. A. 1948. Experiments with lice and fleas. I. The baby mouse. *Parasitology* **39**, 119–24.

Buxton, P. A. 1955. *The natural history of tsetse flies.* London: H. K. Lewis.

Cavanaugh, D. C. 1971. Specific effect of temperature upon transmission of the plague bacillus by the oriental rat flea, *Xenopsylla cheopis*. *American Journal of Tropical Medicine and Hygiene* **20**, 264–73.

Challier, A. 1965. Amelioration de la methode de détermination de l'age physiologique des glossines. *Bulletin de la Sociétié Pathologie Exotique et de Ses Filiales* **58**, 250–9.

Challier, A. & C. Laveissiere 1973. Un nouveau piège pour la capture des glossines (*Glossina*: Diptera, Muscidae): description et essais sur le terrain. *Cahiers ORSTOM, Serie Entomologie Medicale et Parasitologie* **11**, 251–62.

Challier, A., M. Eyraud, A. Lafaye & C. Laveissiere 1977. Amelioration due rendement du piege biconique pour glossines (Diptera, Glossinidae) par l'emploi d'un cone inferieur bleu. *Cahiers ORSTOM, Serie Entomologie Medicale et Parasitologie* **15**, 283–6.

Chamberlain, R. W. & W. D. Sudia 1961. Mechanisms of transmission of viruses by mosquitoes. *Annual Review of Entomology* **6**, 371–90.

Chamberlain, R. W., R. K. Sikes, D. B. Nelson & W. D. Sudia 1954. Studies on the N. American arthropod-borne encephalitides. VI. Quantitative determinations of virus–vector relationships. *American Journal of Hygiene* **60**, 278–85.

Chang, Y. H. & C. L. Judson 1977. The role of isoleucine in differential egg production by the mosquito *Aedes aegypti* Linnaeus (Diptera: Culicidae) following feeding on human or guinea pig blood. *Comparative Physiology and Biochemistry* **57A**, 23–8.

Chapman, R. F. 1961. Some experiments to determine the methods used in host finding by tsetse flies, *Glossina medicorum*. *Bulletin of Entomological Research* **52**, 83–97.

Chapman, R. F. 1982. Chemoreception: the significance of receptor numbers. *Advances in Insect Physiology* **16**, 247–356.

Chaudhury, M. F. B. & T. S. Dhadialla 1976. Evidence of hormonal control of ovulation in tsetse flies. *Nature* **260**, 243–4.

Chen, C. C. & B. R. Laurence 1985. An *in vitro* study on the encapsulation of microfilariae. *Developmental and Comparative Immunology* **9**, 174.

Christensen, B. M. 1978. *Dirofilaria immitis*: effects on the longevity of *Aedes trivittatus*. *Experimental Parasitology* **44**, 116–23.

Christensen, B. M. & M. M. LaFond 1986. Parasite induced suppression of the immune response in *Aedes aegypti* by *Brugia pahangi*. *Journal of Parasitology* **72**, 216–9.

Christensen, B. M., K. F. Forton, M. M. LaFond & R. B. Grieve 1987. Surface changes on *Brugia pahangi* microfilariae and their association with immune evasion in *Aedes aegypti*. *Journal of Invertebrate Pathology* **49**, 14–18.

Chung, H. L., L. C. Feng & S. L. Feng 1951. Observations concerning the successful transmission of kala-azar in North China by the bites of naturally infected *Phlebotomus chinensis*. *Bulletin of the Peking Natural History Museum* **19**, 302–26.

Ciurea, I. & G. Dinulescu 1924. Ravages causes par la mouchede Goloubatz en Roumanie; ses attaques contre les animaux et contre l'homme. *Annals of Tropical Medicine and Parasitology* **18**, 323–42.

Clay, T. 1963. A new species of *Haematomyzus* Piaget (Phthiraptera, Insecta). *Proceedings of the Zoological Society of London* **141**, 153–61.

Clifford, C. M., J. F. Bell, G. J. Moore & C. Raymond 1967. Effects of limb disability on lousiness in mice. IV. Evidence of genetic factors in susceptibility to *Polyplax serrata*. *Experimental Parasitology* **20**, 56–67.

Coatney, G. R., W. E. Collins, W. McWilson, & P. G. Contacos 1971. *The primate malarias*. Bethesda, MD: US Department of Health Education and Welfare. National Institute of Allergy and Infectious Diseases.

Cockerell, T. D. A. 1918. New species of North American fossil beetles, cockroaches and tsetse flies. *Proceedings of the US National Museum* **54**, 301–11.

Colless, D. H. & W. T. Chellapah 1960. Effects of body weight and size of blood meal upon egg production in *Aedes aegypti* (Linnaeus) (Diptera, Culicidae). *Annals of Tropical Medicine and Parasitology* **54**, 475–482.

Collins, F. H., R. K. Sakai, K. D. Vernick, S. Paskewitz, D. C. Seeley, L. H. Miller,

W. E. Collins, C. C. Campbell, & R. W. Gwadz 1986. Genetic selection of a refractory strain of the malaria vector *Anopheles gambiae*. *Science* **234**, 607–10.

Coluzzi, M., A. Concetti & F. Ascoli 1982. Effect of cibarial armature of mosquitoes (Diptera: Culicidae) on blood-meal haemolysis. *Journal of Insect Physiology* **28**, 885–8.

Colvin, J., J. Brady & G. Gibson 1989. Visually guided, upwind turning behaviour of free-flying tsetse flies in odour-laden wind: a wind-tunnel study. *Physiological Entomology* **14**, 31–39.

Colyer, C. N. & C. O. Hammond 1968. *Flies of the British Isles*. London: Frederick Warne.

Compton-Knox, P. & K. L. Hayes 1972. Attraction of *Tabanus* spp. (Diptera: Tabanidae) to traps baited with carbon dioxide and other chemicals. *Environmental Entomology* **1**, 323–6.

Cornford, E. M., B. J. Freeman & A. J. MacInnis 1976. Physiological relationships and circadian periodicities in rodent trypanosomes. *Transactions of the Royal Society for Tropical Medicine and Hygiene* **70**, 238–43.

Croft, S. L., J. S. East & D. H. Molyneux 1982. Antitrypanosomal factor in the haemolymph of *Glossina*. *Acta Tropica* **39**, 293–302.

Cupp, E. W. & G. M. Stokes 1976. Feeding patterns of *Culex salinarius* Coquillett in Jefferson parish, Louisiana. *Mosquito News* **36**, 332–5.

Dalmat, H. T. 1952. Longevity and further flight range studies on the blackflies (Diptera: Simuliidae), with the use of dye markers. *Annals of the Entomological Society of America* **45**, 23–37.

Daniel, T. L. & J. G. Kingsolver 1983. Feeding strategy and the mechanics of blood sucking in insects. *Journal of Theoretical Biology* **105**, 661–72.

David, C. T. 1986. Mechanisms of directional flight in wind. In *Mechanisms in insect olfaction*, T. L. Payne, M. C. Birch & C. E. S. Kennedy (eds), Ch. 5, pp. 49–57.

David C. T., J. S. Kennedy, A. R. Ludlow, J. N. Perry & C. Wall 1982. A reappraisal of insect flight towards a distant point source of wind-borne odour. *Journal of Chemical Ecology* **8**, 1207–15.

Davidson, G. & C. C. Draper 1953. Field studies of some of the basic factors concerned in the transmission of malaria. *Transactions of the Royal Society for Tropical Medicine and Hygiene* **47**, 522–35.

Davies, H. 1964. The eradication of tsetse in the Chad river system of Northern Nigeria. *Journal of Applied Ecology* **1**, 387–403.

Davis, E. E. 1984. Regulation of sensitivity in the peripheral chemoreceptor systems for host-seeking behaviour by a haemolymph-borne factor in *Aedes aegypti*. *Journal of Insect Physiology* **30**, 179–83.

Davis, E. E. & P. G. Sokolove 1975. Temperature response of the antennal receptors in the mosquito, *Aedes aegypti*. *Journal of Comparative Physiology* **96**, 223–36.

Davis, E. E. & P. G. Sokolove 1976. Lactic acid sensitive receptors on the antennae of the mosquito, *Aedes aegypti*. *Journal of Comparative Physiology* **105**, 43–54.

Davis, E. E., D. A. Haggart & M. F. Bowen 1987. Receptors mediating host-seeking behaviour in mosquitoes and their regulation by endogenous hormones. *Insect Science and its Applications* **8**, 637–41.

Day, J. F. & J. D. Edman 1983. Malaria renders mice susceptible to mosquito feeding when gametocytes are most infective. *Journal of Parasitology* **69**, 163–70.

Day, J. F. & J. D. Edman 1984. The importance of disease-induced changes in mammalian body temperature to mosquito blood-feeding. *Comparative Physiology and Biochemistry* **77**, 447–52.

Daykin, P. N., F. E. Kellog & R. H. Wright 1965. Host finding and repulsion of *Aedes aegypti*. *Canadian Entomologist* **97**, 239–63.

De Azambuja, P., C. C. Freitas & F. S. Garcia 1986. Evidence and partial characterisation of an inducible antibacterial factor in the haemolymph of *Rhodnius prolixus*. *Journal of Insect Physiology* **32**, 807–12.

De Azambuja, P., J. A. Guimaraes & E. S. Garcia 1983. Haemolytic factor from the crop of *Rhodnius prolixus*: evidence and partial characterisation. *Journal of Insect Physiology* **29**, 833–7.

DeFoliart, G. R., P. R. Grimstad & D. M. Watts 1987. Advances in mosquito-borne arbovirus/vector research. *Annual Review of Entomology* **32**, 479–505.

Dei Cas, E., I. Maurois, E. Dutoit, *et al*. 1980a. Études sur les gametocytes des *Plasmodium* des mammifères: morphologie et infectivité des gametocytes de *Plasmodium inui*. *Cahiers ORSTOM, Série Entomologie Médicale et Parasitologie* **18**, 103–31.

Dei Cas, E., P. Maurois, I. Landau, *et al*. 1980b. Morphologie et infectivité des gametocytes de *Plasmodium inui*. *Annales de Parasitologie Humaine et Comparée* **55**, 621–33.

Dethier, V. G. 1954. Notes on the biting response of tsetse flies. *American Journal of Tropical Medicine and Hygiene* **3**, 160–71.

Dethier, V. G. 1976. *The hungry fly*. Cambridge, Mass.: Harvard University Press.

Detinova, T. S. 1962. *Age grouping methods in diptera of medical importance*. World Health Organisation Monograph, no. 47.

Dickerson, G. & M. M. J. Lavoipierre 1959. Studies on the methods of feeding blood-sucking arthropods. II. The method of feeding adopted by the bed-bug (*Cimex lectularius*) when obtaining a blood meal from the mammalian host. *Annals of Tropical Medicine and Parasitology* **53**, 347–57.

Dindonis, L. L. & J. R. Miller 1980. Host-finding behaviour of onion flies, *Hylemya antiqua*. *Environmental Entomology* **9**, 769–72.

Distelmans, W., F. D'Haeseleer, L. Kaufman & P. Rousseeuw 1982. The suscepti-bility of *Glossina palpalis palpalis* at different ages to infection with *Trypanosoma congolense*. *Annales Societé Belge de Médécine Tropicale* **62**, 41–8.

Dobson, A. P. 1988. Parasite-induced changes in host behaviour. *Quarterly Review of Biology* **63**, 140–65.

Downes, J. A. 1970. The ecology of blood-sucking Diptera: an evolutionary perspective. In *Ecology and physiology of parasites*, A. M. Fallis, (ed.). Toronto: University of Toronto Press.

Duggan, A. J. 1970. An historical perspective. In *The African trypanosomiases*. H. W. Mulligan, (ed.). London: Allen and Unwin.

Dunnet, G. M. 1970. Siphonaptera (fleas). In *The insects of Australia*. Melbourne: Melbourne University Press.

Duval, J., E. Rajaonarivelo & L. Rabenirainy 1974. Ecologie de *Styloconops spinosifrons* (Carter, 1921) (Diptera, Ceratopogonidae) sur les plages de la côte Est de Madagascar. *Cahiers ORSTOM, Série Entomologie Médicale et Parasitologie* **12**, 245–58.

Dwyer, D. M. 1974. Lectin binding saccharides on the parasitic protozoan. *Science, NY* **184**, 471–7.

East, J., D. H. Molyneux, I. Maudlin & P. Dukes 1983. Effect of *Glossina* haemolymph on salivarian trypanosomes *in vitro*. *Annals of Tropical Medicine and Parasitology* **77**, 97–9.

Edman, J., J. Day & E. Walker 1985. Vector–host interplay – factors affecting disease transmission. In *Ecology of mosquitoes: proceedings of a workshop*, L. P.

Lounibos, J. R. Rey & J. H. Frank (eds). New Beach, Florida: Florida Medical Entomology Laboratory.

Edman, J. D., L. A. Webber & H. W. Kale 1972. Effect of mosquito density on the interrelationships of nest behaviour and mosquito feeding success. *American Journal of Tropical Medicine and Hygiene* **21**, 487–91.

Edman, J. D. 1974. Host-feeding patterns of Florida mosquitoes. III. *Culex (Culex)* and *Culex (Neoculex). Journal of Medical Entomology* **11**, 95–104.

Edman, J. D. & H. W. Kale II. 1971. Host behaviour: its influence on the feeding success of mosquitoes. *Annals of the Entomological Society of America* **64**, 513–16.

Edman, J. D. & A. Spielman 1988. Blood-feeding by vectors: physiology, ecology, behaviour, and vertebrate defence. In *The arboviruses:epidemiology and ecology*, Vol 1. T. P. Monath, (ed.). C. R. C. Press

Edman, J. D. & D. J. Taylor 1968. *Culex nigripalpus*: seasonal shift in the bird–mammal feeding ratio in a mosquito vector of human encephalitis. *Science* **161**, 67–8.

Eguchi, M., I. Haneda & A. Iwamoto 1982. Properties of protease inhibitors from the haemolymph of silkworms. *Bombyx mori, Antheraea pernyi* and *Philosamia cynthia ricini. Comparative Physiology and Biochemistry* **71B**, 569–76.

Eichler, D. A. 1973. Studies on *Onchocerca gutterosa* (Neumann, 1910) and its development in *Simulium ornatum* (Meigen, 1818). 3. Factors affecting the development of the parasite in its vector. *Journal of Helminthology* **47**, 73–88.

Elkinton, J. S., C. Schal, T. Ono & R. T. Carde 1987. Pheromone puff trajectory and upwind flight of male gypsy moths in a forest. *Physiological Entomology* **12**, 399–406.

Emmerson, K. C., K. C. Kim & R. D. Price 1973. Lice. In *Parasites of laboratory animals*, R. J. Flynn, (ed.). Ames, Iowa: Iowa State University Press.

Evans, G. O. 1950. Studies on the bionomics of the sheep ked, *Melophagus ovinus* L., in West Wales. *Bulletin of Entomological Research* **40**, 459–78.

Evans, D. A. & D. S. Ellis 1983. Recent observations on the behaviour of certain trypanosomes within their insect hosts. *Advances in Parasitology* **22**, 1–42.

Ewert, A. 1965. Comparative migration of microfilariae and development o Brugia pahangi in various mosquitoes. *American Journal of Tropical Medicine and Hygiene* **14**, 254–9.

Falleroni, D. 1927. Per la soluzione del problema malario Italiano. *Rivista Malariologia* **6**, 344–409.

Fallis, A. M. & J. N. Raybould 1975. Response of two African simuliids to silhouettes and carbon dioxide. *Journal of Medical Entomology* **12**, 349–51.

Fallis, A. M. & S. M. Smith 1964. Ether extract from birds and CO_2 as attractants for some ornithophilic simuliids. *Canadian Journal of Zoology* **42**, 723–30.

Fallis, A. M., G. F. Bennett, G. Griggs & T. Allen 1967. Collecting *Simulium venustum* female in fan traps and on silhouettes with the aid of carbon dioxide *Canadian Journal of Zoology* **45**, 1011–17.

Farkas, S. R. & H. H. Shorey 1972. Chemical trail-following by flying insects: a mechanism for orientation to a distant odour source. *Science* **178**, 67–8.

Farmer, J., S. H. P. Maddrell & J. H. Spring 1981. Absorption of fluid by the midgut of *Rhodnius. Journal of Experimental Biology* **94**, 301–16.

Faust, E. C., P. F. Russel & R. C. Jung 1977. *Craig and Faust's Clinical Parasitology* 8th edn. Philadelphia: Lea & Febiger.

Feingold, B. F. & E. Benjamini 1961. Allergy to flea bites: clinical and experimen tal observations. *Annals of Allergy* **19**, 1274–89.

Ferris, G. R. 1931. The louse of elephants *Haematomyzus elephantis* Piaget (Mallo phaga: Haematomyzidae). *Parasitology* **23**, 112–27.

Foster, W. A. 1976. Male sexual maturation of the tsetse flies *Glossina morsitans* Westwood and *G. austeni* Newstead (Diptera: Glossinidae) in relation to feeding. *Bulletin of Entomological Research* **66**, 389–99.

Fredeen, F. J. H. 1961. A trap for studying the attacking behaviour of blackflies *Simulium articum* Mall. *Canadian Entomologist* **93**, 73–8.

Freier, J. E. & S. Friedman 1976. Effect of host infection with *Plasmodium gallinaceum* on the reproductive capacity of *Aedes aegypti*. *Journal of Invertebrate Pathology* **28**, 161–6.

Friend, W. G. 1965. The gorging response in *Rhodnius prolixus* Stahl. *Canadian Journal of Zoology* **43**, 125–32.

Friend, W. G. 1978. Physical factors affecting the feeding responses of *Culiseta inornata* to ATP, sucrose and blood. *Annals of the Entomological Society of America* **71**, 935–40.

Friend, W. G. & J. J. B. Smith 1971. Feeding in *Rhodnius prolixus*: mouthpart activity and salivation and their correlation with changes of electrical resistance. *Journal of Insect Physiology* **17**, 233–43.

Friend, W. G. & J. J. B. Smith 1975. Feeding in *Rhodnius prolixus*: increasing sensitivity to ATP during prolonged food deprivation. *Journal of Insect Physiology* **21**, 1081–4.

Friend, W. G. & J. J. B. Smith 1977. Factors affecting feeding by blood-sucking insects. *Annual Review of Entomology* **22**, 309–31.

Friend, W. G. & J. G. Stoffolano 1984. Feeding responses of the horsefly, *Tabanus nigrovittatus*, to physical factors, ATP analogues and blood fractions. *Physiological Entomology* **9**, 395–402.

Gad, A. M., W. A. Maier & G. Piekorski 1979. Pathology of *Anopheles stephensi* after infection with *Plasmodium berghei berghei*. I. Mortality rate. *Zeitschrift für Parasitenkunde* **60**, 249–61.

Galun, R. 1966. Feeding stimulants of the rat flea *Xenopsylla cheopis* Roth. *Life Science* **5**, 1335–42.

Galun, R. 1975. The role of host blood in the feeding behaviour of ectoparasites. In *Dynamic aspects of host–parasite relationships*, Vol II, A. Zuckerman, (ed.). New York: Wiley.

Galun, R. 1986. Diversity of phagostimulants used for recognition of blood meal by haematophagous arthropods. In *Host-regulated development mechanisms in vector arthropods*, D. Borovsky & A. Spielman (eds). Florida: IFAS, University of Florida.

Galun, R. 1987. The evolution of purinergic receptors involved in recognition of a blood meal by haematophagous insects. *Memorias Instituto Oswaldo Cruz* **82**, 5–9.

Galun, R. & J. P. Kabayo 1988. Gorging response of *Glossina palpalis palpalis* to ATP analogues. *Physiological Entomology* **13**, 419–23.

Galun, R., Y. Avidor & M. Bar-Zeev 1963. Feeding response in *Aedes aegypti*: stimulation by adenosine triphosphate. *Science* **124**, 1674–5.

Galun, R., W. G. Friend & S. Nudelman 1988. Purinergic reception by culicine mosquitoes. *Journal of Comparative Physiology* **163A**, 665–70.

Galun, R., L. C. Koontz & R. W. Gwadz 1985. Engorgement response of anopheline mosquitoes to blood fractions and artificial solutions. *Physiological Entomology* **10**, 145–9.

Garcia, R. & F. J. Radovsky 1962. Haematophagy by two non-biting muscid flies and its relationship to tabanid feeding. *Canadian Entomologist* **94**, 1110–16.

Garms, R., J. F. Walsh & J. B. Davies 1979. Studies on the re-invasion of the Onchocerciasis Control Programme in the Volta River Basin by *Simulium damnosum* s. l. with emphasis on the southwestern areas. *Tropenmedizin und Parasitologie* **30**, 345–62.

Gatehouse, A. G. 1970. The probing response of *Stomoxys calcitrans* to certain physical and olfactory stimuli. *Journal of Insect Physiology* **16**, 61–74.

Gatehouse, A. G. 1972. Some responses of tsetse flies to visual and olfactory stimuli. *Nature* **236**, 63–4.

Gee, J. C. 1975. Diuresis in the tsetse fly *Glossina austeni*. *Journal of Experimental Biology* **63**, 381–90.

Gellissen, G. 1983. Lipophorin as the plasma coagulogen in *Locusta migratoria*. *Naturwissenschaften* **70**, 45–6.

Gibson, G. & J. Brady 1985. Anemotactic flight paths of tsetse flies in relation to host odour – preliminary video study in nature. *Physiological Entomology* **10**, 395–406.

Gibson, G. & J. Brady 1988. Flight behaviour of tsetse flies in host odour plumes: the initial response to leaving or entering odour. *Physiological Entomology* **13**, 29–42.

Gillett, J. D. 1967. Natural selection and feeding speed in a blood-sucking insect. *Proceedings of the Royal Society London Ser. B* **167**, 316–29.

Gillett, J. D. & J. Connor 1976. Host temperature and the transmission of arboviruses by mosquitoes. *Mosquito News* **36**, 472–7.

Gillies, M. T. 1980. The role of carbon dioxide in host-finding by mosquitoes (Diptera: Culicidae): a review. *Bulletin of Entomological Research* **70**, 525–32.

Gillies, M. T. & T. J. Wilkes 1969. A comparison of the range of attraction of animal baits and carbon dioxide for some West African mosquitoes. *Bulletin of Entomological Research* **59**, 441–56.

Gillies, M. T. & T. J. Wilkes 1970. The range of attraction of single baits for some West African mosquitoes. *Bulletin of Entomological Research* **60**, 225–35.

Gillies, M. T. & T. J. Wilkes 1972. The range of attraction of animal baits and carbon dioxide for mosquitoes. Studies in a freshwater area of West Africa. *Bulletin of Entomological Research* **61**, 389–404.

Glasgow, J. P. 1961. The feeding habits of *Glossina swynnertoni*. *Journal of Animal Ecology* **30**, 77–85.

Glasgow, J. P. 1963. *The distribution and abundance of tsetse*. Oxford: Pergamon.

Goodchild, A. J. P. 1955. Some observations on growth and egg production of the blood-sucking reduviids, *Rhodnius prolixus* and *Triatoma infestans*. *Proceedings of the Royal Entomological Society* **30**, 137–44.

Gooding, R. H. 1968. A note on the relationship between feeding and insemination in *Pediculus humanus*. *Journal of Medical Entomology* **5**, 265–6.

Gooding, R. H. 1972. Digestive processes of haematophagous insects. I. A literature review. *Quaestiones Entomologicae* **8**, 5–60.

Gooding, R. H. 1974. Digestive processes in haematophagous insects. Control of trypsin secretion in *Glossina morsitans morsitans*. *Journal of Insect Physiology* **20**, 957–64.

Gooding, R. H. 1975. Inhibition of diuresis in the tsetse fly (*Glossina morsitans*) by ouabain or acetazolamide. *Experientia* **31**, 938–9.

Gooding, R. H. 1977. Digestive processes of haematophagous insects. XIV Haemolytic activity in the midgut of *Glossina morsitans morsitans* Westwood (Diptera: Glossinidae). *Canadian Journal of Zoology* **55**, 1899–905.

Goodwin, R. F. W. 1973. The cost effectiveness of animal disease control. *Span* **16**, 63–4.

Gordon, R. M., W. Crewe & K. C. Willett 1956. Studies on the deposition migration and development to the blood forms of trypanosomes belonging to the *Trypanosoma brucei* group. I. An account of the process of feeding adopted by the tsetse fly when obtaining a blood meal from the mammalian host, with special reference to the ejection of saliva and the relationship of the feeding process to the deposition of the metacyclic trypanosomes. *Annals of Tropical Medicine* **50**, 426–37.

Gore, T. C. & G. Pittman-Noblet 1978. The effect of photoperiod on the deep body temperature of domestic turkeys and its relationship to the diurnal periodicity of *Leucocytozoon smithi* gametocytes in the peripheral blood of turkeys. *Poultry Science* **57**, 603–7.

Gotz, P. 1986. Mechanisms of encapsulation in dipteran hosts. In *Immune mechanisms in invertebrate vectors*. A. Lackie, (ed.). Zoological Society of London Symposia, no. 56, Oxford: Oxford Science.

Gotz, P. & H. G. Boman 1985. Insect immunity. In *Comprehensive insect physiology, biochemistry and pharmacology*, Vol. 3, G. A. Kerkut & L. I. Gilbert (eds), 453–85. Oxford: Pergamon.

Gotz, P., I. Roettgen & W. Lingg 1977. Encapsulement humoral en tant que réaction de défense chez les Diptères. *Annales de Parasitologie Humaine et Comparée* **52**, 95–7.

Graf, R., A. S. Raikhel, M. R. Brown, A. O. Lea & H. Briegel 1986. Mosquito trypsin: immunocytochemical localisation in the midgut of blood-fed *Aedes aegypti* (L.). *Cell and Tissue Research* **245**, 19–27.

Green, C. H. 1989. The use of two-coloured screens for catching *Glossina palpalis palpalis* (Robineau-Desvoidy) (Diptera: Glossinidae). *Bulletin of Entomological Research* **79**, 81–93.

Green, C. H. & D. Cosens 1983. Spectral responses of the tsetse fly, *Glossina morsitans morsitans*. *Journal of Insect Physiology* **29**, 795–800.

Grimstad, P. R., S. L. Paulson & G. B. Craig Jr 1985. Vector competence of *Aedes hendersoni* (Diptera: Culicidae) for La Crosse virus and evidence of a salivary gland escape barrier. *Journal of Medical Entomology* **22**, 447–53.

Grimstad, P. R., Q. E. Ross & G. B. Craig Jr 1980. *Aedes triseriatus* (Diptera: Culicidae) and La Crosse virus. II. Modification of mosquito feeding behaviour by virus infection. *Journal of Medical Entomology* **17**, 1–7.

Gwadz, R. W. 1969. Regulation of blood meal size in the mosquito. *Journal of Insect Physiology* **15**, 2039–44.

Haarlov, N. 1964. Life cycle and distribution pattern of *Lipoptena cervi* (L.)(Dipt., Hippobosc.) on Danish deer. *Oikos* **15**, 93–129.

Hacker, C. S. & W. L. Kilama 1974. The relationship between *Plasmodium gallinaceum* density and the fecundity of *Aedes aegypti*. *Journal of Invertebrate Pathology* **23**, 101–5.

Hackett, L. W. & A. Missiroli 1931. The natural disappearance of malaria in certain regions of Europe. *American Journal of Hygiene* **13**, 57–78.

Hailman, J. P. 1979. Environmental light and conspicuous colours. In *The behavioural significance of colour*, E. H. Burtt, Jr (ed.). New York: Garland STPM Press.

Hall, D. R., P. S. Beevor, A. Cork, *et al*. 1984. 1-Octen-3-ol: a potent olfactory stimulant and attractant for tsetse isolated from cattle odours. *Insect Science and its Applications* **5**, 335–9.

Halstead, S. B. 1984. *Tropical and geographical medicine*, K. S. Warren & A. A. F. Mahmoud (eds). New York: McGraw-Hill.

Hansens E. J., E. M. Bosler & J. W. Robinson 1971. Use of traps for study and control of slat marsh flies. *Journal of Economic Entomology* **64**, 1481–6.

Haque, A. & A. Capron 1982. Transplacental transfer of rodent microfilariae induces antigen-specific tolerance in rats. *Nature* **299**, 361–3.

REFERENCES

Hargrove, J. W. 1980. Improved estimates of the efficiency of traps for *Glossina morsitans morsitans* Westwood and *G. pallidipes* Austen (Diptera: Glossinidae) with a note on the effect of the concentration of accompanying host odour on efficiency. *Bulletin of Entomological Research* **70**, 579–87.

Harris, J. A., J. E. Hillerton & S. V. Morant 1987. Effect on milk production of controlling muscid flies, and reducing fly-avoidance behaviour by the use of Fenvalerate ear tags during the dry period. *Journal of Dairy Research* **54**, 165–71.

Harris, P., D. F. Riordan & D. Cooke 1969. Mosquitoes feeding on insect larvae. *Science* **164**, 184–5.

Harrison, G. 1978. *Mosquitoes, malaria and man: a history of the hostilities since 1880.* London: John Murray.

Hawkes, C. & T. H. Coaker 1979. Factors affecting the behavioural response of the adult cabbage root fly, *Delia brassica*, to host plant odour. *Entomologia Experimentalis et Applicata* **25**, 45–58.

Hawking, F. 1962. Microfilaria infestation as an instance of periodic phenomena seen in host–parasite relationships. *Annals of the New York Academy of Sciences* **98**, 940–53.

Hawking, F. 1976. Circadian rhythms in *Trypanosoma congolense*. *Transactions of the Royal Society for Tropical Medicine and Hygiene* **70**, 170.

Hawking, F., K. Gammage & M. J. Worms 1972. The asexual and sexual circadian rhythms of *Plasmodium vinckei, P. chabaudi, P. berghei* and *P. gallinaceum*. *Parasitology* **65**, 189–201.

Hawking, F., M. E. Wilson & K. Gammage 1971. Guidance for cyclic development and short-lived maturity in the gametocytes of *Plasmodium falciparum*. *Parasitology* **65**, 549–59.

Hawking, F., M. J. Worms, K. Gammage & P. A. Goddard 1966. The biological purpose of the blood-cycle of the malaria parasite *Plasmodium cynomolgi*. *Lancet* ii, 422–4.

Hawking, F., M. J. Worms & K. Gammage 1968. 24- and 48-hour cycles of malaria parasites in the blood; their purpose, production and control. *Transactions of the Royal Society for Tropical Medicine and Hygiene* **62**, 731–60.

Hawkins, R. I. 1966. Factors affecting blood clotting from salivary glands and crops of *Glossina austeni*. *Nature* **212**, 738–9.

Hecker, H. & W. Rudin 1981. Morphometric parameters of the midgut cells of *Aedes aegypti* L. (Insecta, Diptera) under various conditions. *Cell and Tissue Research* **219**, 619–27.

Helle, T. & J. Aspi 1983. Does herd formation reduce insect harassment among reindeer? A field experiment with animal traps. *Acta Zoologica Fennica* **175**, 129–31.

Hellman, K. & R. I. Hawkins 1964. Anticoagulant and fibrinolytic activities from *Rhodnius prolixus* Stahl. *Nature* **201**, 1008–9.

Henry, V. G. & R. H. Conley 1970. Some parasites of European wild hogs in the southern Appalachians. *Journal of Wildlife Management* **34**, 913–17.

Hill, P., D. S. Saunders & J. A. Campbell 1973. The production of 'symbiont-free' *Glossina morsitans* and an associated loss of female fertility. *Transactions of th Royal Society for Tropical Medicine and Hygiene* **67**, 727–8.

Hinton, H. E. 1958. The phylogeny of the panorpoid orders. *Annual Review c Entomology* **3**, 181–206.

Hocking, B. 1953. The intrinsic range and speed of flight of insects. *Transactions c the Royal Society of London* **104**, 223–345.

Hocking, B. 1957. Louse control through textile fibre size. *Bulletin of Entomologica Research* **48**, 507–14.

260

Hocking, B. 1971. Blood-sucking behaviour of terrestrial arthropods. *Annual Review of Entomology* **16**, 1–26.

Hockmeyer, W. T., B. A. Schieffer, B. C. Redington & B. V. Eldridge 1975. *Brugia pahangi* effects upon the flight capability of *Aedes aegypti*. *Experimental Parasitology* **38**, 1–5.

Hoffman, J. A., A. Porte & P. Joly 1970. Sur la localisation d'une activité phenoloxydasique dans les coagulocytes de *Locusta migratoria* L. (Orthoptere). *Comptes Rendus Hebdomadaires des Séances, Paris* **270D**, 629–31.

Hooke, R. 1664. *Micrographia*.

Hopkins, F. H. E. 1949. The host associations of the lice of mammals. *Proceedings of the Zoological Society of London* **119**, 387–604.

Hosoi, T. 1958. Adenosine 5' phosphates as the stimulating agent in blood for inducing gorging of the mosquito. *Nature* **181**, 1664–5.

Hosoi, T. 1959. Identification of blood components which induce gorging in the mosquito. *Journal of Insect Physiology* **3**, 191–218.

Houseman, J. G., A. E. R. Downe & P. E. Morrison 1985. Similarities in digestive proteinase production in *Rhodnius prolixus* (Hemiptera: Reduviidae) and *Stomoxys calcitrans* (Diptera: Muscidae). *Insect Biochemistry* **15**, 471–4.

Howcroft, T. K. & R. D. Karp 1987. Demonstration of cell mediated cytotoxicity to allogeneic and xenogenic tissue in the American cockroach *Periplaneta americana*, using a combination *in vivo/in vitro* assay. *Transplantation* **44**, 129–35.

Howe, M. A. & M. J. Lehane 1986. Post-feed buzzing in the tsetse, *Glossina morsitans morsitans*, is an endothermic mechanism. *Physiological Entomology* **11**, 279–86.

Huang, C. T. 1971. Vertebrate serum inhibitors of *Aedes aegypti* trypsin. *Insect Biochemistry* **1**, 27–38.

Hudson, A. 1970. Notes on the piercing mouthparts of three species of mosquitoes (Diptera: Culicidae) viewed with the scanning electron microscope. *Canadian Entomologist* **102**, 501–9.

Hudson, B. W., B. F. Feingold & L. Kartman 1960. Allergy to flea bites. II. Investigations of flea bite sensitivity in humans. *Experimental Parasitology* **9**, 264–70.

Huff, C. G. 1929. Ovulation requirements of *Culex pipiens* Linn. *Biological Bulletin Marine Biological Laboratory (Woods Hole)* **56**, 347–50.

Huff, C. 1931. A proposed classification of disease transmission by arthropods. *Science* **74**, 456–7.

Humphries, D. A. 1967. Function of combs in ectoparasites. *Nature* **215**, 319.

Hunter, D. M. & D. W. Moorhouse 1976. The effects of *Austrosimulium pestilens* on the milk production of dairy cattle. *Australian Veterinary Journal* **52**, 97–9.

Huq, M. 1961. African horse sickness. *Veterinary Record* **73**, 123.

Ibrahim, E. A. R., G. A. Ingram & D. H. Molyneux 1984. Haemagglutinins and parasite agglutinins in haemolymph and gut of *Glossina*. *Tropenmedizin und Parasitologie* **35**, 151–6.

James, M. T. & R. F. Harwood 1969. *Herm's medical entomology*. London: Macmillan.

Janse, C. J., B. Mons, R. J. Rouwenhorst *et al*. 1985. *In vitro* formation of ookinetes and functional maturity of *Plasmodium berghei* gametocytes. *Parasitology* **91**, 19–29.

Jefferies, D. 1984. Transmission of disease by haematophagous arthropods: host–parasite interactions and fluid mechanics. Ph.D. Thesis, University of Salford.

Jenkins, D. W. 1964. Advances in medical entomology using radio-isotopes. *Experimental Parasitology* **3**, 474–90.

Jenni, L., D. H. Molyneux, J. L. Livesey & R. Galun 1980. Feeding behaviour of tsetse flies infected with salivarian trypanosomes. *Nature* **283**, 383–5.

Jobling, B. 1976. On the fascicle of blood-sucking Diptera. In addition a description of the maxillary glands in *Phlebotomus papatasi*, together with the musculature of the labium and pulsatory organ of both the latter species and also of some other Diptera. *Journal of Natural History* **10**, 457–61.

Jones, J. C. & D. R. Pillitt 1973. Blood-feeding behaviour of adult *Aedes aegypti* mosquitoes. *Biological Bulletin* **145**, 127–39.

Jordan, A. M. 1958. The mating behaviour of females of *Glossina palpalis* (R.-D) in captivity. *Bulletin of Entomological Research* **49**, 35–43.

Jordan, A. M. 1974. Recent developments in the ecology and methods of control of tsetse flies (*Glossina* spp.) (Dipt., Glossinidae) – a review. *Bulletin of Entomological Research* **63**, 361–99.

Jordan, A. M. 1986. *Trypanosomiasis control and African rural development*. London: Longman.

Jordan, A. M. & C. F. Curtis 1968. The performance of *Glossina austeni* when fed on lop-eared rabbits and goats. *Transactions of the Royal Society for Tropical Medicine and Hygiene* **62**, 123–4.

Jordan, A. M. & C. F. Curtis 1972. Productivity of *Glossina morsitans morsitans* Westwood maintained in the laboratory, with particular reference to the sterile-insect release method. *Bulletin of the World Health Organisation* **46**, 33–8.

Jordan, K. 1962. Notes on the *Tunga caecigena* (Siphonaptera: Tungidae). *Bulletin of the British Museum Natural History*. (Entomology) **12**, 353–64.

Kaaya, G. P. & N. A. Ratcliffe 1982. Comparative study of haemocytes and associated cells of some medically important dipterans. *Journal of Morphology* **173**, 351–65.

Kaaya, G. P., C. Flyg & H. G. Boman 1987. Insect immunity: induction of cecropin and attacin-like antibacterial factors in the haemolymph of *Glossina morsitans morsitans*. *Insect Biochemistry* **17**, 309–15.

Kangwangye, T. N. 1977. Reactions of large mammals to biting flies in Rwenzori National Park, Uganda. In *Proceedings 1st East African Conference on Entomological Pest Control*, C. P. F. Lima (ed.).

Kartman, L. 1953. Factors influencing infection of the mosquito with *Dirofilaria immitis* (Leidy, 1856). *Experimental Parasitology* **2**, 27–78.

Kellogg, F. E. 1970. Water vapour and carbon dioxide receptors in *Aedes aegypti*. *Journal of Insect Physiology* **16**, 99–108.

Kellogg, F. E. & R. H. Wright 1962. The guidance of flying insects. V. Mosquito attraction. *Canadian Entomologist* **94**, 1009–16.

Kennedy, J. S. 1940. The visual responses of flying mosquitoes. *Proceedings of the Zoological Society of London* **109**, 221–42.

Kennedy, J. S. 1983. Zigzagging and casting as a programmed response to wind-borne odour: a review. *Physiological Entomology* **8**, 109–20.

Kettle, D. S. 1984. *Medical and veterinary entomology*. London: Croom Helm.

Khan, A. A. & H. I. Maibach 1966. Quantitation of effect of several stimuli on landing and probing by *Aedes aegypti*. *Journal of Economic Entomology* **59**, 902–5.

Khan, A. A. & H. I. Maibach 1970. A study of the probing response of *Aedes aegypti*. I. Effect of nutrition on probing. *Journal of Economic Entomology* **63**, 974–6.

Khan, A. A. & H. I. Maibach 1971. A study of the probing response of *Aedes aegypti*. 2. Effect of desiccation and blood feeding on probing to skin and an artificial target. *Journal of Economic Entomology* **64**, 439–42.

Kilama, W. L. & G. B. Craig 1969. Monofactorial inheritance of susceptibility to
Plasmodium gallinaceum in *Aedes aegypti. Annals of Tropical Medicine and
Parasitology* **63**, 419–32.
Killick-Kendrick R. & D. H. Molyneux 1981. Transmission of leishmaniasis by the
bite of phlebotomid sandflies: possible mechanisms. *Transactions of the Royal
Society for Tropical Medicine and Hygiene* **75**, 152–4.
Killick-Kendrick, R., R. Leaney, P. D. Ready & D. H. Molyneux 1977. *Leishmania*
in phlebotomid sandflies. IV. The transmission of *Leishmania mexicana amaz-
onensis* to hamsters by the bite of experimentally infected *Lutzomyia longipalpis.
Proceedings of the Royal Society London B* **196**, 105–15.
Kim, K. C. 1985. Evolution and host association of Anoplura. In *Coevolution of
parasitic arthropods and mammals.* K. C. Kim, (ed). New York: Wiley.
Kim, K. C. & H. W. Ludwig 1982. Parallel evolution, cladistics and classification
of parasitic Psocodea. *Annals of the Entomological Society of America* **75**, 537–48.
Kim, K. D. & P. H. Adler 1985. Evolution and host associations of Anoplura. In
Coevolution of parasitic arthropods and mammals, K. C. Kim, (ed). New York:
Wiley.
Kingsolver, J. G. 1987. Mosquito host choice and the epidemiology of malaria.
American Naturalist **130**, 811–27.
Klein, T. A., B. A. Harrison, R. G. Andre *et al.* 1982. Detrimental effects of
Plasmodium cynomolgi infections on the longevity of *Anopheles dirus. Mosquito
News* **42**, 265–71.
Klowden, M. J. 1986. Distension-mediated egg maturation in the mosquito.
Journal of Insect Physiology **33**, 83–87.
Klowden, M. J. & A. O. Lea 1979. Abdominal distention terminates subsequent
host-seeking behaviour of *Aedes aegypti* following a blood meal. *Journal of Insect
Physiology* **25**, 583–5.
Klowden, M. J., E. E. Davis & M. F. Bowen 1987. Role of the fat body in the
regulation of host-seeking behaviour in the mosquito, *Aedes aegypti. Journal of
Insect Physiology* **33**, 643–6.
Komano, H., D. Mizuno & S. Natori 1980. Purification of a lectin induced in the
haemolymph of *Sarcophaga peregrina* larvae on injury. *Journal of Biological
Chemistry* **255**, 2919–24.
Kramer, L. D., J. L. Hardy, S. B. Presser & E. G. Houk 1981. Dissemination
barriers for western equine encephalomyelitis virus in *Culex tarsalis* infected
after ingestion of low viral doses. *American Journal of Tropical Medicine and
Hygiene* **30**, 190–7.
Krynski, S., A. Kuchta & E. Becla 1952. Research on the nature of the noxious
action of guinea-pig blood on the body louse. [In Polish]. *Bulletin Institute of
Marine Medicine, Gdansk* **4**, 104–7.

La Breque, G. C., D. W. Meifert & J. Rye 1972. Experimental control of stable flies,
Stomoxys calcitrans (Diptera: Muscidae), by the release of chemosterilized
adults. *Canadian Entomologist* **104**, 885–7.
Laarman, J. J. 1958. The host-seeking behaviour of anopheline mosquitoes.
Tropical and Geographical Medicine **10**, 293–305.
Lackie, A. M. 1986. Evasion of insect immunity by helminth larvae. In *Immune
mechanisms in invertebrate vectors,* A. M. Lackie (ed.), Zoological Society of
London Symposia, no. 56, Oxford: Oxford Science.
Lackie, A. M. 1988. Immune mechanisms in insects. *Parasitology Today* **4**, 98–105.
Lackie, A. M & G. R. Vasta 1988. The role of galactosyl-binding lectin in the
cellular immune response of the cockroach *Periplaneta americana* (Dictyoptera).
Immunology **64**, 353–7.

263

LaFond, M. M., B. M. Christensen & B. A. Lasee 1985. Defence reactions of mosquitoes to filarial worms: potential mechanisms for avoidance of the response by *Brugia pahangi* microfilaria. *Journal of Invertebrate Pathology* **46**, 26–30.

Lainson, R. & J. J. Shaw 1979. The role of animals in the epidemiology of South American leishmaniasis. In *Biology of the Kinetoplastidae*, W. H. R. Lumsden & D. A. Evans (eds). New York: Academic Press.

Lainson, R. & J. J. Shaw 1987. Evolution, classification and geographical distribution. In *The leishmaniases in biology and medicine*, W. Peters & R. Killick-Kendrick (eds). New York: Academic Press.

Lall, S. B. 1969. Phagostimulants of haematophagous tabanids. *Entomologia Experimentalis et Applicata* **12**, 325–36.

Lane, N. J. & J. B. Harrison 1979. An unusual cell surface modification: a double plasma membrane. *Journal of Cell Science* **39**, 355–72.

Langley, P. A. 1970. Post-teneral development of thoracic flight musculature in the tsetse flies *Glossina austeni* and *G. morsitans*. *Entomologia Experimentalis et Applicata* **13**, 133–40.

Langley, P. A. 1977. Physiology of tsetse flies (*Glossina* spp.)(Diptera: Glossinidae): a review. *Bulletin of Entomological Research* **67**, 523–74.

Langley, P. A. & H. Maly 1969. Membrane feeding technique for tsetse flies (*Glossina* spp.). *Nature* **221**, 855–6.

Langley, P. A., M. J. R. Hall, T. Fetton & M. Ceesay 1988. Determining the age of tsetse flies, *Glossina* spp. (Diptera: Glossinidae): an appraisal of the pteridine fluorescence technique. *Bulletin of Entomological Research* **78**, 387–95.

Langley, P. A., R. W. Pimley & D. A. Carlson 1975. Sex recognition pheromone in tsetse fly *Glossina morsitans*. *Nature* **254**, 51–3.

Larrivee, D. H., E. Benjamini, B. F. Feingold & M. Shimuzu 1964. Histologic studies of guinea pig skin: different stages of allergic reactivity to flea bites. *Experimental Parasitology* **15**, 491–502.

Laurence, B. R. 1966. Intake and migration of the microfilariae of *Onchocerca volvulus* (Leukart) in *Simulium damnosum* Theobald. *Journal of Helminthology* **40**, 337–42.

Laurence, B. R. & F. R. N. Pester 1961. The ability of *Anopheles gambiae* Giles to transmit *Brugei patei* (Buckley, Nelson and Heisch). *Journal of Tropical Medicine and Hygiene* **64**, 169–71.

Lavoipierre, M. M. J. & M. Hamachi 1961. An apparatus for observations on the feeding mechanism of the flea. *Nature* **192**, 998–9.

Lavoipierre, M. M. J., G. Dickerson & R. M. Gordon 1959. Studies on the methods of feeding of blood-sucking arthropods. I. The manner in which triatomine bugs obtain their blood meal as observed in the tissues of the living rodent with some remarks on the effects of the bite on human volunteers. *Annals of Tropical Medicine and Parasitology* **53**, 235–50.

Lea, A. O. & E. Van Handel 1982. A neurosecretory hormone-releasing factor from ovaries of mosquito fed blood. *Journal of Insect Physiology* **28**, 503–8.

Lee, R. 1974. Structure and function of the fascicular stylets, and the labral and cibarial sense organs of male and female *Aedes aegypti* (L.). *Quaestiones Entomologica* **10**, 187–215.

Lehane, M. J. 1975. Observations on the structure and function of the gut in *Stomoxys calcitrans* (Insecta: Diptera). Ph.D. Thesis, University of Leeds.

Lehane, M. J. 1976a. Digestive enzyme secretion in *Stomoxys calcitrans* (Diptera: Muscidae). *Cell and Tissue Research* **170**, 275–87.

Lehane, M. J. 1976b. The formation and histochemical structure of the peritrophic membrane in the stable fly, *Stomoxys calcitrans*. *Journal of Insect Physiology* **22**, 1551–7.

Lehane, M. J. 1977a. An hypothesis of the mechanism controlling proteolytic digestive enzyme production levels in *Stomoxys calcitrans*. *Journal of Insect Physiology* **23**, 713–15.

Lehane, M. J. 1977b. Transcellular absorption of lipids in the midgut of the stablefly, *Stomoxys calcitrans*. *Journal of Insect Physiology* **23**, 945–54.

Lehane, M. J. 1985. Determining the age of an insect. *Parasitology Today* **1**, 81–5.

Lehane, M. J. 1987. Quantitative evidence for merocrine secretion in an insect midgut cell. *Tissue and Cell* **19**, 451–561.

Lehane, M. J. 1988. Evidence for secretion by the release of cytoplasmic extrusions from midgut cells of *Stomoxys calcitrans*. *Journal of Insect Physiology* **34**, 949–53.

Lehane, M. J. 1989. The intracellular pathway and kinetics of digestive enzyme secretion in an insect midgut cell. *Tissue and Cell* **21**, 101–11.

Lehane, M. J. & J. W. Hargrove 1988. Field experiments on a new method for determining age in tsetse flies (Diptera: Glossinidae). *Ecological Entomology* **13**, 319–22.

Lehane, M. J. & B. R. Laurence 1977. Flight muscle ultrastructure of susceptible and refractory mosquitoes parasitized by larval *Brugia pahangi*. *Parasitology* **74**, 87–92.

Lehane, M. J. & T. S. Mail 1985. Determining the age of adult male and female *Glossina morsitans morsitans* using a new technique. *Ecological Entomology* **10**, 219–24.

Lehane, M. J. & A. Msangi (forthcoming). Lectin and peritrophic membrane development in the gut of *Glossina morsitans morsitans* Westwood and a discussion of their role in protecting the fly against trypanosome infection. *Medical and Veterinary Entomology*.

Lehane, M. J. & C. J. Schofield 1981. Field experiments of dispersive flight by *Triatoma infestans*. *Transactions of the Royal Society for Tropical Medicine and Hygiene* **75**, 399–400.

Lehane, M. J. & C. J. Schofield 1982. Flight initiation in *Triatoma infestans* (Klug) (Hemiptera: Reduviidae). *Bulletin of Entomological Research* **72**, 497–510.

Leonard, C., K. Soderhall & N. A. Ratcliffe 1985. Studies of prophenoloxidase and protease activity of *Blaberus craniifer* haemocytes. *Insect Biochemistry* **15**, 803–10.

Lester, H. M. O. & L. Lloyd 1929. Notes on the process of digestion in tsetse flies. *Bulletin of Entomological Research* **19**, 39–60.

Lewis, D. J. 1953. *Simulium damnosum* and its relation to Onchocerciasis in the Anglo-Egyptian Sudan. *Bulletin of Entomological Research* **43**, 597–644.

Lewis, L. F., D. M. Christenson & G. W. Eddy 1967. Rearing the long-nosed cattle louse and cattle biting louse on host animals in Oregon. *Journal of Economic Entomology* **60**, 755–7.

Lewis, T. & L. R. Taylor 1965. Diurnal periodicity of flight by insects. *Transactions of the Royal Entomological Society, London* **116**, 393–479.

Linley, J. R. & J. B. Davies 1971. Sandflies and tourism in Florida and the Bahamas and Caribbean area. *Journal of Economic Entomology* **64**, 264–78.

Linsenmair, K. E. 1973. Die Windorientierurung läufender Insekten. *Fortschritte der Zoologie* **21**, 59–79.

Lodmell, D. L., J. F. Bell, C. M. Clifford *et al.* 1970. Effects of limb disability on lousiness in mice. V. Hierarchy disturbance on mutual grooming and reproductive capacities. *Experimental Parasitology* **27**, 184–92.

Lowther, J. K. & D. M. Wood 1964. Specificity of a black fly, *Simulium euryadmiculum* Davies, towards its host, the common loon. *Canadian Entomologist* **96**, 911–13.

Maa, T. C. & A. G. Marshall 1981. Diptera Pupipara of the New Hebrides (South Pacific): taxonomy, zoogeography, host association and ecology. *Quarterly Journal of the Taiwan Museum* **34**, 213–32.

MacCormack, C. P. 1984. Human ecology and behaviour in malaria control in tropical Africa. *Bulletin of the World Health Organisation* **62**, 81–7.

Macdonald, W. W. 1962a. The genetic basis of susceptibility to infection with semi-periodic *Brugia malayi* in *Aedes aegypti*. *Annals of Tropical Medicine and Parasitology* **56**, 373–82.

Macdonald, W. W. 1962b. The selection of a strain of *Aedes aegypti* susceptible to infection with semi-periodic *Brugia malayi*. *Annals of Tropical Medicine and Parasitology* **56**, 368–72.

Macdonald, W. W. 1963. Further studies on a strain of *Aedes aegypti* susceptible to infection with sub-periodic *Brugia malayi*. *Annals of Tropical Medicine and Parasitology* **57**, 452–60.

Maddrell, S. H. P. 1963. Control of ingestion in *Rhodnius prolixus* Stal. *Nature* **198**, 210.

Maddrell, S. H. P. 1980. Characteristics of epithelial transport in insect malpighian tubules. *Current Topics in Membrane Transport* **14**, 427–63.

Mahon, R. & A. Gibbs 1982. Arbovirus infected hens attract more mosquitoes. In *Viral diseases in South East Asia and the Western Pacific*, J. S. MacKenzie (ed). New York: Academic Press.

Maier, W. A. & O. Omer 1973. Der einfluss von *Plasmodium cathemerium* auf den Aminosauregehalt und die einzahl von *Culex pipiens fatigans*. *Zeitschrift für Parasitenkunde* **42**, 265–78.

Manson, P. 1878. On the development of *Filaria sanguinis hominis*, and on the mosquito considered as a nurse. *Journal of the Linnaean Society Zoology, London* **14**, 304–11.

Mant, M. J. & K. R. Parker 1981. Two platelet aggregation inhibitors in tsetse (*Glossina*) saliva with studies of thrombin and citrate in *in vitro* platelet aggregation. *British Journal of Haematology* **48**, 601–8.

Margalit, J., R. Galun & M. J. Rice 1972. Mouthpart sensilla of the tsetse fly and their function. I. Feeding patterns. *Annals of Tropical Medicine and Parasitology* **66**, 525–36.

Marshall, A. G. 1980. The function of combs in ectoparasitic insects. In *Fleas – Proceedings of the International Conference on fleas (1977)*, R. Traub & H. Starcke (eds). Rotterdam: Balkema.

Marshall, A. G. 1981. *The ecology of ectoparasitic insects*. New York: Academic Press.

Marx, R. 1955. Über die wirtsfindung und die bedeutung des artspezifischen duftstoffes bei *Cimex lectularius* Linne. *Zeitschrift für Parasitenkunde* **17**, 41–72.

Matthysse, J. G. 1946. Cattle lice: their biology and control. *Cornell Agricultural Experimental Station Bulletin* **832**, 1–67.

Mattingley, P. F. 1965. The evolution of parasite–arthropod vector systems. In *Symposium of the British Society for Parasitology*, no. 3, A. E. R. Taylor (ed). Oxford: Blackwell.

Maudlin, I & P. Dukes 1985. Extrachromosomal inheritance of susceptibility to trypanosome infection in tsetse flies. I. Selection of susceptible and refractory strains of *Glossina morsitans morsitans*. *Annals of Tropical Medicine and Parasitology* **79**, 317–24.

Maudlin, I. & D. Ellis 1985. Association between intracellular rickettsia-like infections of midgut cells and susceptibility to trypanosome infections in *Glossina* species. *Zeitschrift für Parasitenkunde* **71**, 683–7.

Maudlin, I., J. P. Kabayo, M. E. T. Flood & D. A. Evans 1984. Serum factors and the maturation of *Trypanosoma congolense* infections in *Glossina morsitans*. *Zeitschrift für Parasitenkunde* **70**, 11–19.

May, R. M. & R. M. Anderson 1983. Epidemiology and genetics in the coevolution of parasites and hosts. *Proceedings of the Royal Society London B* **219**, 281–313.

Mayer, M. S. & J. D. James 1969. Attraction of *Aedes aegypti* (L.): responses to human arms, carbon dioxide, and air currents in a new type of olfactometer. *Bulletin of Entomological Research* **58**, 629–42.

Mayer, M. S. & J. D. James 1970. Attraction of *Aedes aegypti*. II. Velocity of reaction to host with and without additional carbon dioxide. *Entomologia Experimentalis et Applicata* **13**, 47–53.

Mazokhin-Porshnyakov, G. A. 1969. *Insect vision*. New York: Plenum Press.

McCabe, C. T. & E. Bursell 1975. Interrelationships between amino acid and lipid metabolism in the tsetse fly, *Glossina morsitans*. *Insect Biochemistry* **5**, 781–9.

McGreevy, P. B., J. H. Bryan, P. Oothuman & N. Kolstrup 1978. The lethal effects of the cibarial and pharyngeal armatures of mosquitoes on microfilariae. *Transactions of the Royal Society for Tropical Medicine and Hygiene* **74**, 361–8.

McGreevy, P. B., G. A. H. McClelland & M. M. J. Lavoipierre 1974. Inheritance of susceptibility to *Dirofilaria immitis* infection in *Aedes aegypti*. *Annals of Tropical Medicine and Parasitology* **68**, 97–109.

McKeever, S. 1977. Observations of *Corethrella* feeding on tree frogs (*Hyla*). *Mosquito News* **37**, 522.

McKelvey, J. J. 1973. *Man against tsetse: struggle for Africa*. Cornell University Press.

Mead-Briggs, A. R. 1964. The reproductive biology of the rabbit flea *Spilopsyllus cuniculi* (Dale) and the dependence of this species upon the breeding of its host. *Journal of Experimental Biology* **41**, 371–402.

Medvedev, S. I. & V.Ye. Skylar 1974. Beetles (Coleoptera) from nests of small mammals in Donotsk Province. [In Russian.] *Entomologicheskoe Obozrenie* **53**, 561–71.

Mellink, J. J. 1981. Selections for blood-feeding efficiency in colonized *Aedes aegypti*. *Mosquito News* **41**, 119–125.

Mellink, J. J. & W. van den Bovenkamp 1981. Functional aspects of mosquito salivation in blood feeding in *Aedes aegypti*. *Mosquito News* **41**, 110–15.

Mellor, P. S. & J. Boorman 1980. Multiplication of the bluetongue virus in *Culicoides nubeculosus* (Meigen) simultaneously infected with the virus and microfilaria of *Onchocerca cervicalis* (Railliet and Henry). *Annals of Tropical Medicine and Parasitology* **74**, 463–9.

Miall, R. C. 1978. The flicker fusion frequencies of six laboratory insects, and the response of the compound eye to mains fluorescent 'ripple'. *Physiological Entomology* **3**, 99–106.

Miller, N. & M. J. Lehane 1990. *In vitro* perfusion studies on the peritrophic membrane of the tsetse fly *Glossina morsitans morsitans* (Diptera: Glossinidae). *Journal of Insect Physiology* (in press).

Minchella, D. J. 1985. Host life-history variation in response to parasitism. *Parasitology* **90**, 205–16.

Minchella, D. J. & P. T. Loverde 1983. Laboratory comparison of the relative success of *Biomphalaria glabrata* stocks which are susceptible and insusceptible to infection with *Schistosoma mansoni*. *Parasitology* **86**, 335–44.

Minnick, M. F., R. A. Rupp & K. D. Spence 1986. Aggregation pheromone released from the palps of feeding female *Phlebotomus papatasi* (Psychodidae). *Journal of Insect Physiology* **30**, 153–6.

Mitchell B. K. & H. A. Reinouts van Haga-Kelker 1976. A comparison of the feeding behaviour in teneral and post-teneral *Glossina morsitans* (Diptera, Glossinidae) using an artificial membrane. *Entomologia Experimentalis et Applicata* **20**, 105–12.

Mockford, E. L. 1967. Some Psocoptera from the plumage of birds. *Proceedings Entomological Society Washington* **69**, 307–9.

Mockford, E. L. 1971. Psocoptera from the dusky-footed wood rat in southern California (Psocoptera: Atropidae, Psoguillidae, Liposcelidae). *Pan-Pacific Entomologist* **47**, 127–40.

Moffatt, M. & M. J. Lehane 1990. Trypsin is secreted as a zymogen in the midgut of *Stomoxys calcitrans*. *Insect Biochemistry* (in press).

Mohr, C. O. 1943. Cattle droppings as ecological units. *Ecological Monographs* **13**, 275–98.

Mohrig, W., D. Schittek & R. Hanschke 1979. Immunological activation of phagocytic cells in *Galleria mellonella*. *Journal of Invertebrate Pathology* **34**, 84–7.

Moloo, S. K. 1983. Feeding behaviour of *Glossina morsitans morsitans* infected with *Trypanosoma vivax*, *T. congolense* or *T. brucei*. *Parasitology* **86**, 51–6.

Moloo, S. K. & F. Dar 1985. Probing by *Glossina morsitans centralis* infected with pathogenic *Trypanosoma* species. *Transactions of the Royal Society for Tropical Medicine and Hygiene* **79**, 119.

Moloo, S. K. & S. B. Kutuza 1970. Feeding and crop-emptying in *Glossina brevipalpis* Newstead. *Acta Tropica* **27**, 356–77.

Molyneux, D. H. 1984. Evolution of the Trypanosomatidae: considerations of polyphyletic origins of mammalian parasites. *Center National de la Recherche Scientifique – Institut National de la Santé et de la Recherche Médicale* **1986**, 231–40.

Molyneux, D. H. & D. Jefferies 1986. Feeding behaviour of pathogen-infected vectors. *Parasitology* **92**, 721–36.

Molyneux, D. H. & R. Killick-Kendrick 1987. Morphology, ultrastructure and life cycles. In *The leishmaniases in biology and medicine*, W. Peters & R. Killick-Kendrick (eds). New York: Academic Press.

Molyneux, D. H., R. Killick-Kendrick & R. W. Ashford 1975. *Leishmania* in phlebotomid sandflies. III. The ultrastructure of *Leishmania mexicana amazonensis* in the midgut and pharynx of *Lutzomyia longipalpis*. *Proceedings of the Royal Society B* **190**, 341–57.

Msangi, A. 1988. Lectin activity in the gut of *Glossina morsitans morsitans* Wiedemann. M.Sc. Thesis, University of Wales.

Mukerji, D. & P. Sen-Sarma 1955. Anatomy and affinity of the elephant louse, *Haematomyzus elephantis* Piaget (Insecta: Rhyncophthiraptera). *Parasitology* **45**, 5–30.

Mukwaya, L. G. 1977. Genetic control of feeding preference in the mosquitoes *Aedes (Stegomyia) simpsoni* and *aegypti*. *Physiological Entomology* **2**, 133–45.

Mullens, B. A. & R. R. Gerhardt 1979. Feeding behaviour of some Tennessee Tabanidae. *Environmental Entomology* **8**, 1047–51.

Mumcuoglu, Y. & R. Galun 1987. Engorgement response of human body lice *Pediculus humanus* (Insecta: Anoplura) to blood fractions and their components. *Physiological Entomology* **12**, 171–4.

Murlis, J. 1986. The structure of odour plumes. In *Mechanisms in insect olfaction*, T. L. Payne, M. C. Birch & C. E. J. Kennedy (eds). Oxford: Clarendon Press.

Murray, M. D. 1957. The distribution of the eggs of mammalian lice on their hosts. II. Analysis of the oviposition behaviour of *Damalinia ovis*. *Australian Journal of Zoology* **5**, 19–29.

Murray, M. D. 1963. Influence of temperature on the reproduction of *Damalinia equi* (Denny). *Australian Journal of Zoology* **11**, 183–9.

Murray, M. D. 1987. Effects of host grooming on louse populations. *Parasitology Today* **3**, 276–8.

REFERENCES

Murray, M. D. & D. G. Nicholls 1965. Studies on the ectoparasites of seals and penguins. I. The ecology of the louse *Lepidophthirus macrorhini* Enderlein on the southern elephant seal, *Mirounga leonina* (L.). *Australian Journal of Zoology* **13**, 437–54.

Napier Bax, S. 1937. The senses of smell and sight in *Glossina swynnertoni*. *Bulletin of Entomological Research* **28**, 539–82.

Nasci, R. S. 1982. Differences in host choice between the sibling species of treehole mosquitoes *Aedes triseriatus* and *Aedes hendersoni*. *American Journal of Tropical Medicine and Hygiene* **31**, 411–15.

Nelson, R. L. 1965. Carbon dioxide as an attractant for *Culicoides*. *Journal of Medical Entomology* **2**, 56–7.

Nelson, W. A. 1987. Other blood-sucking and myiasis-producing arthropods. In *Immune responses in parasitic infections: immunology, immunopathology and immunoprophylaxis*, Vol. IV, E. J. L. Soulsby (ed). Boca Raton, FL: CRC Press.

Nelson, W. A. & G. C. Kozub 1980. *Melophagus ovinus* (Diptera: Hippoboscidae): evidence of local mediation in acquired resistance of sheep to keds. *Journal of Medical Entomology* **17**, 291–7.

Nelson, W. A., J. F. Bell, C. M. Clifford & J. E. Keirans 1977. Interaction of ectoparasites and their hosts. *Journal of Medical Entomology* **13**, 389–428.

Nelson, W. A., J. E. Keirans, J. F. Bell & C. M. Clifford 1975. Host–ectoparasite relationships. *Journal of Medical Entomology* **12**, 143–66.

Newson, R. M. & R. G. Holmes 1968. Some ectoparasites of the coypu (*Myocastor coypus*) in eastern England. *Journal of Animal Ecology* **37**, 471–81.

Nogge, G. 1976. Sterility in tsetse flies (*Glossina morsitans* Westwood) caused by loss of symbionts. *Experientia* **32**, 995.

Nogge, G. 1978. Aposymbiotic tsetse flies, *Glossina morsitans morsitans* obtained by feeding on rabbits immunized specifically with symbionts. *Journal of Insect Physiology* **24**, 299–304.

Nogge, G. 1981. Significance of symbionts for the maintenance of an optimal nutritional state for successful reproduction in haematophagous arthropods. *Parasitology* **82**, 101–4.

Nogge, G. & R. Ritz 1982. Number of symbionts and its regulation in tsetse flies, *Glossina* spp. *Entomologia Experimentalis et Applicata* **31**, 249–54.

Nuttal, G. H. F. 1899. On the role of insects, arachnids, and myriapods as carriers in the spread of bacterial and parasitic disease of man and animals. A critical and historical study. *Johns Hopkins Hospital Reports* **8**, 1–154.

Ogston, C. W. & W. T. London 1980. Excretion of hepatitis B surface antigen by the bedbug *Cimex hemipterus* Fabr. *Transactions of the Royal Society for Tropical Medicine and Hygiene* **74**, 823–5.

Ogston, D. 1983. *The physiology of hemostasis*. London: Croom Helm.

O'Meara, G. F. 1979. Variable expressions of autogeny in three mosquito species. *International Journal of Invertebrate Reproduction* **1**, 253–61.

O'Meara, G. F. 1985. Ecology of autogeny in mosquitoes. In *Ecology of mosquitoes*, L. P. Lounibos, J. R. Rey and J. H. Frank (eds). Florida: Florida Medical Laboratory.

O'Meara, G. F. 1987. Nutritional ecology of blood-feeding Diptera. In *Nutritional ecology of insects, mites, spiders, and related invertebrates*, F. Slansky & J. G. Rodriguez (eds). New York: Wiley.

O'Meara, G. F. & J. D. Edman 1975. Autogenous egg production in the salt marsh mosquito, *Aedes taeniorrhynchus*. *Biological Bulletin* **149**, 384–96.

O'Meara, G. F. & D. G. Evans 1973. Blood-feeding requirements of the mosquito: geographical variation in *Aedes taeniorrhynchus*. *Science* **180**, 1291–3.

O'Meara, G. F. & D. G. Evans 1976. The influence of mating on autogenous egg development in the mosquito, *Aedes taeniorrhynchus*. *Journal of Insect Physiology* **22**, 613–17.

Omer, S. M. & M. T. Gillies 1971. Loss of response to carbon dioxide in palpectomized female mosquitoes. *Entomologia Experimentalis et Applicata* **14**, 251–2.

Osbrink, L. A. & M. A. Rust 1985. Cat flea (Siphonaptera: Pulicidae): factors influencing host-finding behaviour in the laboratory. *Annals of the Entomology Society of America* **78**, 29–34.

Overal, W. L. 1980. Biology and behaviour of North American *Trichobius* batflies (Diptera: Streblidae). Unpublished doctoral dissertation, University of Kansas.

Overal, W. L. & L. R. Wingate 1976. The biology of the batbug *Strictimex antennatus* (Hemiptera: Cimicidae) in South Africa. *Annals of the Natal Museum* **22**, 821–8.

Owaga, M. L. & A. Challier 1985. Catch composition of the tsetse *Glossina pallidipes* Austen in revolving and stationary traps with respect to age, sex ratio and hunger stage. *Insect Science and its Applications* **6**, 711–18.

Pant, C. P., V. Houba & H. D. Engers 1987. Bloodmeal identification in vectors. *Parasitology Today* **3**, 324–6.

Pappas, L. G., C. D. Pappas & G. L. Grossman 1986. Hemodynamics of human skin during mosquito (Diptera: Culicidae) blood feeding. *Journal of Medical Entomology* **23**, 581–7.

Parker, K. R. & R. H. Gooding 1979. Effects of host anaemia, local skin factors and circulating antibodies upon biology of laboratory reared *Glossina morsitans morsitans* (Diptera: Glossinidae). *Canadian Journal of Zoology* **57**, 2393–401.

Parker, K. R. & M. J. Mant 1979. Effects of tsetse salivary gland homogenate on coagulation and fibrinolysis. *Thrombosis and Haemostasis* **42**, 743–51.

Patton, W. S. & F. W. Cragg 1913. A new species of *Philaematomyia*, with some remarks on the genus. *Indian Journal of Medical Research* **1**, 26–33.

Payne, T. L., M. C. Birch & C. E. J. Kennedy (eds) 1986. *Mechanisms in insect olfaction*. Oxford: Clarendon Press.

Peacock, A. J. 1981. Distribution of (Na$^+$ and K$^+$)-ATPase activity in the mid and hind guts of adult *Glossina morsitans* and *Sarcophaga nodosa* and the hind gut of *Bombyx mori* larvae. *Comparative Physiology and Biochemistry* **69A**, 133–6.

Peacock, A. J. 1982. Effects of sodium transport inhibitors on diuresis and mid gut (Na$^+$ and K$^+$) ATPase in the tsetse fly *Glossina morsitans*. *Journal of Insect Physiology* **28**, 553–8.

Pearman, J. V. 1960. Some African Psocoptera found on rats. *Entomologist* **93**, 246–50.

Pell, P. E. & D. I. Southern 1976. Effect of the coccidiostat, sulphaquinoxline, on symbiosis in the tsetse fly, *Glossina* species. *Microbios Letters* **2**, 203–11.

Pereira, M. E. A., A. F. B. Andrade & J. M. C. Ribeiro 1981. Lectins of distinct specificity in *Rhodnius prolixus* interact selectively with *Trypanosoma cruzi*. *Science* **211**, 597–9.

Peschken, D. P. & A. J. Thorsteinson 1965. Visual orientation of blackflies (Simuliidae: Diptera) to colour, shape and movement of targets. *Entomologia Experimentalis et Applicata* **8**, 282–8.

Peters, W. 1968. Vorkommen, Zusammensetzung und feinstruktur peritrophischer membranen im tierreich. *Zeitschrift für Morphologie und Ökologie der Tiere* **64**, 21–58.

Peters, W., H. Kolb & V. Kolb-Bachofen 1983. Evidence for a sugar receptor (lectin) in the peritrophic membrane of the blowfly larva, *Calliphora erythrocephala* Mg. (Diptera). *Journal of Insect Physiology* **29**, 275–80.

270

REFERENCES

Peters, W., U. Zimmermann & B. Becker 1973. Investigations on the transport function and structure of peritrophic membranes. IV. Anisotropic cross bands in peritrophic membranes of Diptera. *Journal of Insect Physiology* **19**, 1067–77.

Peterson, D. G. & A. W. A. Brown 1951. Studies of the responses of female *Aedes* mosquito. III. The response of *Aedes aegypti* (L.) to a warm body and its radiation. *Bulletin of Entomological Research* **42**, 535–41.

Phelps, R. J. & G. A. Vale 1976. Studies on the local distribution and on the methods of host location of some Rhodesian Tabanidae (Diptera). *Journal of the Entomological Society of South Africa* **39**, 67–81.

Phipps, J. 1974. The vampire wasps of British Columbia (Hym., Vespidae). An example of haematophagy by *Vespula* sp. *Bulletin of the Entomological Society of Canada* **6**, 134.

Piot, P. & C. J. Schofield 1986. No evidence for arthropod transmission of AIDS. *Parasitology Today* **2**, 294–5.

Politzar, H. & P. Merot 1984. Attraction of the tsetse fly *Glossina morsitans submorsitans* to acetone, 1 octen–3-ol, and the combination of these compounds in West Africa. *Revue d'Élevage et de Médecine Vétérinaire des Pays Tropicaux* **37**, 468–73.

Ponnudurrai, T., P. F. Billingsley & W. Rudin 1988. Differential infectivity of *Plasmodium* for mosquitoes. *Parasitology Today* **4**, 319–21.

Port, G. R., P. F. L. Bateham & J. H. Bryan 1980. The relationship of host size to feeding by mosquitoes of the *Anopheles gambiae* complex (Diptera: Culicidae). *Bulletin of Entomological Research* **70**, 133–44.

Pospisil, J. & J. Zdarek 1965. On the visual orientation of the stable fly (*Stomoxys calcitrans* L.) to colours. *Acta Entomologica Bohemoslovaca* **62**, 85–91.

Price, C. D. & N. A. Ratcliffe 1974. A reappraisal of insect haemocyte classification by the examination of blood from fifteen insect orders. *Zeitschrift für Zellforschung und Mikroskopische Anatomische* **147**, 537–80.

Price, G. D., N. Smith & D. A. Carlson 1979. The attraction of female mosquitoes (*Anopheles quadrimaculatus* Say) to stored human emanations in conjunction with adjusted levels of relative humidity, temperature and carbon dioxide. *Journal of Chemical Ecology* **5**, 383–95.

Price, R. D. 1975. The *Menacanthus eurysternus* complex (Mallophaga: Menoponidae) of the Passeriformes and Piciformes (Aves). *Annals of the Entomological Society of America* **68**, 617-22.

Pringle, G. 1965. A count of sporozoites in an oocyst of *Plasmodium falciparum*. *Transactions of the Royal Society for Tropical Medicine and Hygiene* **59**, 289–90.

Ratcliffe, N. A. 1986. Insect cellular immunity and the recognition of foreignness. In *Immune mechanisms in invertebrate vectors*, A. M. Lackie (ed.), Zoological Society of London Symposia, no. 56. Oxford: Oxford Science.

Ratcliffe, N. A. & S. J. Gagen 1977. Studies on the *in vivo* cellular reactions of insects: an ultrastructural analysis of nodule formation in *Galleria mellonella*. *Tissue and Cell* **9**, 73-85.

Ratcliffe, N. A. & J. B. Walters 1983. Studies on the *in vivo* cellular reactions of insects: clearance of pathogenic and non-pathogenic bacteria in *Galleria mellonella* larvae. *Journal of Insect Physiology* **29**, 407–15.

Ready, P. D. 1978. The feeding habits of laboratory bred *Lutzomyia longipalpis* (Diptera: Psychodidae). *Journal of Medical Entomology* **14**, 545–52.

Reid, G. D. F. & M. J. Lehane 1984. Peritrophic membrane formation in three temperate simuliids, *Simulium ornatum*, *S. equinum* and *S. lineatum* with respect to the migration of onchocercal microfilariae. *Annals of Tropical Medicine and Parasitology* **78**, 527–39.

271

Reinouts van Haga, H. A. & B. K. Mitchell 1975. Temperature receptors on tarsi of the tsetse fly *Glossina morsitans* West. *Nature* **255**, 225–6.

Renwrantz, L. 1986. Lectins in molluscs and arthropods: their occurrence, origin and roles in immunity. In *Immune mechanisms in invertebrate vectors*, A. M. Lackie (ed.), Zoological Society of London Symposia, no. 56. Oxford: Oxford Science.

Renwrantz, L. & A. Stahmer 1983. Opsonizing properties of an isolated hemolymph agglutinin and demonstration of lectin-like recognition molecules at the surface of hemocytes from *Mytilus edulis. Journal of Comparative Physiology* **149**, 535–46.

Ribeiro, J. M. C. 1982. The anti-serotonin and antihistamine activities of salivary secretion of *Rhodnius prolixus. Journal of Insect Physiology* **28**, 69–75.

Ribeiro, J. M. C. 1987. Role of saliva in blood-feeding by arthropods. *Annual Review of Entomology* **32**, 463–78.

Ribeiro, J. M. C. & E. S. Garcia 1981a. Platelet anti-aggregating activity in the salivary secretion of the blood sucking bug *Rhodnius prolixus. Experientia* **37**, 384–5.

Ribeiro, J. M. C. & E. S. Garcia 1981b. The role of saliva in feeding in *Rhodnius prolixus. Journal of Experimental Biology* **94**, 219–30.

Ribeiro, J. M. C. & J. J. F. Sarkis 1982. Anti-thromboxane activity in *Rhodnius prolixus* salivary secretion. *Journal of Insect Physiology* **28**, 655–60.

Ribeiro, J. M. C., G. B. Modi & R. B. Tesh 1989a. Salivary apyrase activity of some old world phlebotomine sandflies. *Insect Biochemistry* **19**, 409–12.

Ribeiro, J. M. C., P. A. Rossignol & A. Spielman 1985. Salivary gland apyrase determines probing time in anopheline mosquitoes. *Journal of Insect Physiology* **31**, 689–92.

Ribeiro, J. M. C., A. Vachereau, G. B. Modi & R. B. Tesh 1989b. A novel vasodilatory peptide from the salivary glands of the sand fly, *Lutzomyia longipalpis. Science* **243**, 212–14.

Ribeiro, J. M. C., J. A. Vaughan & A. F. Azad 1989c. Characterisation of the salivary apyrase activity of three rodent flea species. *Comparative Physiology and Biochemistry* B.

Rice, M. J., R. Galun & J. Margalit 1973. Mouthpart sensilla of the tsetse fly and their function. II. Labial sensilla. *Annals of Tropical Medicine and Parasitology* **67**, 101–7.

Rizki, R. M. & T. M. Rizki 1984. Selective destruction of a host blood cell type by a parasitoid wasp. *Proceedings of the National Academy of Sciences* **81**, 6154–8.

Roberts, L. W. 1981. Probing by *Glossina morsitans morsitans* and transmission of *Trypanosoma (Nannomonas) congolense. American Journal of Tropical Medicine and Hygiene* **30**, 948–51.

Roberts, R. H. 1972. Relative attractiveness of CO_2 and a steer to Tabanidae, Culicidae, and *Stomoxys calcitrans. Mosquito News* **32**, 208–11.

Roberts, R. H. 1977. Attractancy of two black decoys and CO_2 to tabanids (Diptera: Tabanidae). *Mosquito News* **37**, 169–72.

Robinson, A. 1939. The mouthparts and their function in the female mosquito, *Anopheles maculipennis. Parasitology* **31**, 212–42.

Roitt, I. M. 1985. *Immunology.* London: Gower Medical.

Ross, R. 1898. Report on the cultivation of Proteosoma, Labbe, in grey mosquitoes. *Indian Medical Gazette* **33**, 401–8.

Rossignol, P. A. & A. M. Rossignol 1988. Simulations of enhanced malaria transmission and host bias induced by modified vector blood location behaviour. *Parasitology* **97**, 363–72.

Rossignol, P. A., J. M. C. Ribeiro, M. Jungery *et al.* 1985. Enhanced mosquito blood-finding success on parasitaemic hosts: evidence for vector–parasite mutualism. *Proceedings of the National Academy of Science* **82**, 7725–7.

Rossignol, P. A., J. M. C. Ribeiro & A. Spielman 1984. Increased intradermal probing time in sporozoite-infected mosquitoes. *American Journal of Tropical Medicine and Hygiene* **33**, 17–20.

Rossignol, P. A., J. M. C. Ribeiro & A. Spielman 1986. Increased biting rate and reduced fertility in sporozoite-infected mosquitoes. *American Journal of Tropical Medicine and Hygiene* **35**, 277–9.

Rothschild, M. 1975. Recent advances in our knowledge of the Siphonaptera. *Annual Review of Entomology* **20**, 241–59.

Rothschild, M. & T. Clay 1952. *Fleas, flukes and cuckoos*. New York: Philosophical Library.

Rothschild, M & B. Ford 1973. Factors influencing the breeding of the rabbit flea (*Spilopsyllus cuniculi*): a spring-time accelerator and a kairomone in nestling rabbit urine (with notes on *Cediopsylla simplex*, another hormone bound species). *Journal of Zoology* **170**, 87–137.

Rothschild, M., Y. Schlein, K. Parker & S. Sternberg 1972. Jump of the oriental rat flea *Xenopsylla cheopis* (Roths.) *Nature* **239**, 45–8.

Rowland, M. & E. Boersma 1988. Changes in the spontaneous flight activity of the mosquito *Anopheles stephensi* by parasitization with the rodent malaria *Plasmodium yoelii*. *Parasitology* **97**, 221–7.

Rowland, M. W. & S. L. Lindsay 1986. The circadian flight activity of *Aedes aegypti* parasitized with the filarial nematode *Brugia pahangi*. *Physiological Entomology* **11**, 325–34.

Roy, D. N. 1936. On the role of blood in ovulation in *Aedes aegypti*, Linn. *Bulletin of Entomological Research* **27**, 423–9.

Rudin, W. & H. Hecker 1979. Functional morphology of the midgut of *Aedes aegypti* L. (Insecta; Diptera) during blood digestion. *Cell and Tissue Research* **200**, 193–203.

Rudin, W. & H. Hecker 1982. Functional morphology of the midgut of a sandfly as compared to other haematophagous nematocera. *Tissue and Cell* **14**, 751–8.

Sabelis, M. W. & P. Schippers 1984. Variable wind direction and anemotactic strategies of searching for an odour plume. *Oecologia* **63**, 225–8.

Sacks, D. L. 1989. Metacyclogenesis in *Leishmania* promastigotes. *Experimental Parasitology* **69**, 100–3.

Samuel, W. M. & D. O. Trainer 1972. *Lipoptena mazamae* Rondani, 1878 (Diptera: Hippoboscidae) on white-tailed deer in southern Texas. *Journal of Medical Entomology* **9**, 104–6.

Sangiorgi, G. & D. Frosini 1940. Di un principio emolitico (Cimicina) nella saliva del *Cimex lectularius*. *Pathologica* **32**, 189–91.

Sarkis, J. J. F., J. A. Guimaraes & J. M. C. Ribeiro 1986. Salivary apyrase of *Rhodnius prolixus*: kinetics and purification. *Biochemistry Journal* **233**, 885–91.

Sasaki, T. & K. Kobayashi 1984. Isolation of two novel protein inhibitors from haemolymph of silkworm larva, *Bombyx mori*. Comparison with human serum protein inhibitors. *Journal of Biochemistry* **95**, 1009–17.

Saunders, D. S. 1962. The ovulation cycle in *Glossina morsitans* Westwood (Diptera: Muscidae) and a possible method of age determination for female tsetse flies by the examination of their ovaries. *Transactions of the Royal Entomological Society of London* **112**, 221–38.

Schiefer, B. A., R. A. Ward & B. F. Eldridge 1977. *Plasmodium cynomolgi*: effects of malaria infection on laboratory flight performance of *Anopheles stephensi* mosquitoes. *Experimental Parasitology* **41**, 397–404.

Schlein, Y. 1977. Lethal effect of tetracycline on tsetse flies following damage to bacteroid symbionts. *Experientia* **33**, 450–1.

Schlein, Y., A. Warburg, L. F. Schnur & J. Shlomai 1983. Vector compatibility of *Phlebotomus papatasi* dependent on differentially induced digestion. *Acta Tropica* **40**, 65–70.

Schlein, Y., B. Yuval & A. Warburg 1984. Aggregation pheromone released from the palps of feeding female *Phlebotomus papatasi* (Psychodidae). *Journal of Insect Physiology* **30**, 153–6.

Schmit, A. R. & N. A. Ratcliffe 1977. The encapsulation of foreign tissue implants in *Galleria mellonella* larvae. *Journal of Insect Physiology* **23**, 175–84.

Schofield, C. J. 1982. The role of blood intake in density regulation of populations of *Triatoma infestans*. *Bulletin of Entomological Research* **72**, 617–29.

Schofield, C. J. 1985. Population dynamics and control of *Triatoma infestans*. *Annales Société Belge de Médecine Tropicale* **65**, 149–64.

Schofield, C. J. 1988. Biosystematics of the Triatominae. In *Biosystematics of haematophagous insects*, M. W. Service (ed). Oxford: Clarendon Press.

Senghor, J. E. & E. M. Samba 1988. Onchocerciasis control program – the human perspective. *Parasitology Today* **4**, 332–3.

Service, M. W. 1986. *Lecture notes on medical entomology*. Oxford: Blackwell Scientific.

Service, M. W. (ed.) 1988. *Biosystematics of haematophagous insects*. Oxford: Clarendon Press.

Shahan, M. S. & L. T. Giltner 1945. A review of the epizootiology of equine encephalomyelitis in the United States. *Journal of the American Veterinary and Medical Association* **107**, 279–88.

Sippel, W. L. & A. W. A. Brown 1953. Studies on the responses of the female *Aedes* mosquito. Part V. The role of visual factors. *Bulletin of Entomological Research* **43**, 567–74.

Smit, F. G. A. M. 1972. On some adaptive structures in Siphonaptera. *Folia Parasitologica* **19**, 5–17.

Smith, C. N., N. Smith, H. K. Gouck et al. 1970. L-Lactic acid as a factor in the attraction of *Aedes aegypti* to human hosts. *Annals of the Entomological Society of America* **63**, 760–70.

Smith, J. J. B. 1979. Effect of diet viscosity on the operation of the pharyngeal pump in the blood-feeding bug *Rhodnius prolixus*. *Journal of Experimental Biology* **82**, 93–104.

Smith, J. J. B. 1984. Feeding mechanisms. In *Comprehensive insect physiology, biochemistry and pharmacology*, G. A. Kerkut & L. I. Gilbert (eds). Oxford: Pergamon.

Smith, J. J. B. & W. G. Friend 1970. Feeding in *Rhodnius prolixus*: responses to artificial diets as revealed by changes in electrical resistance. *Journal of Insect Physiology* **16**, 1709–20.

Smith, J. J. B. & W. G. Friend 1982. Feeding behaviour in response to blood fractions and chemical phagostimulants in the blackfly, *Simulium venustum*. *Physiological Entomology* **7**, 219–26.

Smith, K. G. V. (ed.) 1973. *Insects and other arthropods of medical importance*. London: British Museum (Natural History).

Smith, V. J. & K. Soderhall 1983a. β 1,3-glucan activation of crustacean hemocytes in vitro and in vivo. *Biological Bulletin* **164**, 299–314.

Smith, V. J. & K. Soderhall 1983b. Induction of degranulation and lysis of haemocytes in the freshwater crayfish, *Astacus astacus* by components of the prophenoloxidase activating system in vitro. *Cell and Tissue Research* **233**, 295–303.

274

Snodgrass, R. E. 1944. The anatomy of the Mallophaga. *Occasional Papers of the California Academy of Sciences* **6**, 145–229.

Snodgrass, R. E. 1946. *The skeletal anatomy of fleas (Siphonaptera)*. Smithsonian Miscellaneous Collections, Vol. 104. Washington, DC: Smithsonian Institute.

Soulsby, E. J. L. 1982. *Helminths, arthropods and protozoa of domesticated animals.* London: Baillière Tindall.

Southwood, T. R. E., S. Khalaf & R. E. Sinden 1975. The micro-organisms of tsetse flies. *Acta Tropica* **32**, 259–66.

Southworth, G. C., G. Mason & J. R. Seed 1968. Studies in frog trypanosomiasis. I. A 24 hour cycle in the parasitaemia level of *Trypanosoma rotatorium* in *Rana clamitans* from Louisiana. *Journal of Parasitology* **54**, 255–8.

Spates, G. E. 1981. Proteolytic and haemolytic activity in the midgut of the stablefly *Stomoxys calcitrans* (L.): partial purification of the haemolysin. *Insect Biochemistry* **11**, 143–7.

Spates, G. E., R. D. Stipanovic, H. Williams & G. M. Holman 1982. Mechanisms of haemolysis in a blood-sucking dipteran, *Stomoxys calcitrans*. *Insect Biochemistry* **12**, 707–12.

Spielman, A. & J. Wong 1974. Dietary factors stimulating oogenesis in *Aedes aegypti*. *Biological Bulletin Marine Biological Laboratory (Woods Hole)* **147**, 433–42.

Sroka, P. & S. B. Vinson 1978. Phenoloxidase activity in the hemolymph of parasitized and unparasitized *Heliothis virescens*. *Insect Biochemistry* **8**, 399–402.

Stange, G. 1981. The ocellar component of flight equilibrium control in dragonflies. *Journal of Comparative Physiology* **141**, 335–47.

Steelman, C. D. 1976. Effects of external and internal arthropod parasites on domestic livestock production. *Annual Review of Entomology* **21**, 155–78.

Stojanovich, C. J. 1945. The head and mouthparts of the sucking lice (Insecta: Anoplura). *Microentomology* **10**, 1–46.

Stoltz, D. B. & D. I. Cook 1983. Inhibition of host phenoloxidase activity by parasitoid Hymenoptera. *Experientia* **30**, 1022–4.

Storey, H. H. 1933. Investigations of the mechanism of the transfer of plant viruses by insect vectors. *Proceedings of the Royal Society London B* **113**, 463–85.

Strangways-Dixon, J. & R. Lainson 1966. The epidemiology of dermal leishmaniasis in British Honduras. Part III. The transmission of *Leishmania mexicana* to man by *Phlebotomus passoanus*, with observations on the development of the parasite in different species of *Phlebotomus*. *Transactions of the Royal Society for Tropical Medicine and Hygiene* **60**, 192–207.

Stys, P. & M. Daniel 1957. *Lyctocoris campestris* F. (Heteroptera: Anthocoridae) as a human facultative ectoparasite. *Acta Societatis Entomologicae Cechosloveniae* **54**, 1–10.

Sutcliffe, J. F. 1978. Feeding behaviour in *Simulium venustum* (Diptera: Simuliidae), sensory aspects and mouthpart mechanics. Ph.D. Thesis, University of Toronto.

Sutcliffe, J. F. 1986. Black fly host location: a review. *Canadian Journal of Zoology* **64**, 1041–53.

Sutcliffe, J. F. 1987. Distance orientation of biting flies to their hosts. *Insect Science and its Applications* **8**, 611–16.

Sutcliffe, J. F. & S. B. McIver 1975. Artificial feeding of simuliids (*Simulium venustum*), factors associated with probing and gorging. *Experientia* **31**, 694–5.

Sutherland, D. R., B. M. Christensen & B. A. Lasee 1986. Midgut barrier as a possible factor in filarial worm vector competency in *Aedes trivittatus*. *Journal of Invertebrate Pathology* **47**, 1–7.

Sutton, O. G. 1947. The problem of diffusion in the lower atmosphere. *Quarterly Journal of the Royal Meteorological Society* **73**, 257–81.

Sutton, O. G. 1953. *Micrometeorology*. New York: McGraw-Hill.

Swellengrebel, N. H. 1929. La dissociation des fonctions sexuelles de nutritives (dissociation gonotrophique) d'*Anopheles maculipennis* comme cause du paludisme dans les Pays-Bas et ses rapports avec l'infection domiciliare. *Annales Institute Pasteur* **43**, 1370–89.

Takahashi, H., H. Komano & S. Natori 1986. Expression of the lectin gene in *Sarcophaga peregrina* during normal development and under conditions where the defence mechanism is activated. *Journal of Insect Physiology* **32**, 771–9.

Tashiro, H. & H. H. Schwardt 1953. Biological studies of horse flies in New York. *Journal of Economic Entomology* **46**, 813–22.

Tawfik, M. S. 1968. Feeding mechanisms and the forces involved in some blood-sucking insects. *Quaestiones Entomologicae* **4**, 92–111.

Teesdale, C. 1955. Studies on the bionomics of *Aedes aegypti* L. in its natural habitats in a coastal region of Kenya. *Bulletin of Entomological Research* **46**, 711–42.

Tempelis, C. H. & R. K. Washino 1967. Host feeding patterns of *Culex tarsalis* in the Sacramento Valley, California, with notes on other species. *Journal of Medical Entomology* **4**, 315–18.

Terra, W. R. 1988. Physiology and biochemistry of insect digestion: an evolutionary perspective. *Brazilian Journal of Medical and Biological Research* **21**, 675–734.

Theodor, O. 1967. *An illustrated catalogue of the Rothschild collection of Nycteribiidae (Diptera) in the British Museum (Natural History)*. London: British Museum (Natural History).

Thompson, B. H. 1976. Studies on the attraction of *Simulium damnosum* s. l. (Diptera: Simuliidae) to its hosts. I. The relative importance of sight, exhaled breath and smell. *Tropenmedizin und Parasitologie* **27**, 455–73.

Thompson, W. H. & B. J. Beatty 1977. Venereal transmission of La Crosse (California encephalitis) arbovirus in *Aedes triseriatus* mosquitoes. *Science*, **196**, 530–1.

Thorsteinson, A. J. & G. K. Bracken 1965. The orientation behavior of horse flies and deer flies (Tabanidae: Diptera). III. The use of traps in the study of orientation of tabanids in the field. *Entomologia Experimentalis et Applicata* **8**, 189–92.

Thorsteinson, A. J., G. K. Bracken & W. Tostawaryk 1966. The orientation behaviour of horse flies and deer flies (Tabanidae: Diptera). VI. The influence of the number of reflecting surfaces on attractiveness to tabanids of glossy black polyhedra. *Canadian Journal of Zoology* **44**, 275–9.

Tillyard, R. J. 1935. The evolution of scorpion flies and their derivatives (order Mecoptera). *Annals of the Entomological Society of America* **28**, 37–45.

Titus, R. G. & J. M. C. Ribeiro 1990. The role of vector saliva in transmission of arthropod-borne disease. *Parasitology Today* **6**, 157–60.

Tobe, S. S. & K. G. Davey 1972. Volume relationships during the pregnancy cycle of the tsetse fly *Glossina austeni*. *Canadian Journal of Zoology* **50**, 999–1010.

Torr, S. J. 1987. The host-orientation behaviour in tsetse flies (Diptera, Glossinidae). Ph.D. Thesis, University of London.

Torr, S. J. 1989. The host-orientated behaviour of tsetse flies (*Glossina*): the interaction of visual and olfactory stimuli. *Physiological Entomology* **14**, 325–40.

Traub, R. 1985. Coevolution of fleas and mammals. In *Coevolution of parasitic arthropods and mammals*, K. C. Kim (ed). New York: Wiley.

Trpis, M., R. E. Duhrkopf & K. L. Parker 1981. Non-Mendelian inheritance of mosquito susceptibility to infection with *Brugia malayi* and *Brugia pahangi*. *Science* **211**, 1435–7.

REFERENCES

Turell, M. J., P. A. Rossignol, A. Spielman et al. 1984. Enhanced arboviral transmission by mosquitoes that concurrently ingested microfilaria. *Science* **225**, 1039–41.

Turner, D. A. & J. F. Invest 1973. Laboratory analyses of vision in tsetse flies (Dipt., Glossinidae). *Bulletin of Entomological Research* **62**, 343–57.

Underhill, G. W. 1940. Some factors influencing feeding activity of simuliids in the field. *Journal of Economic Entomology* **33**, 915–17.

Usinger, R. L. 1966. *Monograph of Cimicidae*. Maryland: Thomas Say Foundation.

Vale, G. A. 1974a. New field methods for studying the response of tsetse flies (Diptera, Glossinidae) to hosts. *Bulletin of Entomological Research* **64**, 199–208.

Vale, G. A. 1974b. The response of tsetse flies (Diptera, Glossinidae) to mobile and stationary baits. *Bulletin of Entomological Research* **64**, 545–88.

Vale, G. A. 1977. Feeding responses of tsetse flies (Diptera: Glossinidae) to stationary hosts. *Bulletin of Entomological Research* **67**, 635–49.

Vale, G. A. 1982. The trap-orientated behaviour of tsetse flies (Glossinidae) and other Diptera. *Bulletin of Entomological Research* **72**, 71–93.

Vale, G. A. 1983. The effects of odours, wind direction and wind speeds on the distribution of *Glossina* (Diptera: Glossinidae) and other insects near stationary targets. *Bulletin of Entomological Research* **73**, 53–64.

Vale, G. A. & D. R. Hall 1985a. The role of 1-octen-3-ol, acetone and carbon dioxide in the attraction of tsetse flies, *Glossina* spp. (Diptera: Glossinidae), to ox odour. *Bulletin of Entomological Research* **75**, 209–17.

Vale, G. A. & D. R. Hall 1985b. The use of 1-octen-3-ol, acetone and carbon dioxide to improve baits for tsetse flies, *Glossina* spp. (Diptera: Glossinidae). *Bulletin of Entomological Research* **75**, 219–31.

Vale, G. A., D. R. Hall & A. J. E. Gough 1988. The olfactory responses of tsetse flies, *Glossina* spp. (Diptera: Glossinidae), to phenols and urine in the field. *Bulletin of Entomological Research* **78**, 293–300.

Vale, G. A., J. W. Hargrove, G. F. Cockbill & R. J. Phelps 1986. Field trials of baits to control populations of *Glossina morsitans morsitans* West. and *G. pallidipes* Austen (Diptera: Glossinidae). *Bulletin of Entomological Research* **76**, 179–94.

Van Handel, E. 1984. Metabolism of nutrients in the adult mosquito. *Mosquito News* **44**, 573–9.

Vargaftig, B. B., M. Chignard & J. Benveniste 1981. Present concepts on the mechanism of platelet aggregation. *Biochemistry and Pharmacology* **30**, 263–71.

Venkatesh, K. & P. E. Morrison 1982. Blood meal as a regulator of triacylglycerol synthesis in the haematophagous stable fly, *Stomoxys calcitrans*. *Journal of Comparative Physiology* **147**, 49–52.

Venkatesh, K., P. E. Morrison & V. L. Kallapur 1981. Influence of blood meals on the conversion of D-(U–14C)-glucose to lipid in the fat body of the haematophagous stablefly, *Stomoxys calcitrans*. *Comparative Physiology and Biochemistry* **68B**, 425–9.

Waage, J. K. 1979. The evolution of insect/vertebrate associations. *Biological Journal of the Linnean Society* **12**, 187–224.

Waage, J. K. 1981. How the zebra got its stripes – biting flies as selective agents in the evolution of zebra colouration. *Journal of Entomological Society of South Africa* **44**, 351–8.

Waage, J. K. & J. Nondo 1982. Host behaviour and mosquito feeding success: an experimental study. *Transactions of the Royal Society for Tropical Medicine and Hygiene* **76**, 119–22.

Wakelin, D. 1984. *Immunity to parasites.* London: Edward Arnold.

Walters, J. B. & N. A. Ratcliffe 1983. Studies on the *in vivo* cellular reactions of insects: fate of pathogenic and non-pathogenic bacteria in *Galleria mellonella* nodules. *Journal of Insect Physiology* **29**, 417–24.

Ward, R. A. 1963. Genetic aspects of the susceptibility of mosquitoes to malaria infections. *Experimental Parasitology* **13**, 328–41.

Warnes, M. L. & L. H. Finlayson 1985. Responses of the stable fly, *Stomoxys calcitrans* (L.) (Diptera: Muscidae), to carbon dioxide and host odours. I. Activation. *Bulletin of Entomological Research* **75**, 519–27.

Warnes, M. L. & L. H. Finlayson 1986. Electroantennogram responses of the stable fly, *Stomoxys calcitrans*, to carbon dioxide and other odours. *Physiological Entomology* **11**, 469–73.

Warnes, M. L. & L. H. Finlayson 1987. Effect of host behaviour on host preference in *Stomoxys calcitrans*. *Medical and Veterinary Entomology* **1**, 53–7.

Waterhouse, D. F. 1953. The occurrence and significance of the peritrophic membrane, with special reference to adult Lepidoptera and Diptera. *Australian Journal of Zoology* **1**, 299–318.

Watts, D. M., S. Pantuwatana, G. R. Defoliart *et al.* 1973. Transovarial transmission of La Crosse virus (California encephalitis group) in the mosquito, *Aedes triseriatus*. *Science* **182**, 1140–1.

Weber, L. A. & J. D. Edman 1972. Anti-mosquito behaviour of ciconiiform birds. *Animal Behaviour* **20**, 228–32.

Weitz, B. 1963. The feeding habits of *Glossina*. *Bulletin of the World Health Organisation* **28**, 711–29.

Welburn, S. C., D. S. Ellis & I. Maudlin 1987. *In vitro* cultivation of rickettsia-like organisms from *Glossina* spp. *Annals of Tropical Medicine and Parasitology* **81**, 331–5.

Wells, E. A. 1982. Trypanosomiasis in the absence of tsetse. In *Perspectives in trypanosomiasis research*, J. R. Baker (ed). Research Studies Press.

Wenk, P. 1953. Der Kopf von *Ctenocephalus canis* (Curt.) (Aphaniptera). *Zoologische Jahrbücher Abteilung für Anatomie und Ontogenie der Tiere* **73**, 103–64.

Wenk, P. 1962. Anatomie des Kopfes von *Wilhelmia equina* (Simuliiidae syn. Melusinidae, Diptera). *Zoologische Jahrbücher Abteilung für Anatomie und Ontogenie der Tiere* **80**, 81–134.

Wenk, P. & G. Schlorer 1963. Wirtsorientierung und Kopulation bei blutsaugenden Simuliiden (Diptera). *Zeitschrift für Parasitenkunde* **14**, 177–91.

Weyer, F. 1960. Biological relationships between lice (Anoplura) and microbial agents. *Annual Review of Entomology* **5**, 405–20.

Wharton, R. H. 1957. Studies on filariasis in Malaya: observations on the development of *Wuchereria malayi* in *Mansonia (Mansonioides) longipalpis*. *Annals of Tropical Medicine and Parasitology* **51**, 278–96.

White, G. B. 1974. *Anopheles gambiae* complex and disease transmission in Africa. *Transactions of the Royal Society of Tropical Medicine and Hygiene* **68**, 278–98.

White, G. B. & B. Rosen 1973. Comparative studies on sibling species of the *Anopheles gambiae* Giles complex (Dipt. Culicidae). II. Ecology of species A and B in savanna around Kaduna, Nigeria, during transition from wet to dry season. *Bulletin of Entomological Research* **62**, 613–25.

White, G. B., S. A. Magayuka & P. F. L. Boreham 1972. Comparative studies on sibling species of the *Anopheles gambiae* Giles complex (Dipt. Culicidae): bionomics and vectorial activity of species A and species B at Segera, Tanzania. *Bulletin of Entomological Research* **62**, 295–317.

W.H.O. 1985. *Tropical diseases research.* Seventh Program Report. Avignon, France: Barthelemy.

Wigglesworth, V. B. 1931. Physiology of excretion in a blood-sucking insect, *Rhodnius prolixus* (Hemiptera: Reduviidae). *Journal of Experimental Biology* **8**, 411–27.

Wigglesworth, V. B. 1941. The sensory physiology of the human louse *Pediculus humanus corporis* de Greer (Anoplura). *Parasitology* **33**, 67–109.

Wigglesworth, V. B. & J. D. Gillett 1934. The function of antennae in *Rhodnius prolixus* and the mechanism of orientation of the host. *Journal of Experimental Biology* **11**, 120–39.

Wijers, D. J. B. 1958. Factors that may influence the infection rate of *Glossina palpalis* with *Trypanosoma gambiense*. I. The age of the fly at the time of the infected feed. *Annals of Tropical Medicine and Parasitology* **52**, 385–90.

Willett, K. C. 1966. Development of the peritrophic membrane in *Glossina* (tsetse flies) and its relation to infection with trypanosomes. *Experimental Parasitology* **13**, 290–5.

Wilson, J. J., P. V. Neame & J. G. Kelton 1982. Infection induced thrombocytopaenia. *Seminars in Thrombosis and Haemostasis* **8**, 217–33.

Wilson, M. 1978. The functional organisation of locust ocelli. *Journal of Comparative Physiology* **124**, 297–316.

Woke, P. A. 1937. Comparative effects of the blood of man and of canary on egg production of *Culex pipiens* Linn. *Journal of Parasitology* **23**, 311–13.

Wood, D. M. 1964. Studies on the beetles *Leptinillus validus* (Horn) and *Platypsylus castoris* Rissema (Coleoptera: Leptinidae) from beaver. *Proceedings of the Entomological Society of Ontario* **95**, 33–63.

Wood, S. F. 1942. Observations on Vectors of Chagas' disease in the United States. I. California. *Bulletin of the Californian Academy of Science* **41**, 61–9.

Worms, M. J. 1972. Circadian and seasonal rhythms in blood parasites. In *Behavioural aspects of parasite transmission*, E. U. Canning & C. A. Wright (eds). London: Linnean Society.

Wright, R. H. 1958. The olfactory guidance of flying insects. *Canadian Entomologist* **90**, 81–9.

Wright, R. H. 1968. Tunes to which mosquitoes dance. *New Scientist* **37**, 694–7.

Wright, R. H. & F. E. Kellogg 1962. Response of *Aedes aegypti* to moist convection currents. *Nature* **194**, 402–3.

Yuill, T. M. 1983. The role of mammals in the maintenance and dissemination of La Crosse virus. In *California serogroup viruses*. C. H. Calisher & W. H. Thompson (eds). New York: Alan R. Liss.

Zachary, D. & J. A. Hoffman 1973. The haemocytes of *Calliphora erythrocephala* (Meig.) (Diptera). *Zeitschrift für Zellforschung und Mikroskopische Anatomie* **14**, 55–73.

Index

Acalypterae 209
Adult emergence
  cuticular hardening 27
  feeding 27
  sexual maturation 27
Aedes 75, 211, 213, 215
Aedes aegypti 11, 12, 23, 47, 48, 57, 59, 61,
  109, 128, 142, 158, 159, 173, 175, 183,
  191, 212, 214, 216, 217, 218
Aedes africanus 212
Aedes albopictus 212
Aedes bromeliae 212
Aedes hendersoni 22
Aedes polynesiensis 164
Aedes scutellaris 159
Aedes simpsoni 23
Aedes taeniorhynchus 107, 108, 218
Aedes triseriatus 22, 170
Aedes trivittatus 173, 191
Aedes vexans 218
Africa 1, 2, 3, 21, 23, 199, 205, 212, 218,
  220, 228, 231, 233, 235
African horse sickness 3, 4, 222
Age determination
  fluorescence accumulation 94, 240
  from ovarian changes 93, 239
AIDS 163
Akabane virus 222
Akiba 222
Alberprosenia goyovargasi 202
Allobosca 242
Amblycera 195, 197
Amblyopinini 247
America 1, 4, 22, 140, 200, 205, 212, 218,
  220, 224
Anaplasmosis 227
Anemotaxis
  shear forces 39
Anopheles 211, 212, 215
Anopheles annulipes 156
Anopheles arabiensis 23, 24
Anopheles atroparvus 183
Anopheles bwambae 23
Anopheles dirus 64
Anopheles freeborni 64
Anopheles gambiae 18, 23, 24, 64, 158, 214,
  217
Anopheles gambiae
  as a species complex 23

Anopheles labranchiae atroparvus 187, 195
Anopheles melas 23, 24
Anopheles merus 23
Anopheles pharoensis 218
Anopheles quadriannulatus 23, 24
Anopheles stephensi 64, 175, 183
Anoplura 7, 117, 120, 121, 195, 197,
  198
Anthocoris nemorum 12
Anthrax 227, 231
Anticoagulant 56, 57, 61
  antithrombin 57
  prolixin 57
Appetitive behaviour
  search patterns 29
  waiting 28
Apterygota 194
Apyrase 172
Arachnida 2
Arboviral 23
Arbovirus
  gut barrier 155, 183
  mesenteronal escape barrier 156
  salivary gland escape barrier 156
  salivary gland infection barrier 156
  trans-ovarial transmission 156
  venereal transmission 156
Arboviruses 150, 155, 163, 181, 183, 212,
  222
Argentina 201
Armigeres 211
Arterial worm 228
Arthropoda 194
Ascodipteron 117, 120, 245
Ashiza 209
Asia 201, 205, 212, 213, 231, 246
Atherix 230
Attacin 190
Attraction
  heat 47
  odours 47
Auchmeromyia luteola 94
Auchmeromyia senegalensis 233
Austenina 235
Australia 150, 205, 206, 230, 231
Austroconops 222
Austroleptis 230
Austrosimulium 218
Austrosimulium pestilens 221

Autogeny 105, 142
  effect of mating 108
  facultative 105
  male 108
  obligate 105
  selection pressures on adults 107
  selection pressures on larvae 107
  sugar feeding 108
Avian trypanosomes 218

Bacteria 181, 189, 190, 191, 227
Bahamas 5
Bartonella bacilliformis 225
Bartonellosis 225
Basilia 245
Basilia hispida 246
Bironella 210
Biting flies 4, 35, 65
Black Death 204
Blackflies 63, 65, 68, 72, 80, 93, 94, 105,
  109, 133, 141, 155, 164, 180, 182, 209
Blood feeding
  abdominal stretch receptors 109
  acquired resistance 136
  adult stages only 94
  density dependent effects 135
  dual sense of hunger 109
  efficiency 138, 139, 153
  food availability 135
  larval stages 94
  nymphal and adult stages 94
  rate 100, 177
  safe period 128
Blood meal
  absorption 91
  egg batch size 93
  identification 88
  ingestion 72
  metabolism 91
  protein metabolism 104
  resource allocation 92, 101
  size 82
  size and excretion of excess water 85
  size and maximising manoeuvrability
    88
  size and morphological adaptations 84
  size and physiological adaptations 84
Blood meal digestion
  batch process 91
  continuous process 91
  digestive enzymes 89, 91
  haemolysins 88, 89
  infection with parasites 90
  proteinase inhibitors 90
  rate 90
  regulation of digestive enzymes 89
  time taken in 88
Blood-sucking insects
  discovery that they are vectors 2

  distribution on the host 112
  evolution of the blood-sucking habit 6
  host preferences 14
  hosts exploited 14
  number of sensory receptors 121
  number of species 1
  opportunists 64
  range of mouthparts 64
  seasonal abundance 4
  selection for rapid feeding 126
  selection of refractory strains 158, 159,
    172, 175, 176
  sexual differences in blood feeding
    status 11, 141
  tolerance threshold of host 4
Blowflies 81
Bluetongue virus 3, 157, 222
Bolivia 201
Boreidae 12
Borrelia recurrentis 196, 197
Bovine ephemeral fever 222
Brachycera 209
Brazil 200, 201
Brill-Zinsser's disease 197
Britain 150
Brugia malayi 155, 158, 159, 212
Brugia pahangi 158, 159, 175, 187, 191
Brugia timori 212
Bubonic plague 172

Cabbage root fly 38
California 201
  encephalitis 227
Calliphoridae 233
Calpe eustrigata 12, 246
Calypterae 209
Camargue 5
Caribbean 5, 212
Carrion's disease 225
Ceratopogonidae 222–4
Cecropin 190
Ceratopogonidae 94, 105, 155, 163, 222,
  223
Chagas disease 153, 157, 161, 201
Chagasia 210
Chiclero's ear 224
Chile 201, 224
China 205
Chomadous 133
Chrysops 155, 227, 228, 229
Chrysops dimidiata 228
Chrysops discalis 227
Chrysops distinctipennis 228
Chrysops excitans 91
Chrysops silacea 228, 229
Cibarial armature 88
  pump 72
Cibarial receptors 153
Cimex 75, 108

Cimex hemipterus 199
Cimex lectularius 25, 62, 89, 199
Cimicidae 199
Clegs 227
Cnephia 218
Coagulation cascade
  extrinsic pathway 53
  factor IX 56
  factor V 56
  factor VIII 53, 57
  factor X 53, 57
  factor XI 53
  factor XII 53
  fibrin 56
  intrinsic pathway 53
  kininogen 53
  phospholipid 53
  thromboplastin 53
Coagulogens 191
Cochabamba valley 201
Coleoptera 8, 247
Colorado 235
  beetle 104
Congo 233
  floor maggot 209, 233
Coquillettidia 215, 217
Crustacea 190
Cryptotylus unicolor 150
Ctenocephalides canis 204, 206
Ctenocephalides felis felis 27, 47, 50, 61, 204, 206
Culex annulirostris 156
Culex molestus 215
Culex nigripalpus 22
Culex pipiens 98, 158
Culex pipiens quinquefasciatus 2
Culex quinquefasciatus 142, 164, 212, 216
Culex salinarius 15
Culex tarsalis 12, 22
Culex tritaeniorhynchus 213
Culicidae 210–18
Culicoides 4, 155, 157, 222, 223
Culiseta 211, 215
Culiseta melanura 16
Cutaneous leishmaniasis 224
Cyclorrhapha 209
Cyclo-developmental transmission 151
Cyclo-propagative transmission 151
Cyprus 4

Damalinia equi 114, 129
Damalinia bovis 114, 129, 191
Deerflies 227
Dengue 3, 212
Dermal leishmaniasis 224
Desertification 4, 233
Dipetalogaster maximus 202
Dipetalonema dracunculoides 242
Dipetalonema reconditum 207

Diptera 11, 51, 70, 79, 88, 104, 110, 162, 180, 184, 189, 208–46
Dipylidium caninum 206
Dirofilaria immitis 155, 158, 173, 212
Dirofilaria repens 155, 212
Dirofilaria roemeri 228
Distribution on host's surface
  host sickness 130
  feathers 114
  hair 116
  hairs 114
  insolation 112
  microclimate 112
  protection 116
  skin thickness 114
  temperature 112
Dog heartworm 212
Duttonella 153

Eastern equine encephalitis 4, 16, 212
Echidnophaga 120, 207
Echidnophaga gallinacea 204
Echidnophaga myrmecobii 206
Elaeophora schneideri 228
Elephantiasis 3, 212
Endopterygota 194
Entomological information
  Entomology Abstracts 193
  Review of Applied Entomology 193
Entomophagous insects 11, 12
Eoctenes 49
Epidemic relapsing fever 196
  typhus 196
Equine infectious anaemia virus 227, 231
Espundia 224
Ethiopia 23, 24
Eucampsipoda aegyptica 96
Europe 18, 196
Evolution of the blood-sucking habit
  entomophagy 11
  host dung 10
  morphological adaptation 11
  nests 6
  phoresy 8
  plant feeders 12
Exopterygota 194

Fannia 162
Fannia benjamini 64
Filaria 23, 24, 88, 151, 155, 158, 162, 163, 173, 175, 182, 191, 192, 207, 212, 218, 222, 227, 242
Fleas 12, 50, 51, 59, 64, 65, 72, 79, 88, 94, 99, 101, 117, 118, 124, 143, 150, 153, 155, 156, 170, 172, 204, 206, 244
Flicker fusion frequency 44
Florida 5, 107
Flour beetles 207
Following swarm 235

Forcipomyia 222
Forest flies 241
Fowl pox 231
France 5
Francisella tularensis 143, 205, 227

Geometrid 247
Glossina 58, 154, 235, 240
Glossina austeni 17, 98, 170, 241
Glossina brevipalpis 235, 241
Glossina fusca 235, 241
Glossina fuscipes 240
Glossina longipalpis 241
Glossina longipennis 30, 235, 241
Glossina morsitans 30, 31, 57, 108, 241
Glossina morsitans centralis 195
Glossina morsitans morsitans 17, 35, 98,
    170, 181
Glossina nigrofusca 235
Glossina pallidipes 30, 31, 35, 241
Glossina palpalis 235, 240
Glossina swynnertoni 82, 241
Glossina tabaniformis 241
Glossina tachinoides 240
Glossinidae 233–41
Gonotrophic concordance 92, 93
    dissociation 94
Great Britain 206

Habronema majus 231
Haemagogus 211, 212, 215
Haematobia 231
Haematobia atripalpis 162, 164
Haematobia irritans 10
Haematobia irritans exigua 231
Haematobia irritans irritans 231
Haematomyzus elephantis 7, 14
Haematomyzus hopkinsi 15
Haematopinus 130
Haematopinus asini 114, 129, 198
Haematopinus eurysternus 114, 129, 196, 198
Haematopota 227, 228, 229
Haematosiphon inodorus 200
Haemogogus 212
Haemolysins 89
Haemosporidia 154
Headfly 231
Helminths 151, 206
Hemimetabola 194
Hemiptera 79, 82, 89, 108, 199–204
    Cimicidae 199
    evolution 89
    Polyctenidae 203
    Reduviidae 200
Hepatocystis 155, 222
Hippobosca 94, 120, 121, 242, 243
Hippobosca equina 243
Hippobosca longipennis 242
Hippoboscidae 117, 241–3, 245

Hippo-flies 227
Holland 18
Holometabola 194
Hornfly 231, 232
Horseflies 48, 94, 227
Host behavioural defences
    aggregation behaviour 132
    aggregation sites 133
    anting 135
    dust baths 135
    efficiency 131
    host sickness 133, 177
    insect feeding success 130
    intensity of attack 130, 131, 139
    mud bathing 135
    permanent ectoparasites 128
    stampeding 133
    temporary ectoparasites 130
Host, blood viscosity and temperature 77
Host choice
    ecological factors 18
    physiological factors 16
    species complexes 23
    effects on insect fecundity 17
    effects on vector capacity of the insect 18
    factors determining 15
    genetic factors 23
    geographical considerations 21
    host abundance 20
    host availability 22, 142
    host behavioural defences 16
    housing quality 20
    locomotory abilities 14
    morphological characters of the host 18
    seasonal variations in 22
Host feathers 111
Host grooming
    ectoparasite distribution 128
    effect of host size 129
    host sickness 130
    moulting 129
    mutual 129
Host
    haematocrit 75
    homeostasis 111
    hormones 99
    skin 64, 111
    sickness 48
Host immune response 121
    acquired resistance 136, 138, 178
    fleas 204
    irritability 128
    louse infestation 196
    population regulation 138
    pruritis 121, 126, 204
    sequence of responses 124
    spectrum of reactivity 126
    Type I response 124
    Type IV response 124

Host location 25
  activation and orientation 26, 30
  anemotaxis 26, 37
  appetitive searching 26, 27
  attractants from feeding insects 40
  attraction 26, 46
  circadian periodicity 27
  distance 40
  heat 26, 47
  height of approach to the host 39
  host fevers 48
  hunger 27
  infrared radiation 46
  number of receptors 25
  olfaction 26, 30
  optomotor anemotaxis 37, 38
  optomotor responses 26
  orientation 46
  sound 46
  the behavioural framework 25
  vision 26, 40, 46
  water vapour 48
Human immunodeficiency virus 163
Hydrotaea 65, 231
Hydrotaea armipes 64
Hydrotaea irritans 231
Hymenolepis diminuta 207
Hymenolepis nana 207

India 4, 201, 205, 213
Insect anti-haemostatic factors
  apyrase 58, 59, 173
  factor VIII 58
  histamine 58, 61
  immune response 56
  inflammation 61
  platelets 58, 60
  serotonin 58
  thromboxane 58
  vasoconstriction 58, 61
  vasodilation 59, 60
Insect defence mechanisms
  antitrypanosome factor 191
  avoidance 191
  blood cells 184
  encapsulation 186
  energetic costs 159
  immunosuppression 192
  lectins 159, 185, 190
  nodule formation 189
  peritrophic membrane 181
  phagocytosis 185
  prophenoloxidase cascade 186
  self/non-self recognition 185
  wounding response 191
Insect repellants 16
International Code of Zoological
  Nomenclature 195
Ischnocera 195

Ivermectin 219

Japan 213
Japanese encephalitis virus 213
Jigger flea 204, 207

Kairomones 100
Kala-azar 224

La Crosse virus 133, 170
Languriidae 247
Latin America 212
Lectins 153, 185, 190
Leeches 57
Leishmania 151, 160, 161
Leishmania enhancing factor 160
Leishmania major 160
Leishmania tropica 90
Leishmaniases 3, 151, 224
Lepidophthirus macrorhini 113
Lepidoptera 246
Leptinillus 247
Leptininae 247
Leptinus 247
Leptocimex boueti 199
Leptoconops 222, 223
Leptoconops spinosifrons 223
Leucocytozoon 218, 222
Leucocytozoon smithi 167
Lice 64, 65, 84, 94, 113, 117, 121, 137, 141,
  178, 206
Linognathidae 120
Linognathus ovillus 113
Linognathus pedalis 113, 196
Lipophorin 191
Lipoptena 49, 117, 242, 243
Lipoptena cervi 115, 116
Loa loa 155, 227, 228
Loa loa papionis 228
Locusta 181
Loiasis 229
Lutzomyia 224, 225
Lutzomyia longipalpis 64
Lyctocoris campestris 12

Mal de Caderas 227, 231
Malaria 2, 3, 24, 133, 151, 155, 157, 158,
  161, 162, 163, 167, 172, 173, 182, 212,
  225
  origins 162
  retreat from Europe 18
Mallophaga 7, 195, 197, 198
Mansonella ozzardi 222
Mansonia 211, 212, 215, 217
Mansonoides 215
Mecoptera 12
Megistopoda aranea 244
Melophagus 242
Melophagus ovinus 113, 117, 120, 129, 241,
  243

Menacanthus eurysternus 15
Menacanthus stramineus 7
Mesozoic 7
Mexico 224
Microtus arvalis 129
Middle East 205
Mirounga leonina 113
Mosquito 1, 2, 4, 11, 12, 15, 16, 18, 19, 20,
    22, 24, 27, 32, 33, 34, 36, 39, 41, 44, 45,
    46, 47, 48, 57, 58, 61, 62, 63, 64, 65, 69,
    72, 79, 80, 88, 89, 92, 93, 94, 98, 105,
    107, 108, 109, 121, 126, 127, 130, 131,
    133, 135, 138, 141, 142, 150, 155, 156,
    157, 158, 159, 163, 164, 167, 168, 170,
    173, 175, 177, 180, 182, 183, 184, 187,
    191, 192, 209, 247
Moths 12, 34, 39, 117
Mouthparts 7, 64
    blackflies 68
    blood ingestion 72
    blood ingestion, haematocrit and
        parasites 77
    bugs 65
    fleas 72
    lice 65
    mosquitoes 69
    muscoid Diptera 70
    non-blood feeding 7
    tabanids 70
Movement between hosts 49
    blood meal size 50
    jumping 50
    morphological adaptations 49
    wing loss 49
Mozambique 21
Mucocutaneous leishmaniasis 224
Murine typhus 206
Mus musculus 133, 168
Musca 162
Musca lusoria 162
Musca planiceps 231
Musca xanthomelas 162
Muscidae 64, 72, 80, 94, 109, 150, 155, 180,
    230–2
Mycetome 96
Myotis velifer 48
Myxomatosis 150, 156, 206, 218

Nagana 2, 3, 21, 233
Nannomonas 154
Nematocera 209, 219
Nemorhina 235
Neosomy 204, 245
Niger 218
Noctuid 12, 246
Nosopsyllus fasciatus 208
Nutrition 94, 104
    differences between hosts 98
    host health 97

host hormones 100
life history pattern 94
maximal protein intake 97
symbionts 94, 198
Nycteribii 245
Nycteribiidae 94, 117, 120, 121, 243, 245–6
Nycterophilia 244

Ocelli 41
Olfaction 46
    acetone 30, 35
    butanone 30
    carbon dioxide 30, 32, 34, 40
    deterrence 36
    lactic acid 30, 33, 35
    octenol 30, 35
    odour plume 30, 34, 36, 37
    phenolic compounds 30, 35
    pulsed and continuous sources 34
    receptors 36
    response to mixtures 35
    synergism 35, 36
Onchocerca 163
Onchocerca gibsoni 222
Onchocerca volvulus 155, 164, 218, 219
Onchocerciasis 219
    Control Program 219
Onion fly 38
Opsonins 185
Oriental sore 224
Orinoco 1
Ornithocoris toledo 200
Oroya fever 225
Oryctolagus cuniculus 100
Overgrazing 233

Pappataci fever 225
Parafilaria bovicola 162
Parafilaria multipapillosa 162, 164
Paraguay 201
Parahaemoproteus 222
Parasite transmission
    arboviruses 155
    biological transmission 150
    blockage of the intestine 170
    circadian variation in infection 167
    contacting the vector 163
    cyclo-developmental transmission 151
    cyclo-propagative transmission 151
    degree of susceptibility 159
    feeding rates 172
    filarial worms 155, 164
    high parasitaemia 163
    HIV 163
    host sickness 168
    infection rates 159
    invasion of vector 178
    lectins 151
    leishmania 151

INDEX

Parasite transmission *cont.*
  malaria 154, 164, 167
  mechanical transmission 143
  offspring congregate at particular site
    164
  periodicity 164
  propagative transmission 150
  routes 143
  selection pressures on the vector 172
  trypanosomes 153, 166
  vector feeding behaviour 170
  vector feeding time 168
Parasite strategies for contacting a vector
  163
Parasite-induced vector pathology 173
  death 173
  familial infections 175
  impaired flight 175
  nutrient depletion 175
  reduced longevity 173
  reduced reproduction 173
Parasitoid wasps 191
Pediculus 75, 196
Pediculus capitis 196, 198
Pediculus humanus 48, 94, 120, 173, 196, 198
Pediculus humanus capitis 196
Pediculus humanus humanus 196
Pediculus serrata 126
Penicillidia 245
Periodic ectoparasite
  definition of term xiv
Peritrophic membrane 151, 154, 179
  bulk filtration 87
  function 81, 181
  protection against invaders 181
  structure 79, 179
  Type I 180
  Type II 180
Permanent ectoparasites
  attachment to the host 120
  burrowing in skin 117
  comb function 118
  combs 117
  convergent evolution 118
  definition of term xiv
  egg attachment 120
  morphological specializations 116
  neosomy 117, 120
  sensory organs 121
  shape 117, 120
  size 116
  water loss 119
  wing loss 117
Peru 201
Phagostimulants 63
  nucleotides 63
  solution tonicity 64
Pharyngeal pump 72
Philaematomyia lineata 65

Philoliche zonata 31
Phlebotominae 224
Phlebotomus 224
Phlebotomus papatasi 90, 226
Phormia regina 109
Phthiraptera 195–9
Phthirus pubis 196, 198
Pigeon fly 241
Piroplasms 3
Plague 204
Plasmodiidae 167
Plasmodium 162
Plasmodium berghei 183, 212
Plasmodium chabaudi 168
Plasmodium corporis 196
Plasmodium cynomolgi 158, 175, 212
Plasmodium falciparum 18, 155, 167, 183
Plasmodium gallinaceum 158, 173, 183, 212
Plasmodium inui 164
Plasmodium knowlesi 212
Plasmodium malariae 18
Plasmodium vivax 18
Plasmodium yoelii 164, 175
Platelet 58
Polio virus 231
Polyctenidae 117, 120, 121, 203
Polyplax serrata 128, 137
Population regulation
  acquired resistance 138
  density dependent mechanisms 131,
    139, 141
Potamonautes 220
Probing stimulants 61
  ATP 62
  heat 62
  substrate thickness 62
  various 62
Propagative transmission 150
Prophenoloxidase cascade 186
Prosimulium 218
Protocalliphora 233
Pseudolynchia canariensis 241
Psocids 6
Psorophora 211, 215
Psychodidae 224–7
Pterygota 194
Pulex 15
Pulex irritans 204
Pupipara 209, 243, 244, 245
Pycnomonas 154
Pyralidae 117, 247

Quediini 247

Rabbits 98
Rana clamitans 166
Rattus norveigicus 129
Reduviidae 63, 179, 200
Relapsing fever 197

Resilin 51
Rhagionidae 11, 209, 230
Rhodnius 65, 86, 108
Rhodnius pallescens 195
Rhodnius prolixus 48, 57, 58, 59, 61, 62, 63, 72, 73, 75, 77, 85, 89, 91, 190, 201
Rhynchophthirina 7, 195
Rickettsia prowazekii 173, 196, 197
Rickettsia typhi (mooseri) 173, 206
Rickettsiae 196
Rickettsiae-like organisms 159
Rift Valley Fever virus 3, 168
Rinderpest 227
River blindness 3, 218
Rochalimaea quintana 196
Romania 4, 218

Sabethes 211, 212
Sabethes chloropterus 212
Sahara 21, 23
St Louis encephalitis 22
Saliva
  anaesthetic 61
  digestive enzymes 56, 89
  lubricant 56
  skin penetration 61
  surface tension and mouthparts 61
Salivaria 153
Salmonella enteritidis 206
Sand flea 204
Sandflies 59, 64, 79, 80, 93, 94, 105, 151, 153, 160, 170, 180, 209, 224
Sandfly fever 225
Scarabaeid beetles 10
Schistosoma mansoni 159, 177
Scottish highlands 5
Scramble competition 140
Seal 113
Semliki forest 23
Sensu lato – definition 195
Sensu stricto – definition 195
Sexual maturation
  feeding 27
  following adult emergence 27
Sheep 113
Sheep ked 241, 243
Shizophora 209
Simuliidae 44, 45, 46, 62, 163, 180, 182, 218–22
Simulium colombaschense 4
Simulium damnosum 23, 195, 219, 220, 221
  species complex 23
Simulium equinum 180
Simulium euryadminiculum 36, 47, 221
Simulium lineatum 180
Simulium neavei 220
Simulium ochraceum 220
Simulium ornatum 180
Simulium rugglesi 221

Simulium venustum 48, 63, 221
Sinbis virus 168
Siphonaptera 121, 204–8
Sleeping sickness 2, 3, 21, 233
Somaliland 2
Souma 3
South American sleeping sickness 153, 201
Spaniopsis 230
Species complexes 195
Spilopsyllus cuniculi 99, 150, 156, 206, 208
Sporozoite rate 24
Stableflies 33, 44, 62, 70, 79, 89, 94, 109, 132, 162, 230, 231, 232
Staphylinidae 117
Stephanofilaria stilesi 162, 231
Stercorarian (posterior station) trypanosomes 153, 161
Sticktight fleas 204, 207
Stomoxys 132, 231
Stomoxys calcitrans 33, 44, 62, 70, 79, 89, 94, 109, 132, 162, 230, 231, 232
Stomoxys inornata 231
Stomoxys nigra 231
Stomoxys omega 231
Streblidae 15, 50, 94, 117, 120, 121, 207, 243–5
Sugar feeding
  crop 110
  dual sense of hunger 109
  egg production 108
  proteinase inhibitors 110
  regulation 109
  sugar sources 109
  water conservation 110
Summer mastitis 231
Suprapylaria 151
Surra 3, 227, 231
Symbiotic micro-organisms 94
  distribution among sexes 96
  regulation of numbers 97
  vitamins 96
Symphoromyia atripes 230
Symphoromyia sackeni 230

Tabanidae 4, 43, 44, 45, 46, 63, 65, 70, 72, 80, 93, 105, 109, 116, 150, 153, 155, 180, 206, 209, 231, 227–9
Tabanus 32, 227, 228, 229
Tabanus calens 229
Tabanus fuscicostatus 150
Tabanus nigrovittatus 43
Taeniorhynchus 211
Tanders 133
Tapeworm 207
Temporary ectoparasites
  definition of term xiv
Texas cattle fever 2
Thelazii gulosa 162
Theobaldia 211

Three day fever 225
Tick 2, 124, 197
Transmission routes 143
Trench fever 196
Triatoma 65
Triatoma infestans 25, 116, 140, 192, 195, 201, 202, 203
Triatoma protracta 202
Triatoma rubrofasciata 201
Triatominae 94, 200
Trichobius 244
Trichobius major 48
Trichobius yunkeri 244
Trypanosoma 161
Trypanosoma brucei 154, 170, 172, 191
Trypanosoma brucei evansi 143
Trypanosoma congolense 154, 166, 170, 191
Trypanosoma cruzi 153, 161, 192, 201, 202
Trypanosoma dionisii 191
Trypanosoma duttoni 166
Trypanosoma equinum 150, 227, 231
Trypanosoma equiperdum 150
Trypanosoma evansi 150, 227, 231
Trypanosoma hippicum 150
Trypanosoma lewisi 153, 166
Trypanosoma major 244
Trypanosoma melophagium 242
Trypanosoma minasense 166
Trypanosoma rotatorium 166
Trypanosoma simiae 227
Trypanosoma suis 154
Trypanosoma theileri 153, 227, 242
Trypanosoma venezuelense 150
Trypanosoma vivax 18, 150, 153, 191
Trypanosoma vivax viennei 143, 227
Trypanosomatidae 22, 151, 153, 159, 160, 161, 163, 170, 225, 227, 231, 233
Trypanozoon 154
Tsetse 2, 11, 26, 30, 31, 35, 36, 37, 38, 39, 40, 41, 43, 44, 45, 46, 47, 50, 57, 58, 62, 63, 70, 79, 80, 84, 85, 86, 87, 88, 89, 91, 94, 96, 97, 98, 101, 102, 104, 105, 131, 135, 138, 140, 159, 170, 173, 175, 178, 180, 181, 182, 190, 191, 233, 244, 246
blood meal size and manoeuvrability 50, 82
body temperature 50
following swarm 45
Tularaemia 205, 227
Tunga penetrans 120, 204, 207
Tungidae 204

Uganda 23
Unitred Kingdom 18
Uranotaenia 215
Uropsylla tasmanica 120
Uruguay 201

Usambara mountains 1
Uta 224

Vector/parasite relationships
evolutionary routes 160
longevity of the association 161
Vector/parasite specificity 156
genetics 158
host choice 157
non-Mendelian inheritance 159
physiological factors 157
Venezuelan equine encephalitis 133, 212, 218
Vertebrate blood
nutritional value 7
proteinase inhibitors 90
Vertebrate haemostasis 52
ADP 53
coagulation 53
coagulation cascade 53
collagen 52
inhibition of coagulation 52
platelet plug 53
platelets 53
serotonin 53
thromboxane A2 53
vasoconstriction 52
Vesicular stomatitis virus 225
Viannia 151
Vinchucas 201
Visceral leishmaniasis 224
Vision
colour 41, 43, 44, 46
detection of movement 44
distance 40
intensity-contrast 41, 43, 44
pattern discrimination 44
shape 46
shape discrimination 41, 46
trap size 46
ultra-violet sensitivity 41
Volta 218

Wasp 246
Western equine encephalitis 4, 212, 227
Wuchereria bancrofti 2, 155, 158, 164, 212
Wyeomyia smithii 107, 211

Xenopsylla brasiliensis 205
Xenopsylla cheopis 64, 99, 170, 205, 206

Yellow fever 3, 212
Yersinia pestis 150, 170, 205
Yugoslavia 4

Zanzibar 23
Zebras 44

Blood-sucking insects are agents in the transmission of many of the most debilitating diseases of man including malaria, sleeping sickness, filariasis, yellow fever, typhus and plague. In addition, these insects cause major economic losses in agriculture both by direct damage to livestock and because of the veterinary diseases, such as the various trypanosomiases, which they transmit. *Biology of blood-sucking insects* is a unique, 'topic-led' commentary on the biological themes common to the lives of haematophagous insects, dealing with subjects as diverse as physiology and biochemistry, behavioural biology, economic impact and population dynamics. The book opens with a brief outline of the medical, social and economic impact of blood-sucking insects. Further chapters cover the evolution of the blood-sucking habit, feeding preferences, host location, the ingestion of blood and the various physiological adaptations for dealing with the blood meal. Discussions on host-insect interactions and the transmission of parasites by blood-sucking insects are followed by a final taxonomic chapter which is designed as a useful quick-reference section covering the different groups of insects referred to in the text.

Biology of blood-sucking insects is written in a clear, concise fashion and is well illustrated throughout with a variety of specially prepared line illustrations and photographs. The text provides a summary of knowledge in this important group of insects and will be of direct interest to advanced undergraduate and postgraduate students in medical and veterinary parasitology and entomology.

Mike Lehane graduated from the University of London and has a Ph.D. in Entomology from the University of Leeds. After a Wellcome Fellowship at the London School of Hygiene and Tropical Medicine he moved to the University of Wales, Bangor where he is currently a Senior Lecturer in Parasitology.

ISBN 0-04-445410-0

ISBN 0 04 4454090
ISBN 0 04 4454104

9 780004 454106